Kentucky BARNS

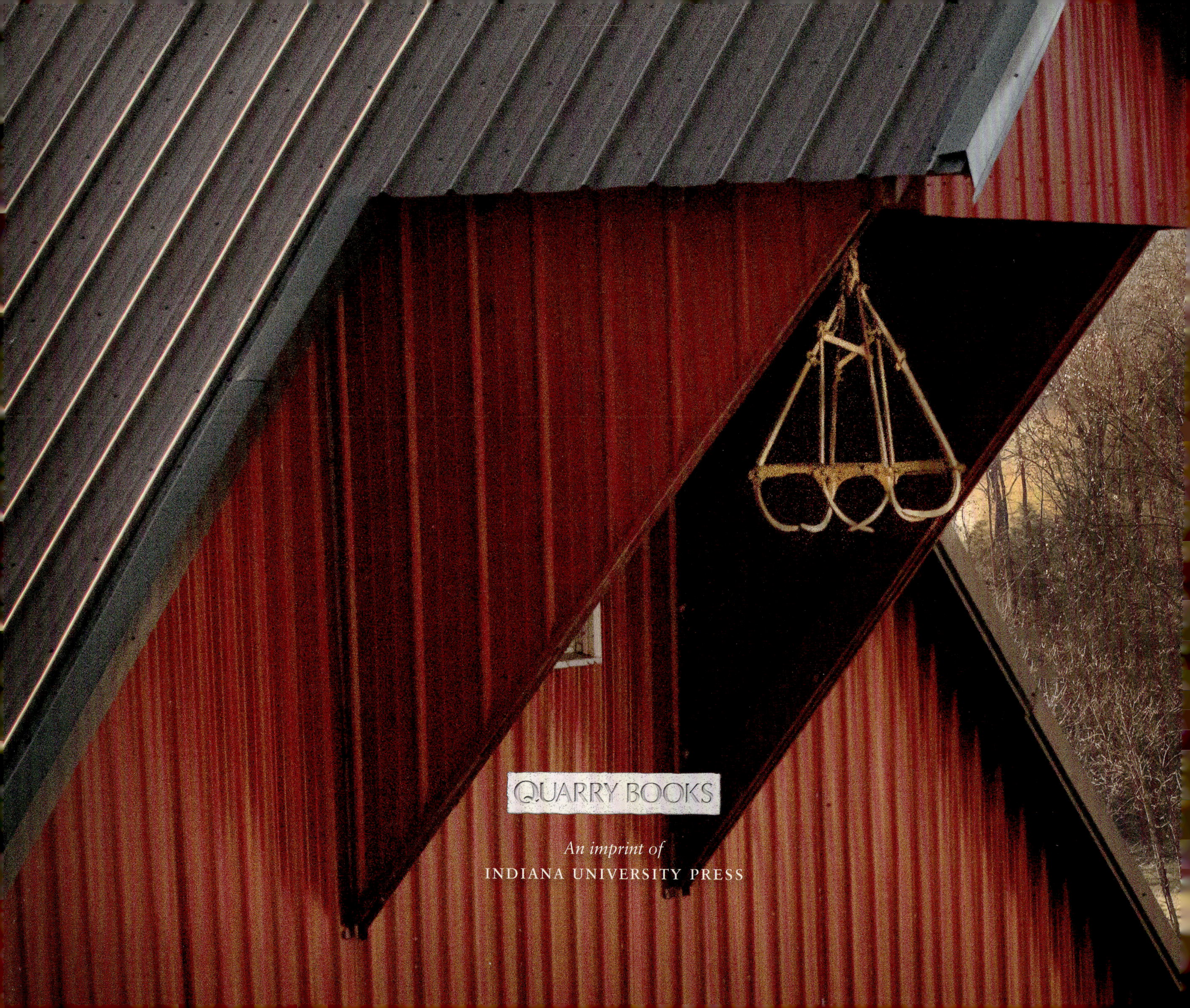
QUARRY BOOKS
An imprint of
INDIANA UNIVERSITY PRESS

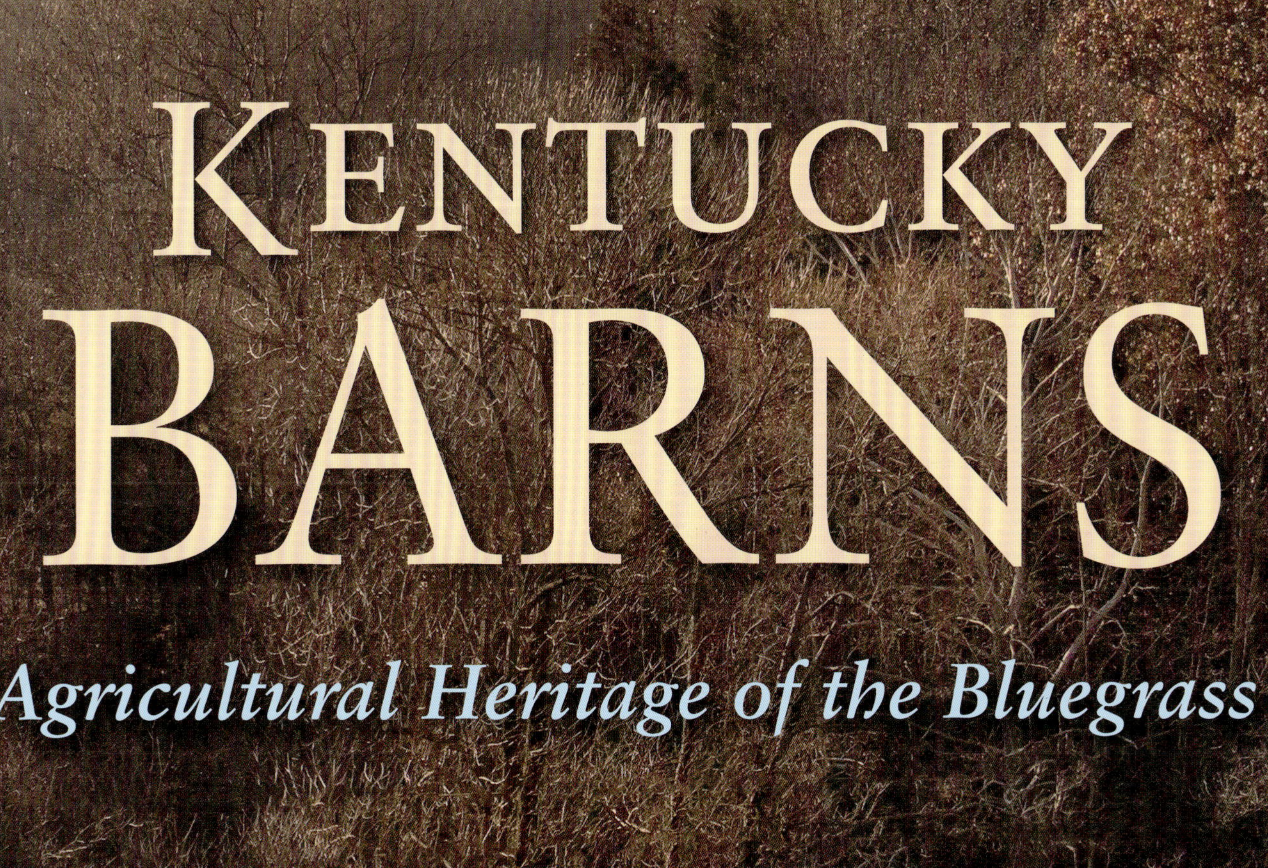

KENTUCKY BARNS

Agricultural Heritage of the Bluegrass

CAROL PEACHEE

Foreword by Mary Berry
Introduction by Janie-Rice Brother

This book is a publication of

QUARRY BOOKS

an imprint of
Indiana University Press
Office of Scholarly Publishing
Herman B Wells Library 350
1320 East 10th Street
Bloomington, Indiana 47405 USA

iupress.indiana.edu

This book is printed on acid-free paper.

Manufactured in China

Library of Congress Cataloging-in-Publication Data

Names: Peachee, Carol, photographer.
Title: Kentucky barns : agricultural heritage of the
   bluegrass / Carol Peachee ; foreword by Mary
   Berry ; introduction by Janie-Rice Brother.
Description: Bloomington, Indiana : Quarry Books, [2019]
   | Includes index. | Identifiers: LCCN 2018061357 (print)
   | LCCN 2019000054 (ebook) | ISBN 9780253042750
   (ebook) | ISBN 9780253042743 (cloth : alk. paper)
Subjects: LCSH: Barns—Kentucky—Pictorial works.
   | Vernacular architecture—Kentucky—Pictorial
   works. | Architectural photography.
Classification: LCC NA8230 (ebook) | LCC NA8230
   .P432 2019 (print) | DDC 725/.37209769—dc23
LC record available at https://lccn.loc.gov/2018061357

1 2 3 4 5   24 23 22 21 20 19

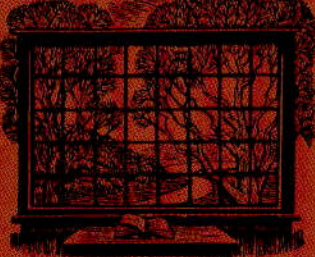

THE BERRY CENTER

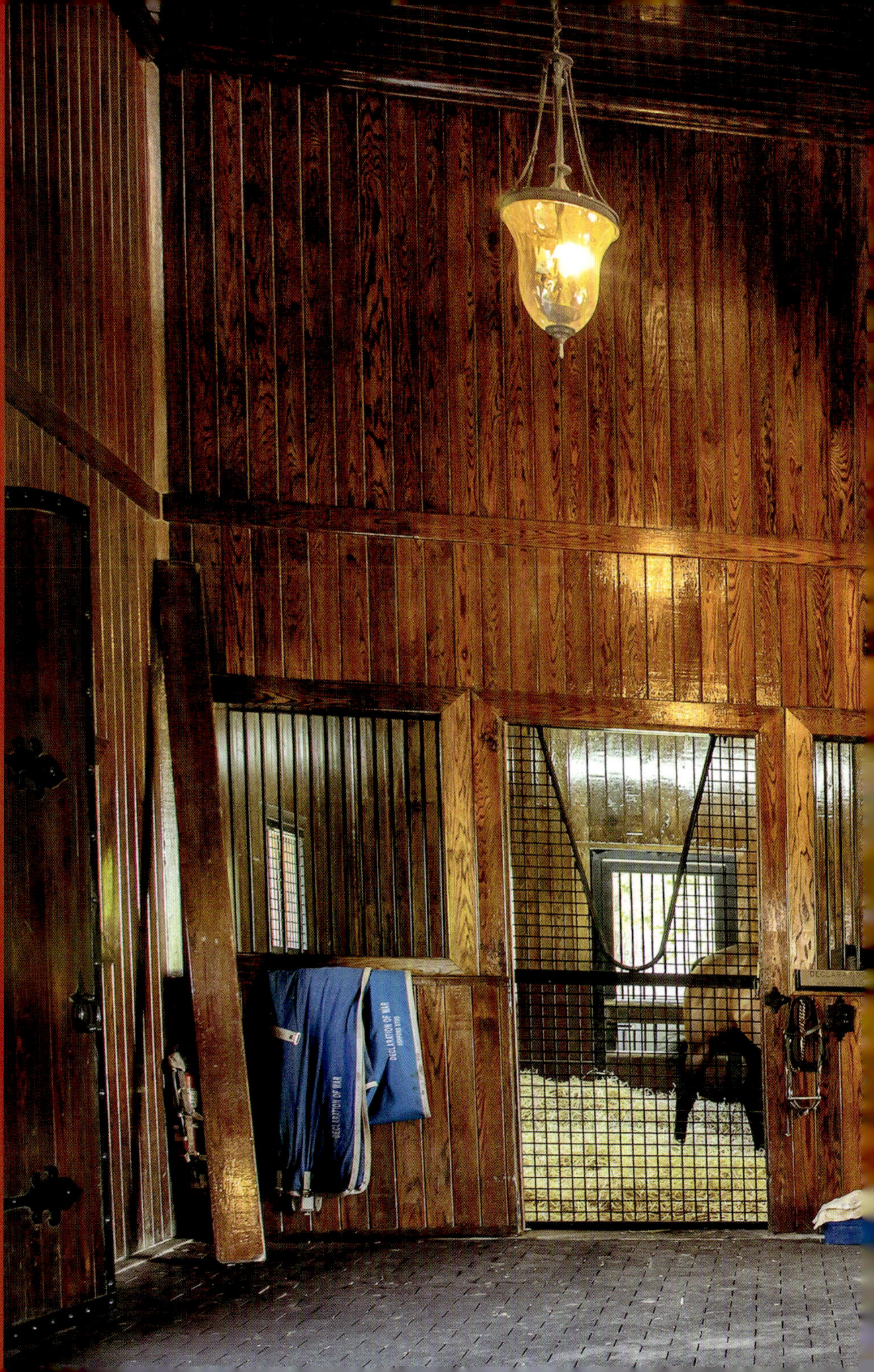

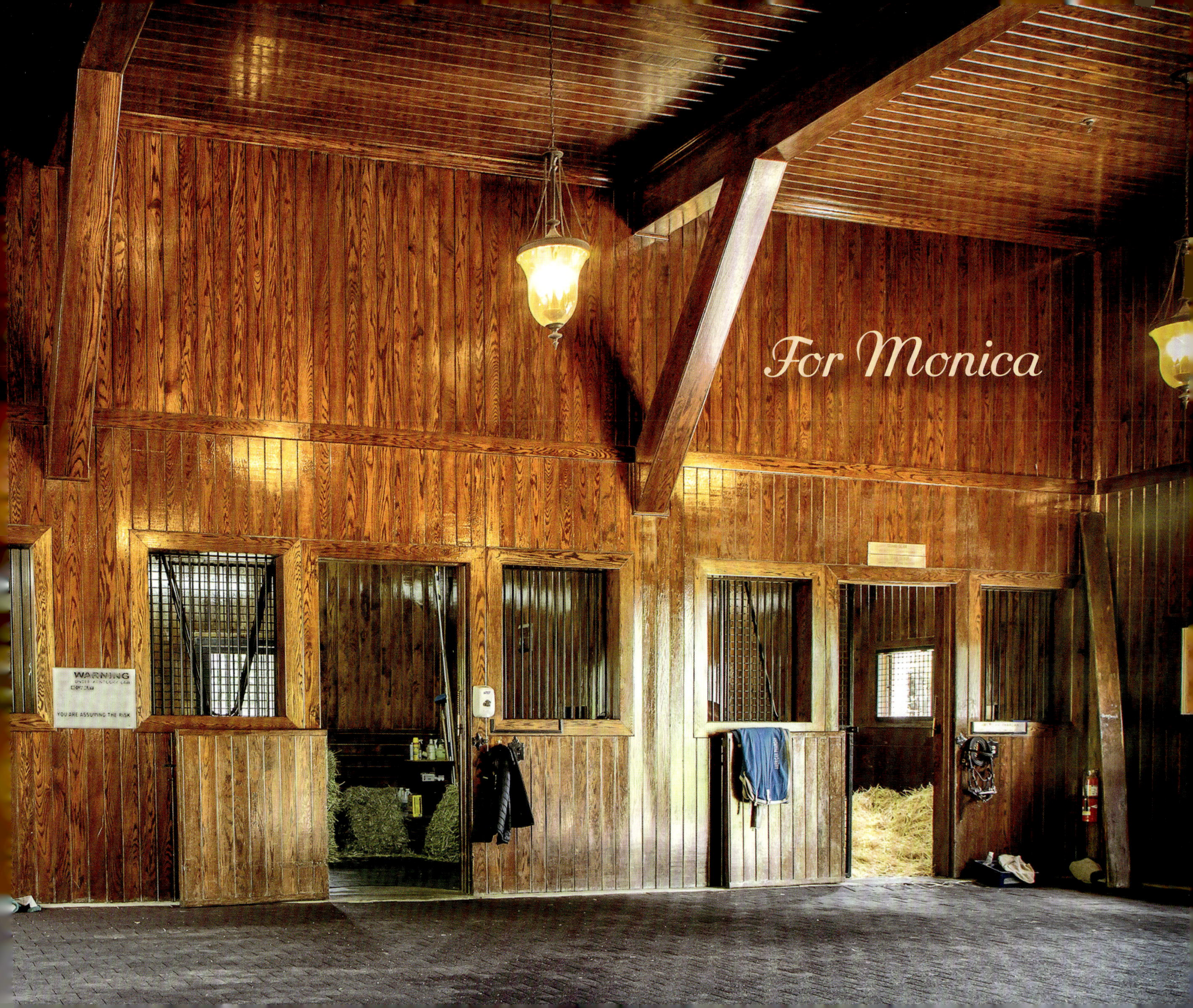
For Monica
WARNING
YOU ARE ASSUMING THE RISK

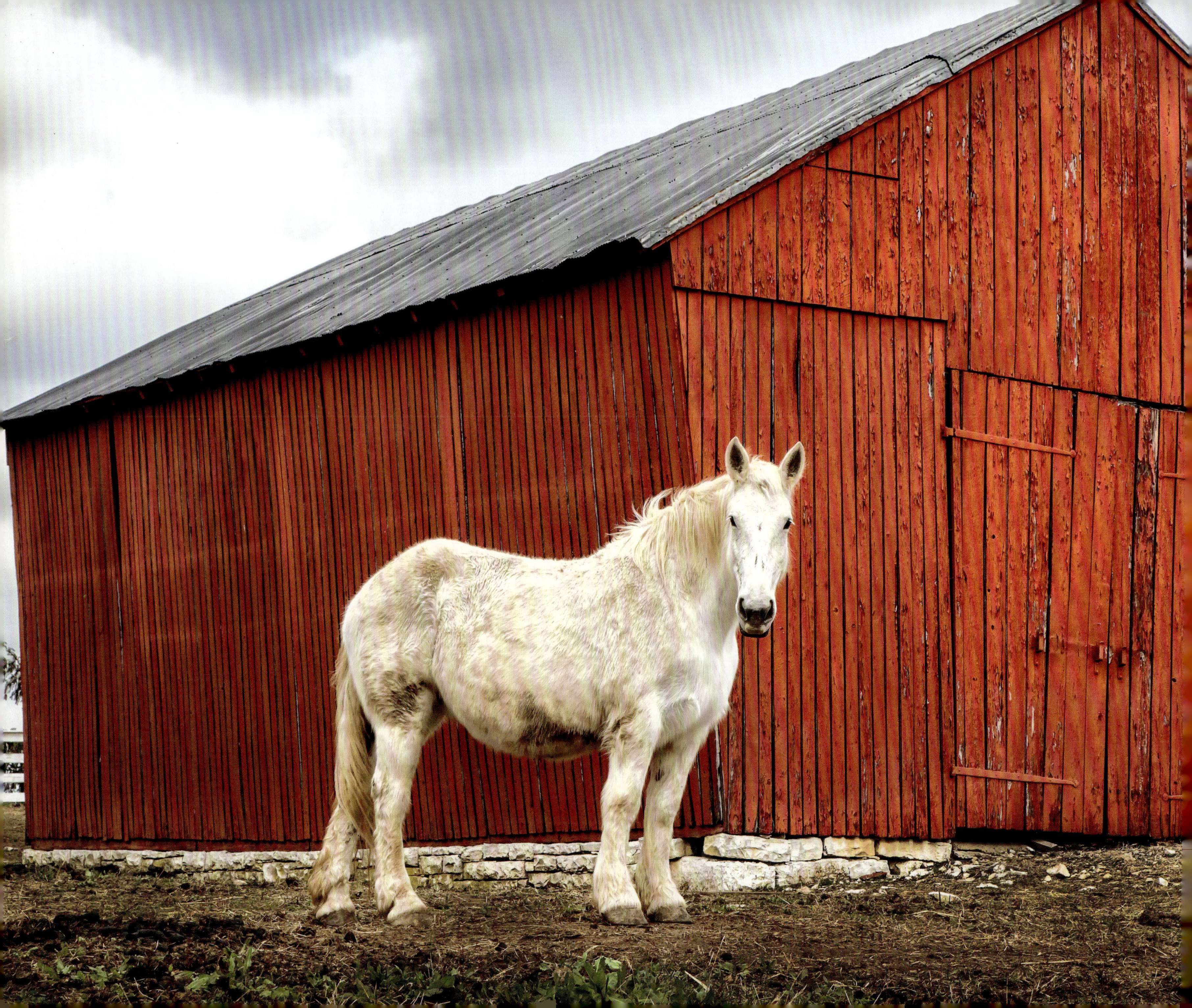

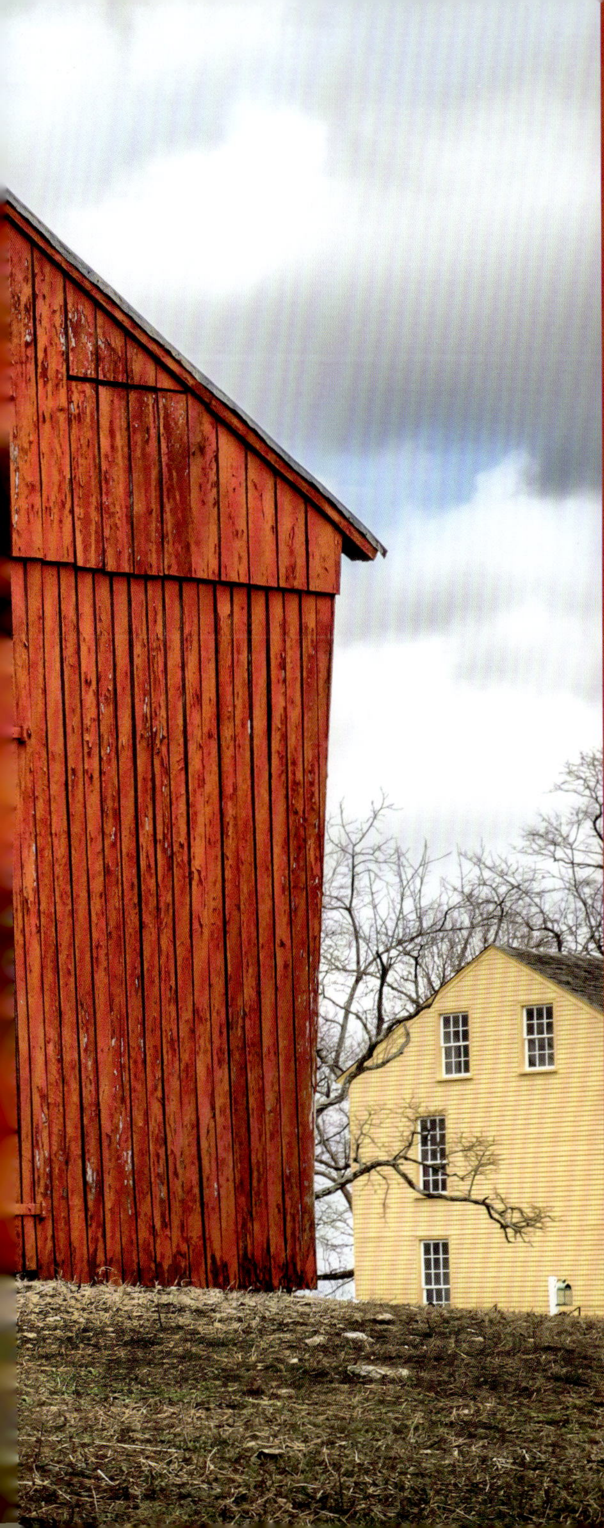

Contents

Foreword

MARY BERRY

I have lived and farmed in Henry County, Kentucky, all of my life—first by birth and then by choice. The country that I grew up in was populated by people who lived by farming. The agriculture in Henry County had diminished a good deal between my father Wendell Berry's childhood and my own, but it was still highly diversified and supported the communities and small towns in our county. People's pride was in their homes, and it showed. I grew up working on the same farms and in the same barns as my father. This is more than nostalgia. There was the seamless passing down of the essential knowledge of good work and good land use.

My children grew up in this culture. I remember thinking that the kind of work my children were doing would be recognizable to seven generations of the farming family that came before them. My youngest daughter is now in her late twenties, and that culture is almost completely gone, along with the farming people and the farmland that had survived until the late nineties. The farmers who are left are dependent on the toxic and erosive grain economy. There isn't a town in Henry County that is not dead or dying. As I look back now, the first sign of this decay was the barns. Farmers didn't need them anymore, or they couldn't afford to fix them.

Carol Peachee's excellent photographs of barns in Kentucky are a tribute to our agrarian past. Her work has given me a welcome opportunity to recall my memories of the barns I played in while growing up and then worked in later. The barns I knew and loved as a child were primarily stock barns and tobacco barns. The work we did there was quiet enough for talk. Much of our conversation was for the benefit of the young people just learning their place in the long lineage of agrarian people. I speak now of the culture of agriculture, which modern industrial agriculture has disrupted with the idea that machines can replace people and that particular places are the same as all places. And the result has been the displacement of millions of people from the countryside to the cities, political discord, and the now pervasive idea that technology will get us out of our dependence on the natural world and our need for each other.

This is an important book and one that I hope will serve not as a collection of ornaments and relics but as a testament to the accomplishments of a people who are firmly placed in a country they know well and who are not in constant economic trouble. This is what we need again—a healthy, prosperous rural landscape dotted with well-kept barns.

There has not been a time in Kentucky's past when some people and some land were not misused, and that must be kept in mind so that we can do better. When many of the barns pictured in this book were built, there was the possibility that a healthy farm culture might flourish. We have barely the sad remnants of that now. But we still have some of the barns and some of the people. My father asked this critical question in *The Unsettling of America* forty years ago, and in my work at the Berry Center, I continue his questioning: Are we or are we not going to take care of our land, our country? These photographs remind us of the capacity of the human imagination when it accepts the necessary limits of form and function. The result is beautiful.

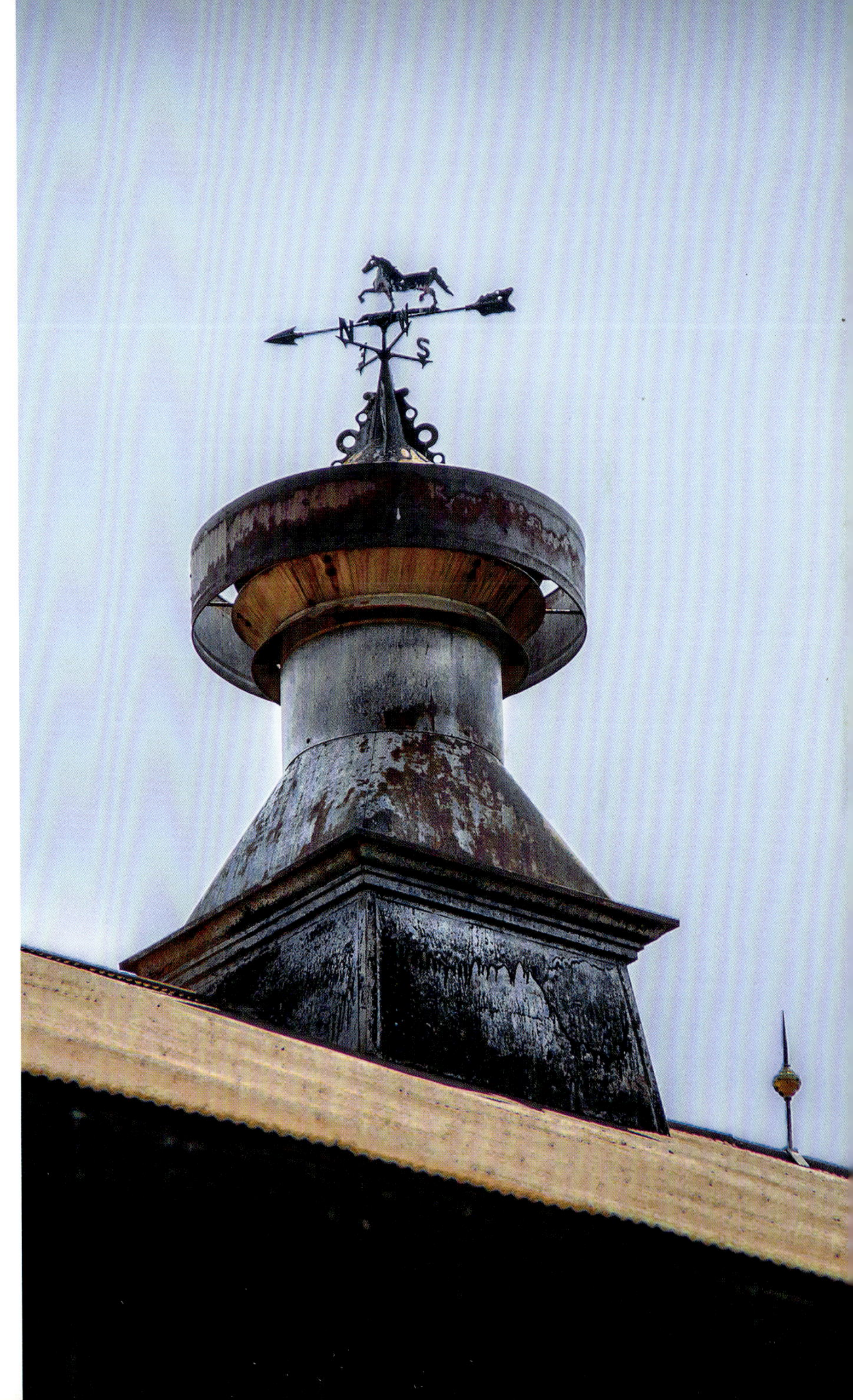

Acknowledgments

To undertake a project like photographing barns statewide in Kentucky is impossible without the expertise, knowledge, and connections of many, many people. I want to thank *all* the people who took time out of their days to drive me around, stand and talk with me, meet with me while I asked questions, and endure my ignorance. There are too many to name, and yet this network of people was incredibly generous and valuable to me.

There are some folks who gave an inordinate amount of attention to this project and to whom I am deeply indebted. First, thank you to Amy Sparrow Potts, who patiently improved my simplistic understanding of barns and helped me formulate what I wanted this project to be about. For their help during my research efforts, I want to offer my gratitude to Marty Perry and the Kentucky Heritage Council folks, who endured my rustling and disorganized searching through their records; to Bill McIntire, who came out of KHC retirement to meet with me and make some suggestions regarding sources and barns; and to Danae Peckler, the National Barn Alliance past president, who was helpful in directing my attention and providing me with an important network.

Many thanks to Tom Eblen at the *Lexington Herald*, who was invaluable in getting the word out to barn owners by writing an article on my search for barns. For her help during the actual photographing, thank you to my friend Sarah Tate, who drove me through old stomping grounds of hers to locate barns; to Dan Thomas in Trigg County, near Cadiz, who walked me through his dark fired tobacco barn, regaling me with great stories; Jerry Zwahlen, who drove me around and provided me with amazing historic details of the Swiss settlement of Ottenheim; and Michael Enzweiler and Whitey Heeb, who unselfishly introduced me to community members and drove me to significant barns in their community of Camp Springs.

I also want to offer enormous thanks to Ag extension agent Matt Futrell, who took an entire day out of his workweek to drive me around Christian and Caldwell Counties, looking for traditional wooden dark fired tobacco barns that were still active. And thanks to Keenan Bishop, another of Kentucky's fine Ag extension agents, who also spent the better part of a day driving me around to significant barns in Franklin and Owen Counties. Thank you also to Orloff Miller, who drove me through northern Kentucky counties and introduced me to vanished towns and historic barns while giving me an archaeologist's perspective. I am deeply in debt to Eddie Gilkison, my go-to guy for Clark, Montgomery, Bourbon, and Fayette Counties; he is a fount of knowledge about farming practices, farming history, and great barns. I

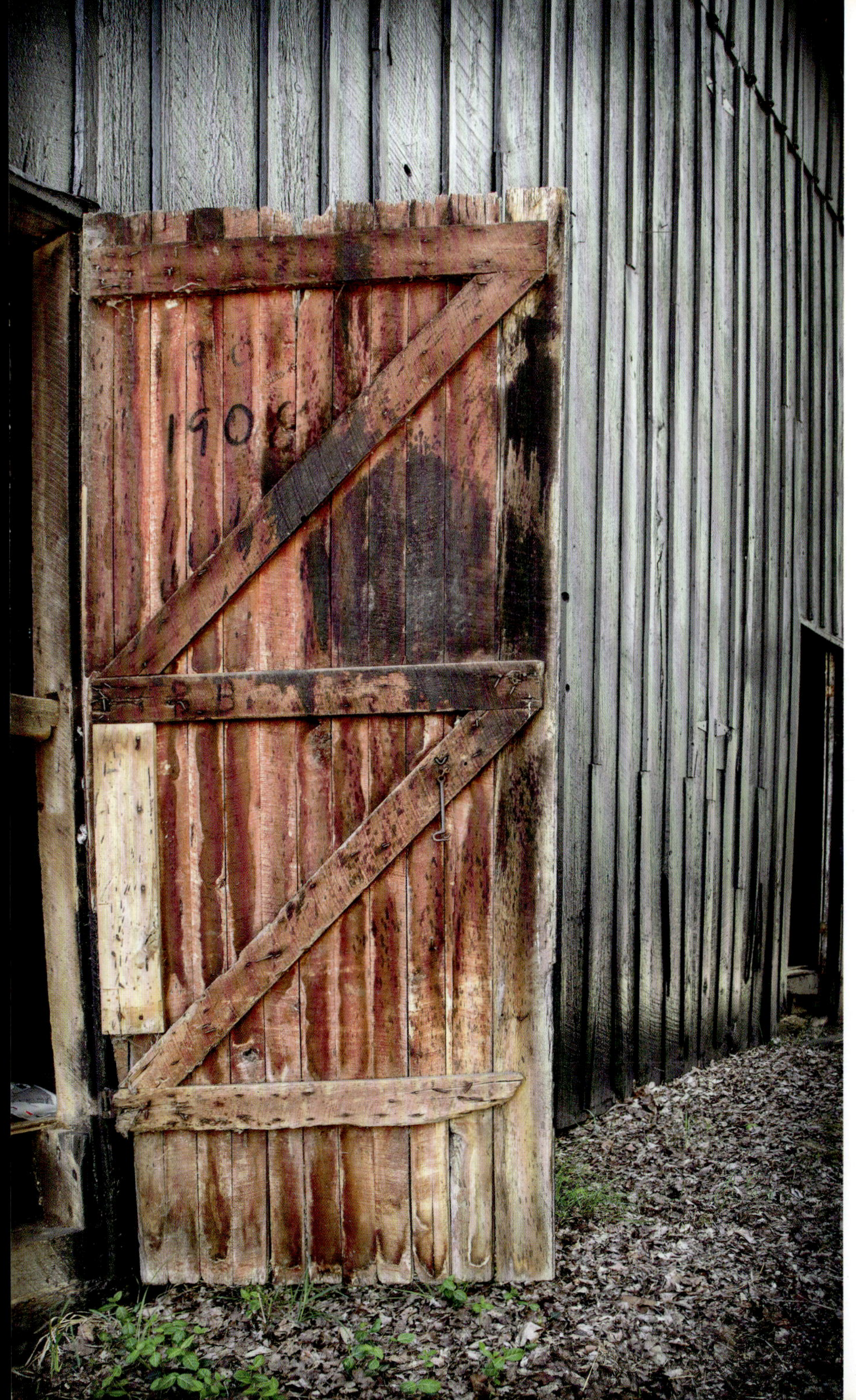

also owe heartfelt thanks to Janie-Rice Brother for her organizational skills, driving ability, historic and architectural knowledge, connections, vast network of cousins, and especially for her willingness to be up to her neck in this.

Finally, thank you to Ashley Runyon, my editor, for her patience and steady compass. And, of course, my deepest gratitude goes to Monica for all that she did to make my life easier, thereby making this project possible.

KENTUCKY
BARNS

Introduction

JANIE-RICE BROTHER

Barns strike a chord with many Kentuckians—and that's not surprising, as Kentucky ranks fourth in the nation in the number of barns built before 1960. Barns are highly recognizable buildings on the Kentucky landscape, and even for those far removed from an agricultural background, they serve as a reminder of rural spaces and our historic farming traditions.

Some of my earliest memories are of barns. The images that play in my head are from the vantage point of someone about knee high, so in addition to visuals of weathered gray wood, dust motes, and old farm equipment, in my memories I also see a lot of horse hooves and my father's legs.

The barn closest to our house, a stock barn used for horses and tobacco, was an escape from the stultifying heat of summer and a blessed break from the wind in the winter. There were kittens in the bales of hay and rusty treasures in every corner. There was usually one tractor parked in the central aisle, and there was always the manure spreader, the receptacle for stall-mucking activities. But the visuals also share space with the smells—dust, wood, tobacco, the sweaty flank of a horse, the sweet smell of hay.

Later, after I moved away from Kentucky for the first time, I realized how distinctive our landscape was, with its stark black barns perched on hillsides and in broad bottoms. Flying in over Lexington, I thrilled at the sight of horse barns, cupolas perched on the roof, flanked by board fences.

Most of the historic barns I've explored across the state were built between 1880 and 1960. After the mid-twentieth century, the practice of standardization in barn building increased. Just like houses, barns evolved as materials became more widely available and as farming practices changed. Log cribs gave way to heavy timber or transitional framed barns, which in turn began to be constructed with nailed and sawn milled lumber in the late nineteenth century.

The catchall barn is the most common historic barn still standing in most places, and Kentucky boasts thousands of frame barns, clad in vertical wood boards on the exterior. There are some brick and stone barns, of course, but those tend to be the exception in the historic rural landscape.

Specialized barns didn't really start showing up in large numbers in Kentucky until after the Civil War; prior to that, most barns were multipurpose buildings that housed livestock, crops, and equipment.

The plans (and functions) of historic barns are just as diverse (and sometimes puzzling) as old houses. Peek past the exterior of an old barn, and you may be surprised by what you find inside. You might even find an old log cabin hidden in the structure of your barn!

Sometimes that might be true—but more often than not, that log structure inside a frame barn is the most common of early Kentucky multipurpose barns. Log cribs stored ear corn or hay, and projecting off the side of the log walls were frame overhangs or frame sheds, which provided shelter for livestock. Over time, those frame extensions were extended to completely hide the original structure.

Barns might not have distinct architectural styles (although I've seen many fancy historic barns, some even sporting gingerbread trim), but they certainly have distinct types, construction methods, and functions. The English barn, for example, can be identified by its plan, which is visible from the exterior by the entry doors on the long side of the barn. Typically this type of barn will have a central aisle, open to the roof, which allows a wagon to be driven into the barn. On either side of the central aisle are side aisles, usually divided into stalls for livestock, with lofts above for storing hay and grain.

A bank barn is built into the side of a hill, with access from two levels. The lower level most often houses livestock, since it provides a more constant temperature than the first level. Bank barns can have an interior plan similar to a transverse frame barn or a plan like the English barn, with entrance doors along the eaves.

Stock barns in Kentucky can often be identified by their lofts, which stored grain for the animals stabled underneath. Kentucky was a major producer of mules before and immediately after the Civil War—the mule industry, along with increased numbers of sheep and cattle, led to a profusion of new stock barn construction. Often stock barns will also sport a hay hood on one gable end, a projection that protected the hayfork from the weather and allowed access into the upper reaches of the loft. Stalls, corncribs, and tack rooms are typically built into stock barns.

Dairy barns, often topped with large gambrel roofs, seem to me to be the queens of stock barns. Sprawling footprints accommodate milking parlors, lofts, and stalls, each division of space reflecting the era in which the barn was built.

Burley tobacco barns, once ubiquitous across Kentucky, tend to be light and airy—the better for drying the hanging tobacco. These barns are longer than they are wide, with openings (usually hinged or sliding doors) on the gable ends. The measurements of a tobacco barn are designed around the tobacco sticks, which are typically around four to four and a half feet long. The long walls are built in bents, usually around ten feet wide—these are the partitions between the vertical posts that support the framing of the barn.

Barns built for fire-cured tobacco are very different from burley tobacco barns. Dark tobacco, as it is known, is grown in far Western Kentucky, and though the growing process is roughly the same as burley tobacco, fire-cured tobacco barns are usually as twice as tall as they are wide. Unlike air-cured burley tobacco, which relies on a barn's air circulation for drying, dark tobacco is cured by the smoke from a smoldering fire laid out along the barn floor.

Although utilitarian, barns are not immune from trends or fads. By the late 1870s, round and octagonal barns were being built across the country, and there are several examples in Kentucky. The octagon made good use of building materials and interior wall space—in other words, if you were a farmer looking to save some money, you got more bang for your buck with a round or octagonal barn. George Washington built a sixteen-sided threshing barn at Mount Vernon in 1793, and if it's good enough for the father of our country . . .

These barn types evolved as a result of available building materials (timber was abundant in Kentucky), the needs of the farmer, and the cultural traditions of the immigrants who transplanted themselves and their farming heritage to Kentucky. For example, the heavy stone foundations of the barns in Camp Springs, Kentucky, can be traced to the German settlers who built the massive structures—and the plans of these barns are very similar to those found in places like Pennsylvania.

There is a power in these rural landscapes—not just in the beauty of the barns or the majesty of the land surrounding them, but in the ability of these symbols of Kentucky to shape our memories. The photographs in this book capture barns that once hummed with activity and were the center of farm life. Some barns are still very much in use—sheep mill about in the barn lot, and horses munch hay in spacious stalls better outfitted than many folks'

bedrooms. Other barns, though, lean into the earth, their timbers and piers sinking each year as the wind teases and slams creaking doors. Barns as billboards, hawking chewing tobacco and tourist attractions, draw our eyes from the road even today, just as the sign painters intended. Oversized quilt squares hang on historic barns across the state, bringing new life to many almost-forgotten barns.

Concrete and asphalt sneak up on some old barns, and before you know it, a field transforms into a parking lot. And I feel that something vital to our sense of being— our sense of ourselves as Kentuckians—has been lost.

Historic houses tend to convey social messages about their occupants—or at least how the residents of that dwelling want to be perceived. I've always thought that to really know how a farm works and to get a sense of the rhythm of that family's daily life—even when looking back 150 years—you need to look at their barns.

The preservation of our barns does not necessarily guarantee the preservation of our farmland or vice versa. The tobacco buyout spelled the death of thousands of burley tobacco barns across Kentucky. But memories are harder to destroy, and I firmly believe in the power of stories across generations. The gorgeous photographs within these pages contain not only stunning examples of light, shadow, and construction techniques but also layers and layers of memories. I hope they inspire us all to hold on to our stories. Perhaps in the process we can take another look at the barns we know and ponder what role they can play in the future.

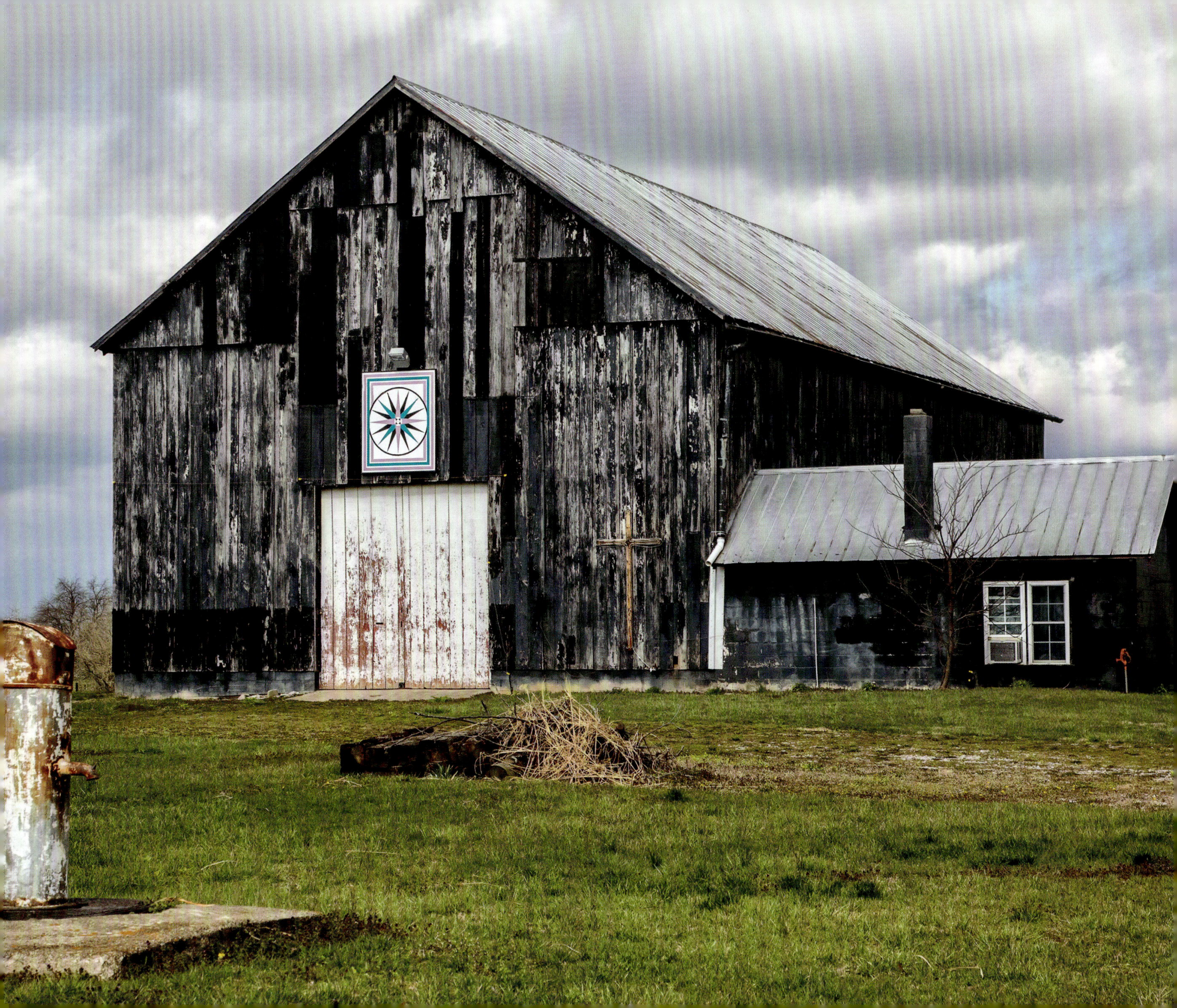

Photographing Kentucky's Agricultural Heritage

As a photographer, what interests me is heritage. I'm interested in practices, characteristics, traditions, qualities, and features of life that an industry or a community has inherited. I want to explore not only what comes from the past to us but also how it has evolved in contemporary life. What have we learned from our past, and what is part of our individual and collective memories? What elements have we inherited that make us who we are and contribute to how we identify ourselves? Is there a way that we can use that understanding to make our lives better today and in the future? To understand ourselves and make those sorts of decisions, I believe we have to study our heritage, and to do that, we need to preserve and conserve our inheritances.

When I decided to explore the agricultural heritage of Kentucky, photographing barns was an obvious place to begin. The barn is an iconic symbol for farms and agriculture. Photographing these historic and heritage buildings, I reasoned, would open up agricultural practices and help me understand the how and why of Kentucky's farming traditions. The agricultural collective memory is strong in this state. I thought exploring the barn would be a key to opening it. What I didn't grasp at the outset is how much of that heritage is on the brink of being lost. As farming practices have shifted from family farms to larger industrial complexes, and as shortsighted policies failed to protect small farming operations, important knowledge that has been handed down for centuries—knowledge necessary to have healthy and truly sustainable food sources—is disappearing.

It is not surprising, then, that the barn, the iconic symbol of farming, is also disappearing. Over the two years it took to create the images in this book, I came to understand that I was photographing not only many barns that would be gone in twenty years but also the vestiges of a farming tradition that is in danger of disappearing. In its place is a nostalgic and idealistic collective memory rather than the true Kentucky agricultural heritage, which is full of history and could play a constructive role in contemporary culture.

I approached this project understanding that there were many ways to choose which barns to photograph and how to present them. I could choose barns for inclusion according to architecture, function, cultural tradition, historic value, notoriety, accessibility, locality, or aesthetics. I sought to include as diverse a variety of barns as I could, and so I used all of those criteria. I located many of the barns in this book because they were surveyed and in the records of the Kentucky Heritage Council as historic. Other barns I chose because they were significant to the agricultural identity of Kentucky. Whenever I could, I included barns with architecture brought to Kentucky by cultures that immigrated here from other countries and other traditions. Some I chose because they were not only historic but open to the public, while others I would never have found if an agricultural agent with intimate knowledge of the area hadn't shown me. The *Lexington Herald* ran an article asking folks to help identify important barns, so their owners brought some of the barns to my attention. Always there were people willing to tell me where a barn might be or drive me to a barn that they knew of. In some instances, such as in far Western Kentucky and northeastern Kentucky, I chose a route and photographed barns from the road; I approached the homes associated with many of these barns and spoke with owners whenever someone would answer the door. Through research; begging; knocking on doors; riding around with agricultural agents, archaeologists, and farmers who knew the area; and just plain luck, I was able to photograph barns in 45 percent of the 120 counties that make up Kentucky.

Included in these images are stock barns for sheep, pigs, cow, dairy, mules, and horses; burley tobacco and dark tobacco firing barns; hay barns; and general-purpose barns. There are barns from Dutch, German, and Swiss colonies that settled here as well as those architectural forms influenced by the English, Irish, and migrations from the northeastern states. I've included barns of brick, stone, and hand-hewn timber, as well as peg barns and milled timber barns. There are fancy stables and log barns included here; there are abandoned barns, active barns, restored barns, and preserved barns. I've included several homesteads with a collection of barns along with solitary barns to show how the barn's place figured into the farm. There are also barns on the Quilt Barn trail, barns painted by the Tobacco Heritage Trail, and Mail Pouch barns. I photographed old equipment, the insides of the barns, and the details of how they were constructed when I could. By the time you get to the end of the book, you should have a pretty good idea of not only the amazing variety of barns but also the long history of Kentucky and the significance of agriculture to its people.

The images in this book represent only a fraction of those actually taken and only a small selection of barns that could have been photographed. I regret that I couldn't have gotten to every county and that I could not have included more barns from each one. The truth is that photographing barns is a massive project of great depth and breadth, and there is a certain urgency to documenting the ones that are still standing; so many will fall, as they are no longer used. I encourage you, if you know a barn, to go out and photograph it. Meanwhile, enjoy the ones I have presented here. Each and every one is an architectural heritage jewel.

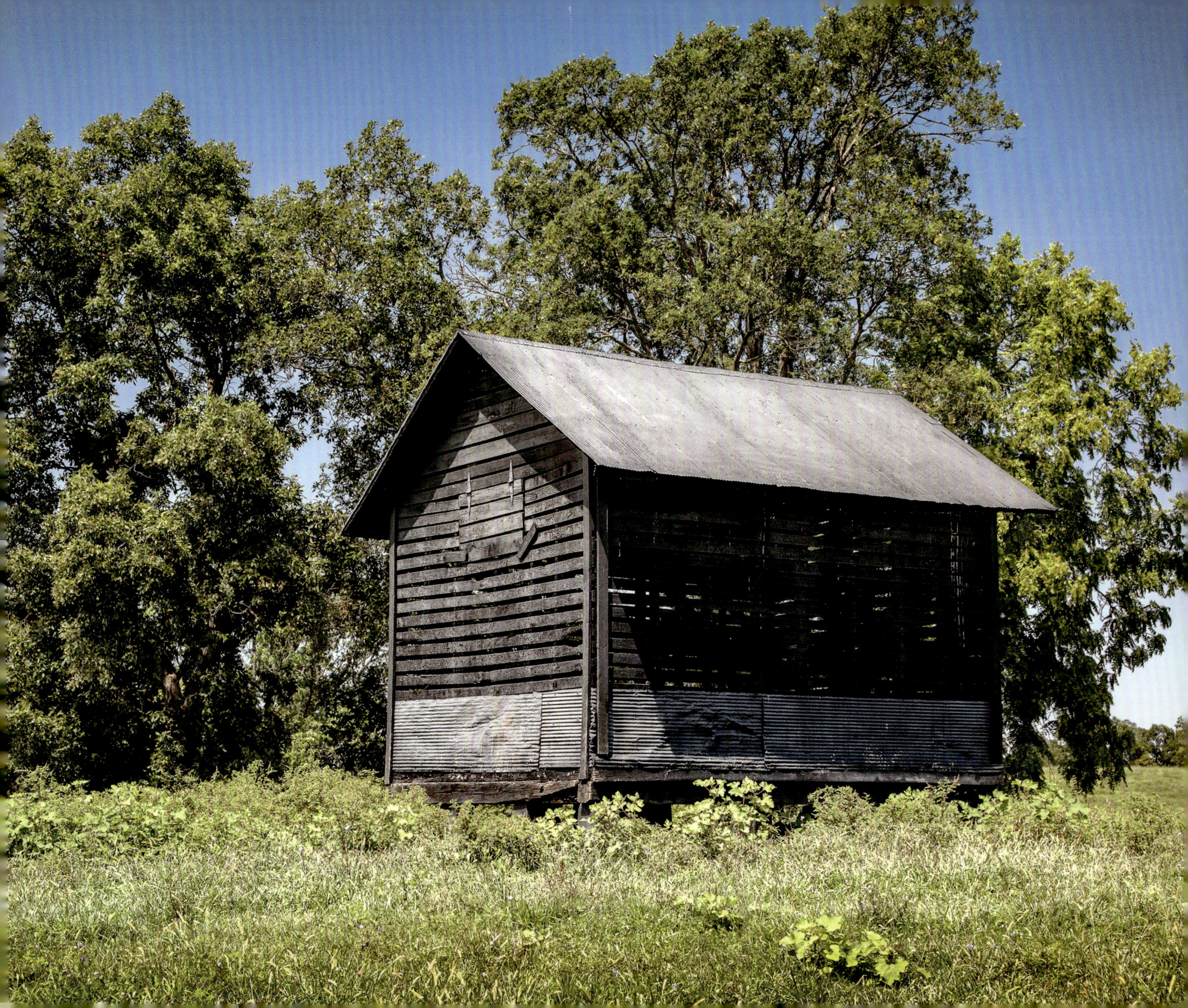

PHOTOGRAPHS

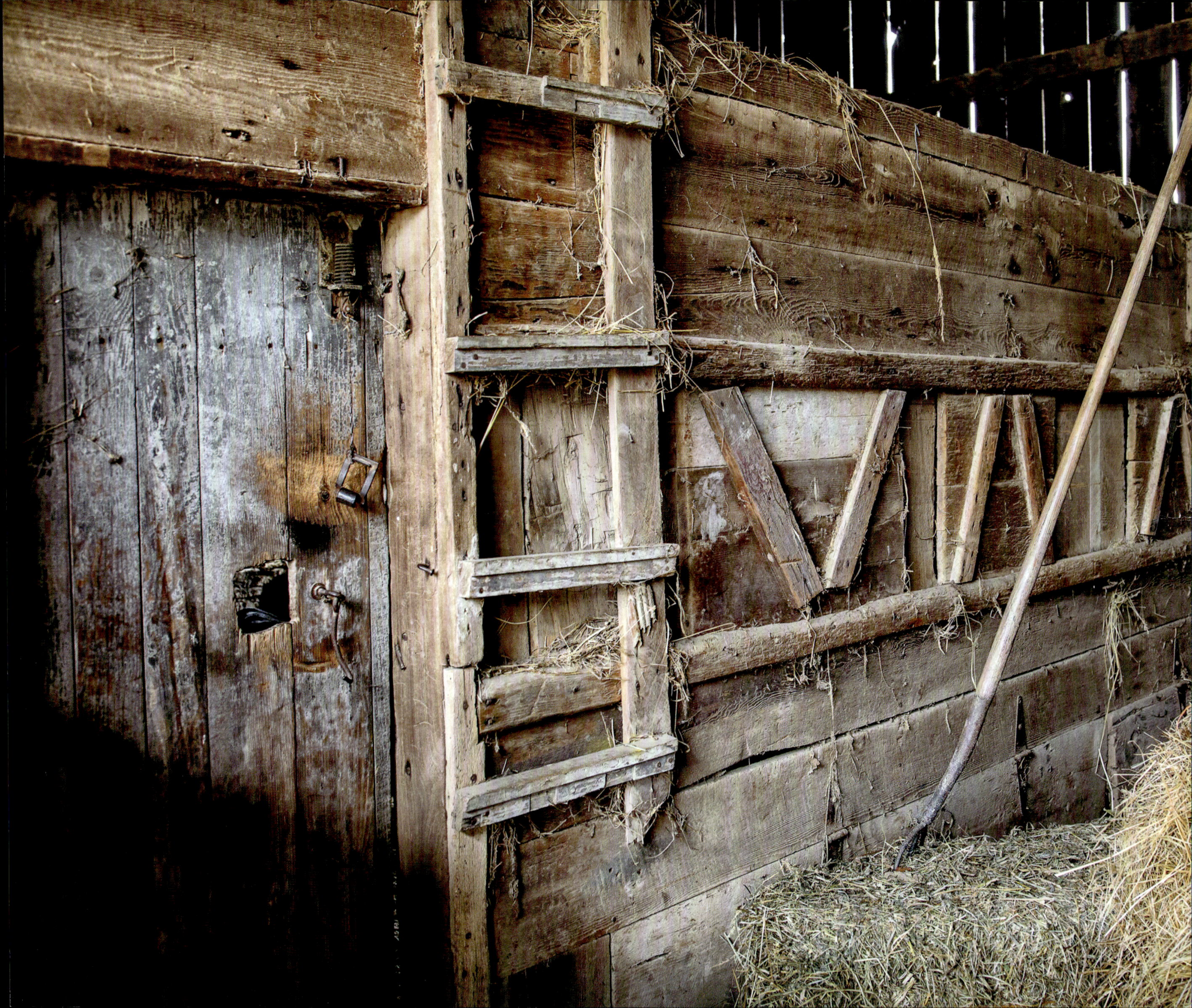

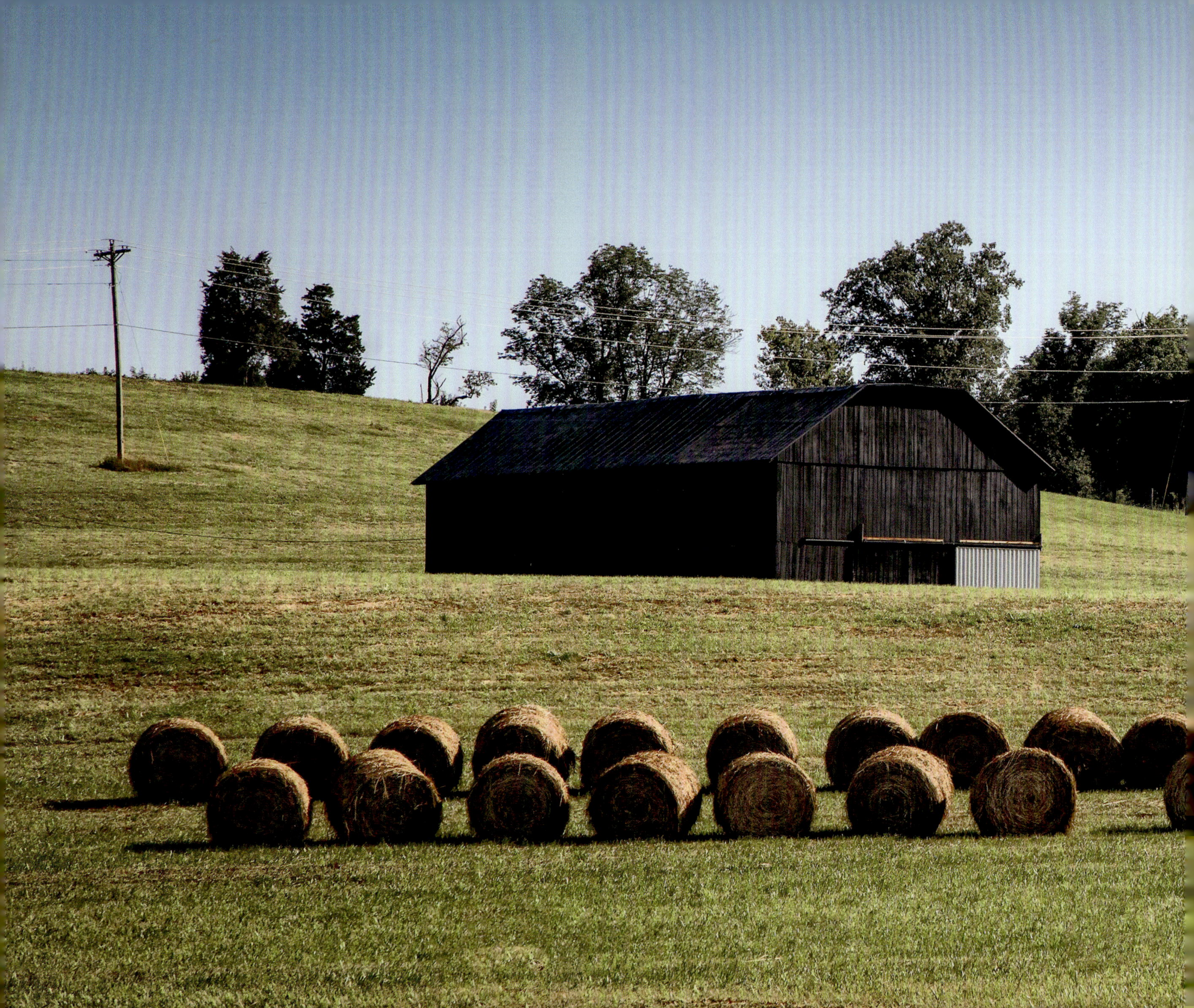

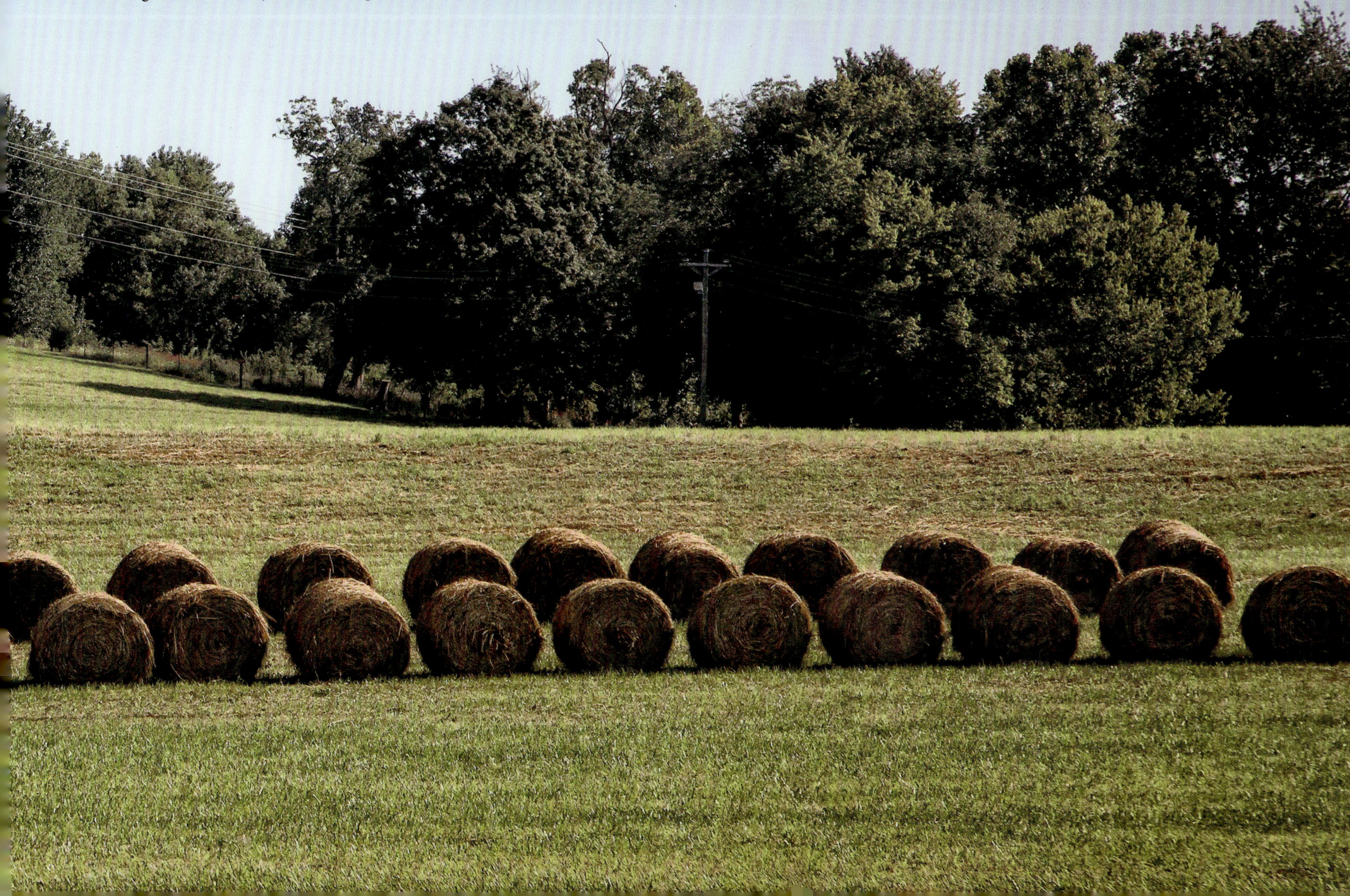
Madison County.

There are at least three other barns in the area with construction similar to this barn. All are built on sections of a Revolutionary War land grant that dates back to the eighteenth century. The Bowling barn, shown here, is on one of the parcels.

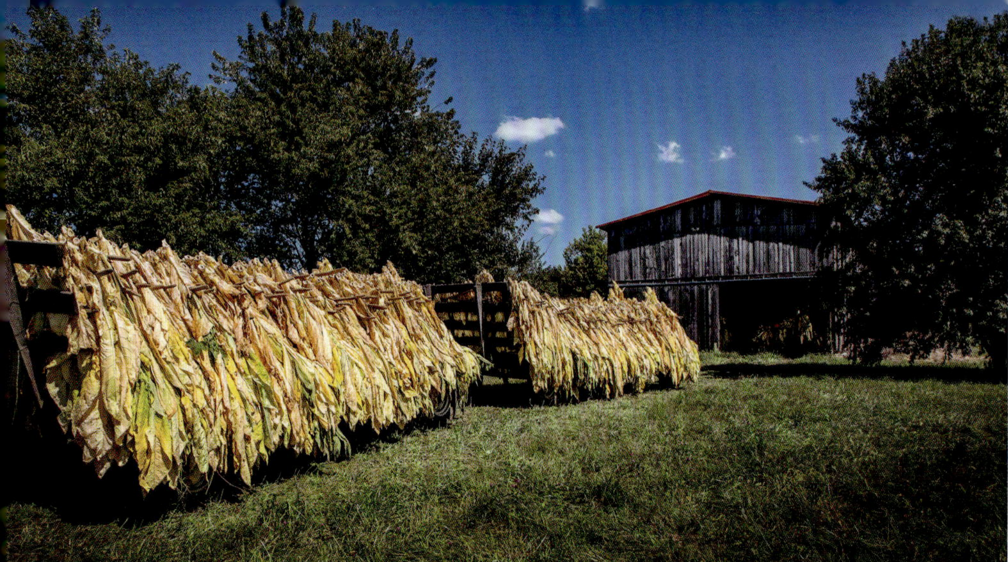

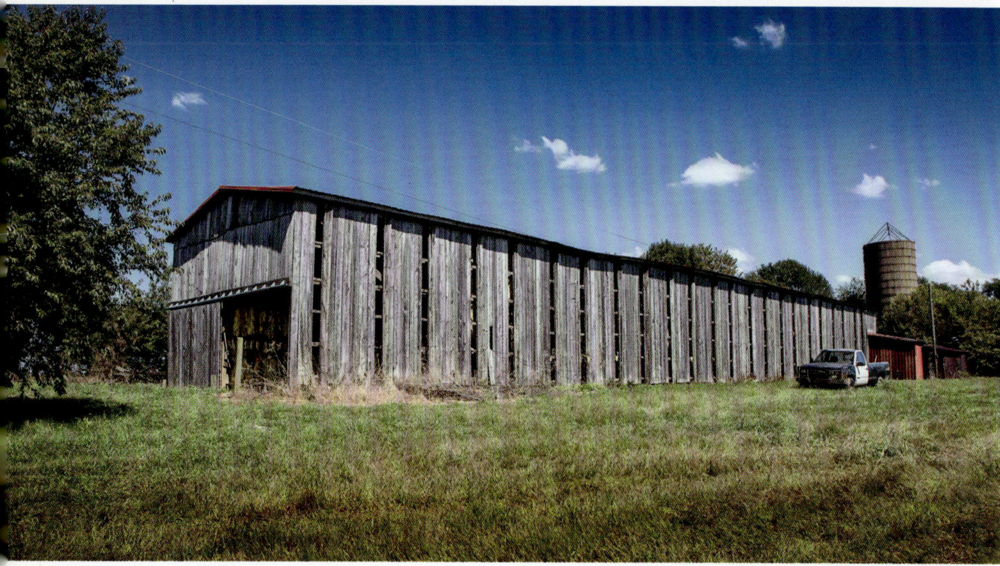

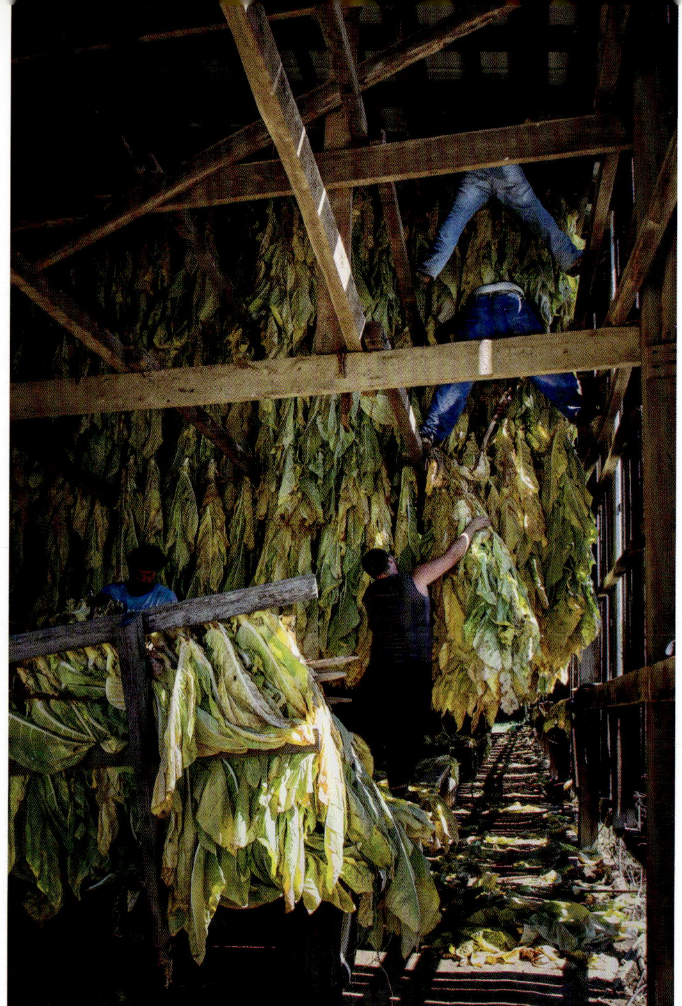

Woodford County.

Burley tobacco is cut in the fields, loaded onto carts, and hung from rafters to cure. Most tobacco acreage has been rented out since the 2004 tobacco buyout.

Woodford County.

Along Leestown Road there are several still-active burley tobacco farms with tobacco hanging. The quilt barn in this image is used for tobacco in the late summer and early fall.

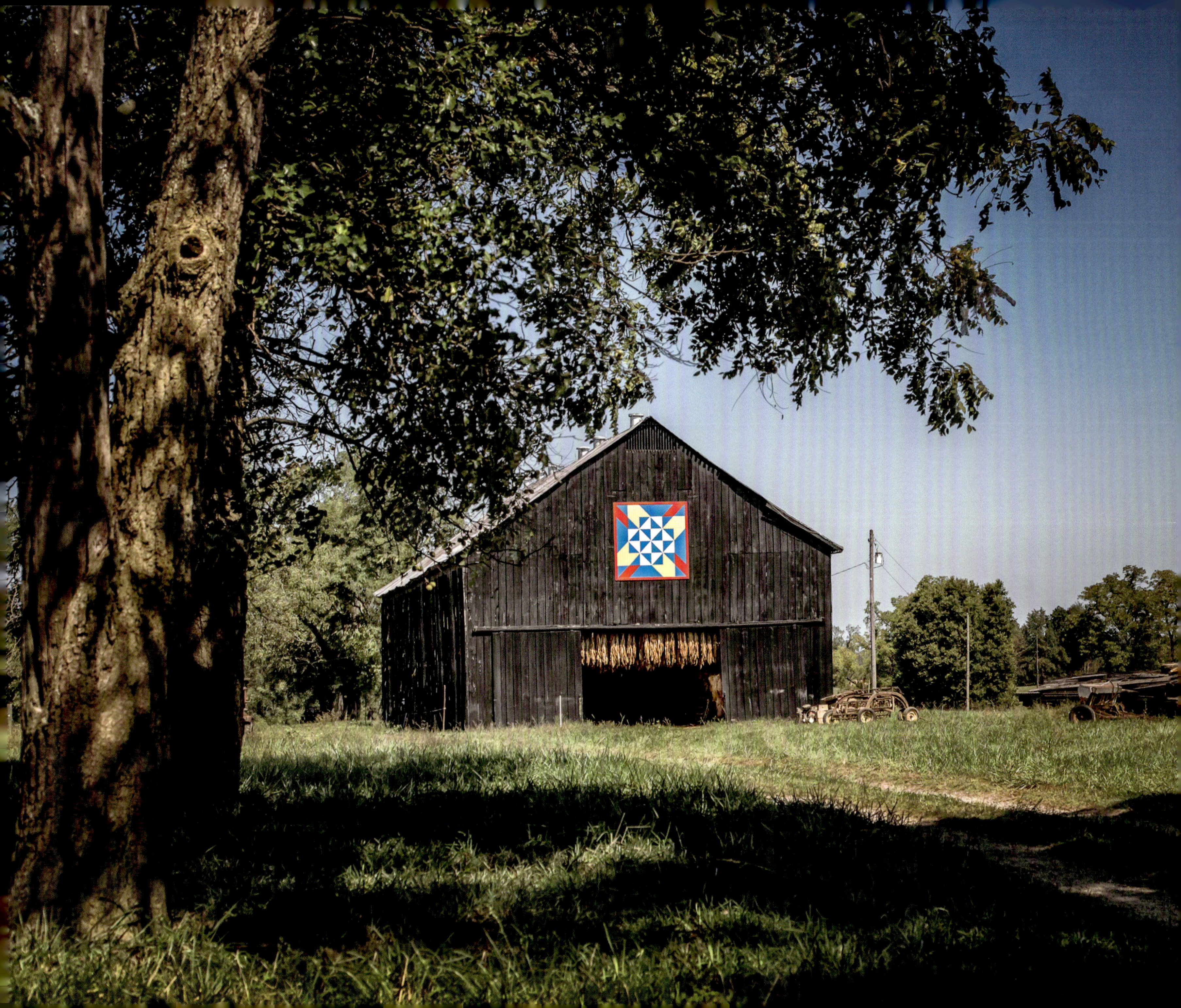

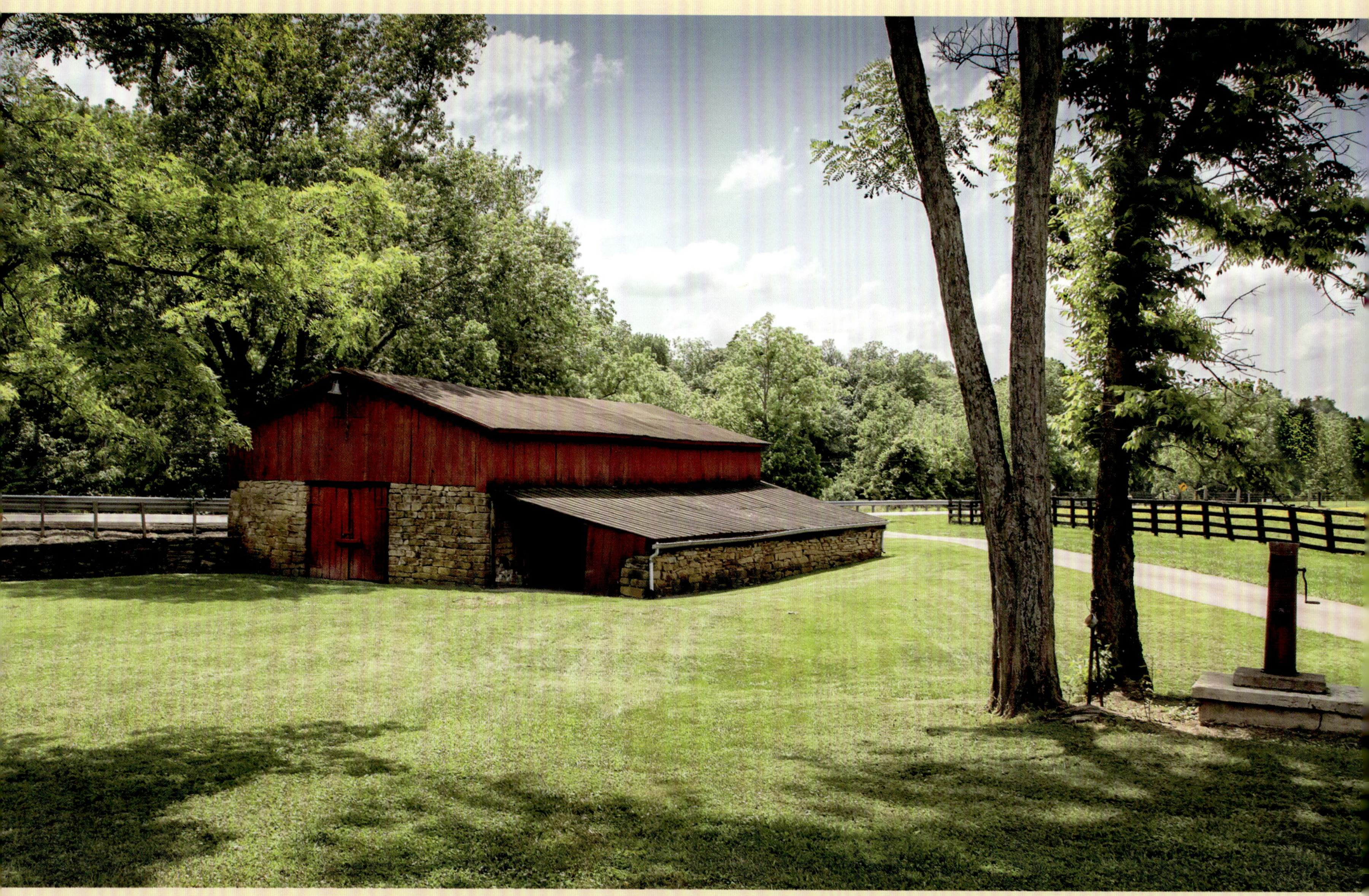

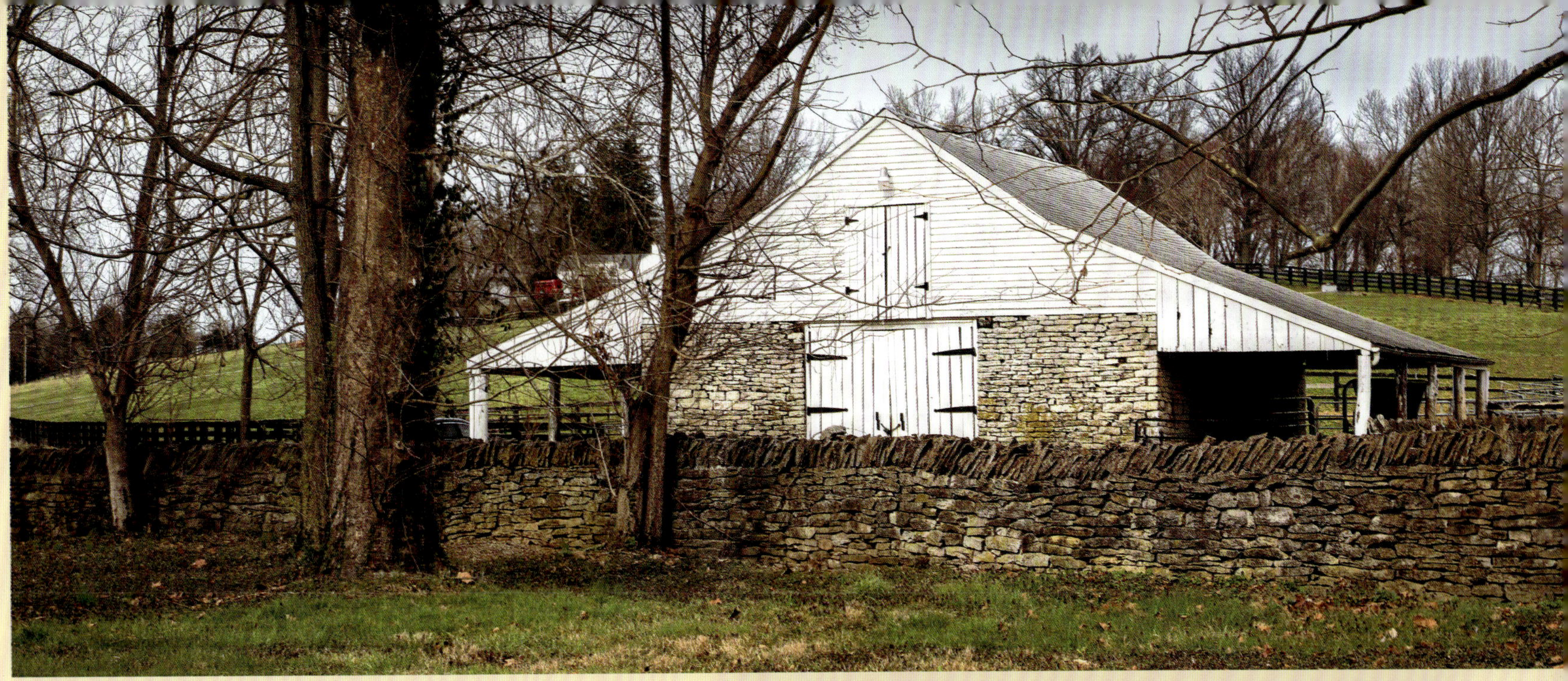

Nicholas County.

The historic home of Kentucky governor Thomas Metcalfe, Forest Retreat was built in 1795, along with this stone stable. Notice the large timbers, with the bark still attached, supporting the wide loft boards.

FACING PAGE

Garrard County.

This stone foundation barn, built around 1825 on the Lucien Perkins property and farm, has been recently restored.

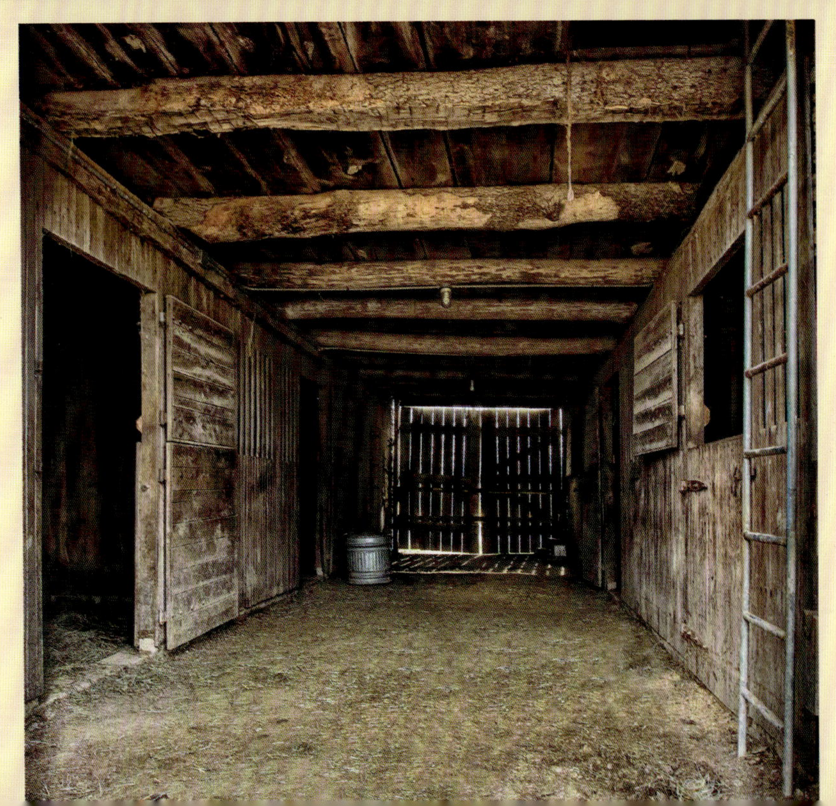

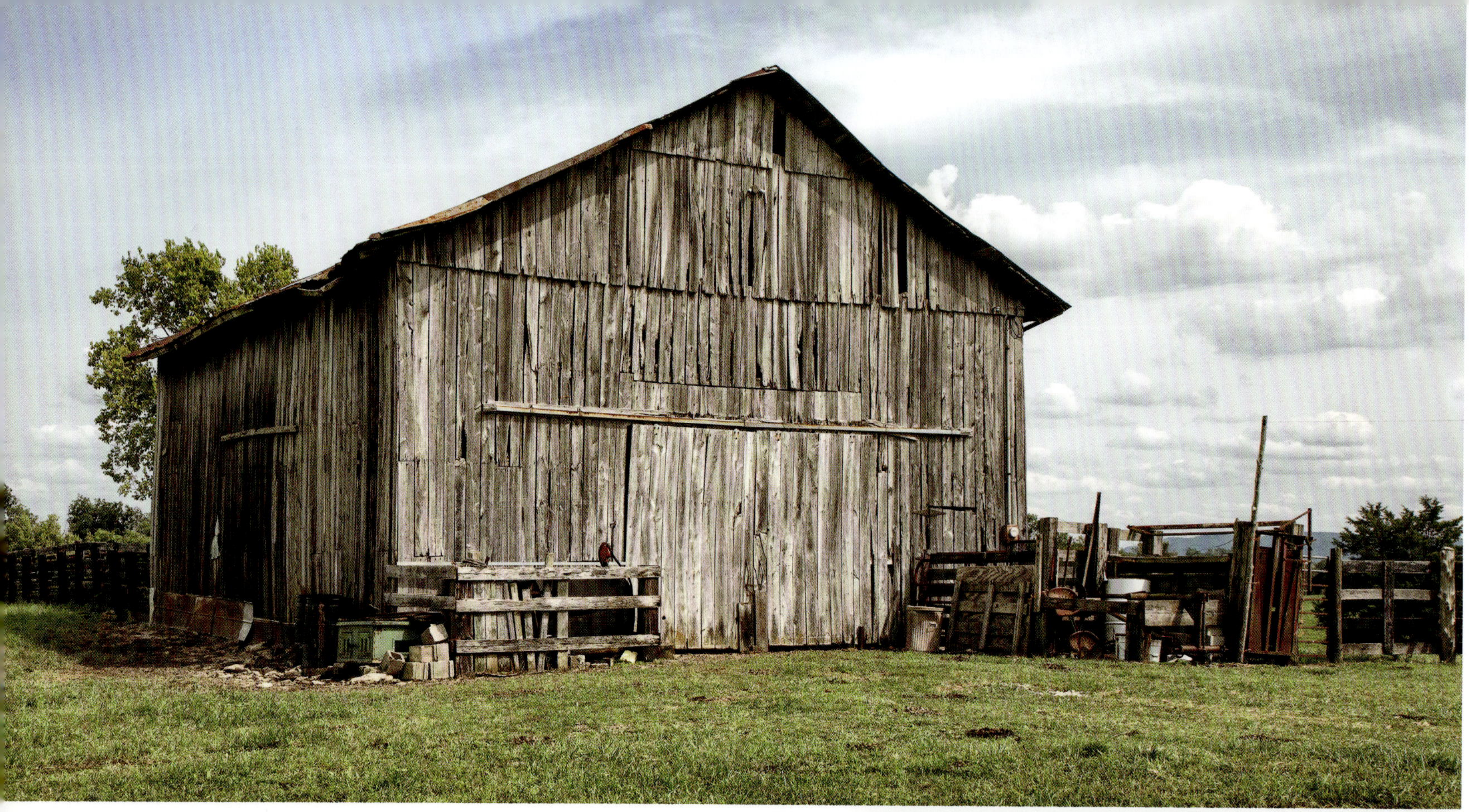

Madison County.

Dan Jenkins's ancestors established one of the earliest African American farms in Madison County on a parcel of the McWilliams Revolutionary War land grant. His uncle and father built this barn over eighty-one years ago. The chain was hand forged.

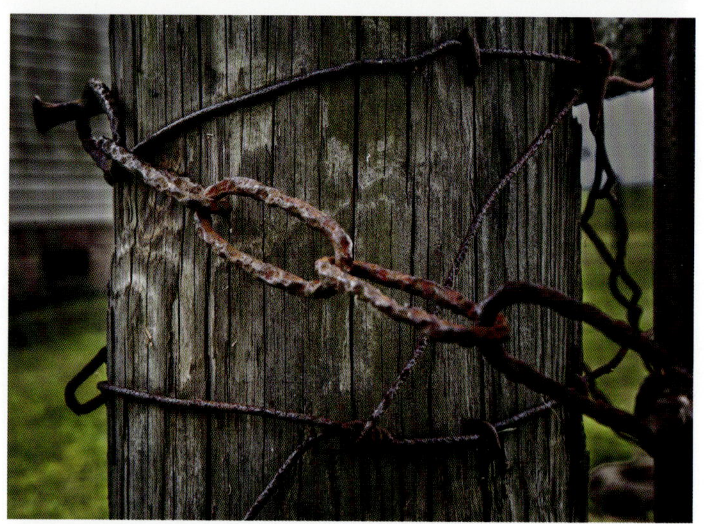

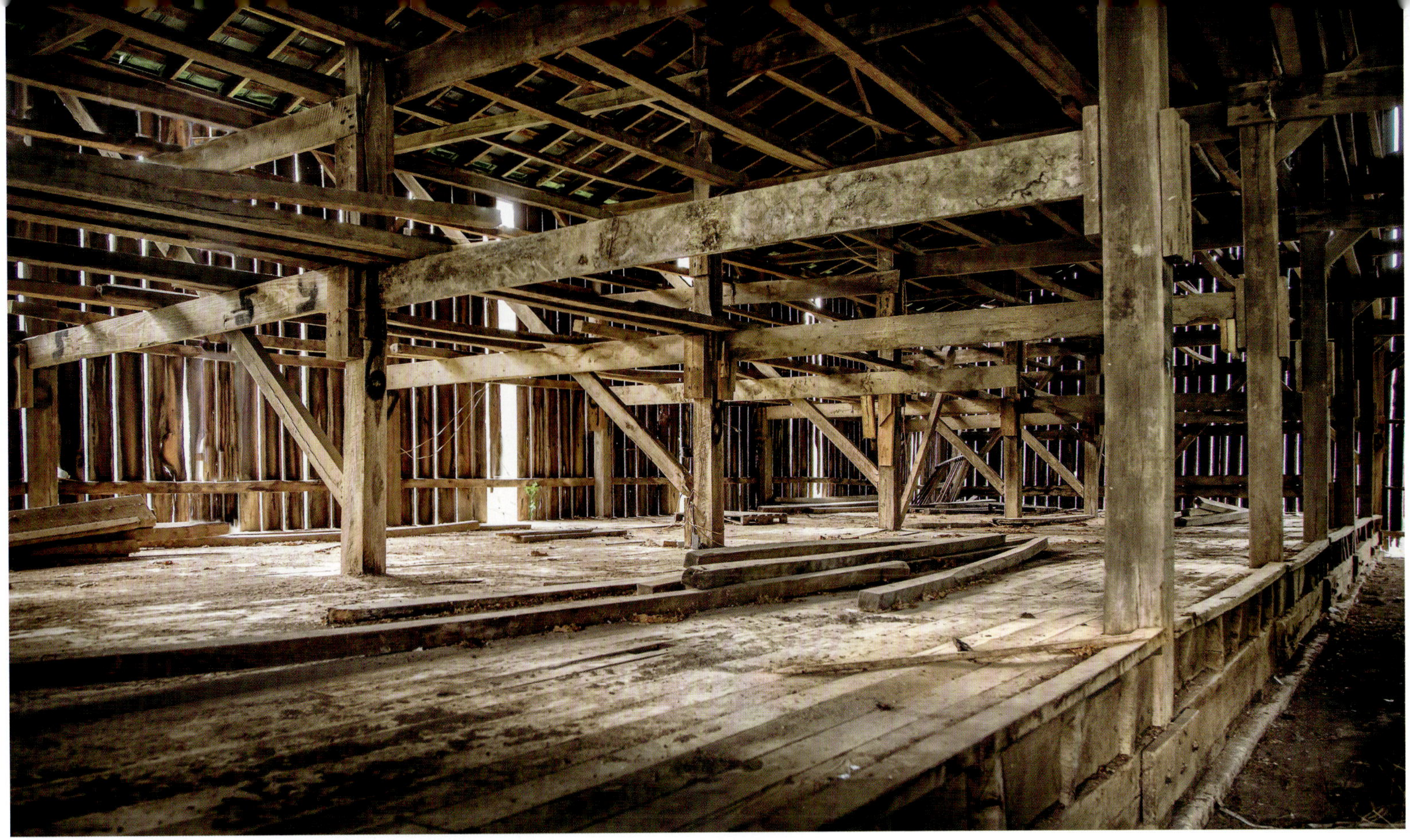

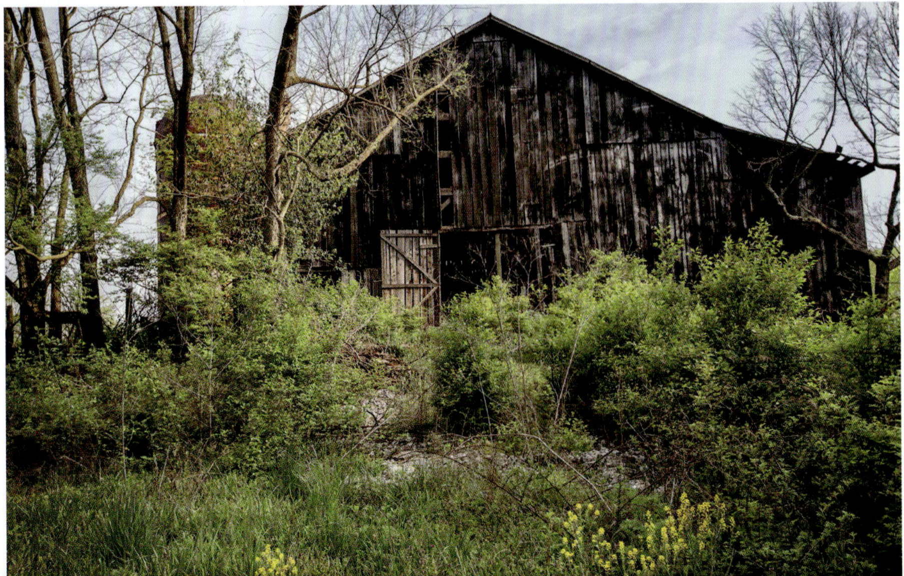

The timbers used to build this barn were once the grandstands for a mile-long dirt racetrack built between 1828 and 1834 to host Thoroughbred flat racing events. Today the racetrack area is covered in asphalt and is a Lexington, Kentucky, street known as Race Street. The barn has the unusual feature of a raised floor on one side of the main aisle.

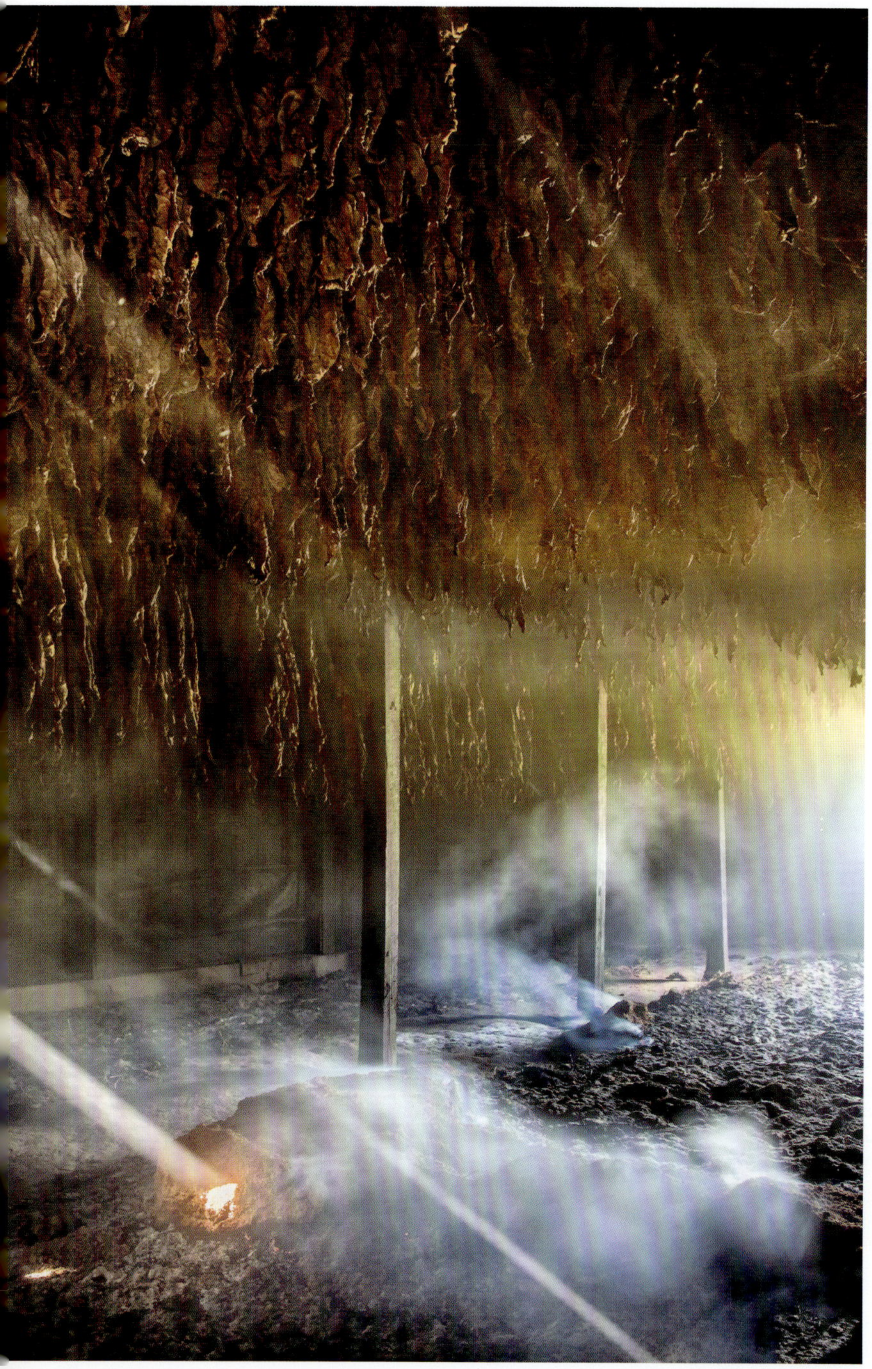

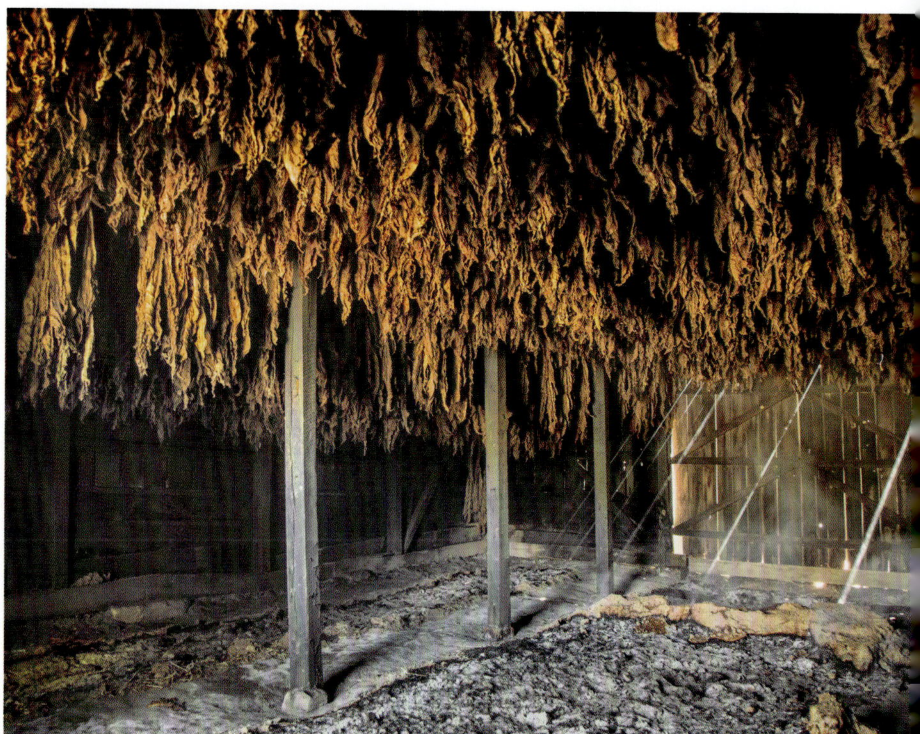

Christian County.

Since most dark tobacco firing barns are now steel, David Jones's farm boasts one of the few remaining wooden firing barns in use. Inside the sealed barn, three rows of hardwood chips smolder for five to six weeks, smoking the leaves before they are stripped from the stalks. The barn has been in Jones's family for more than sixty years.

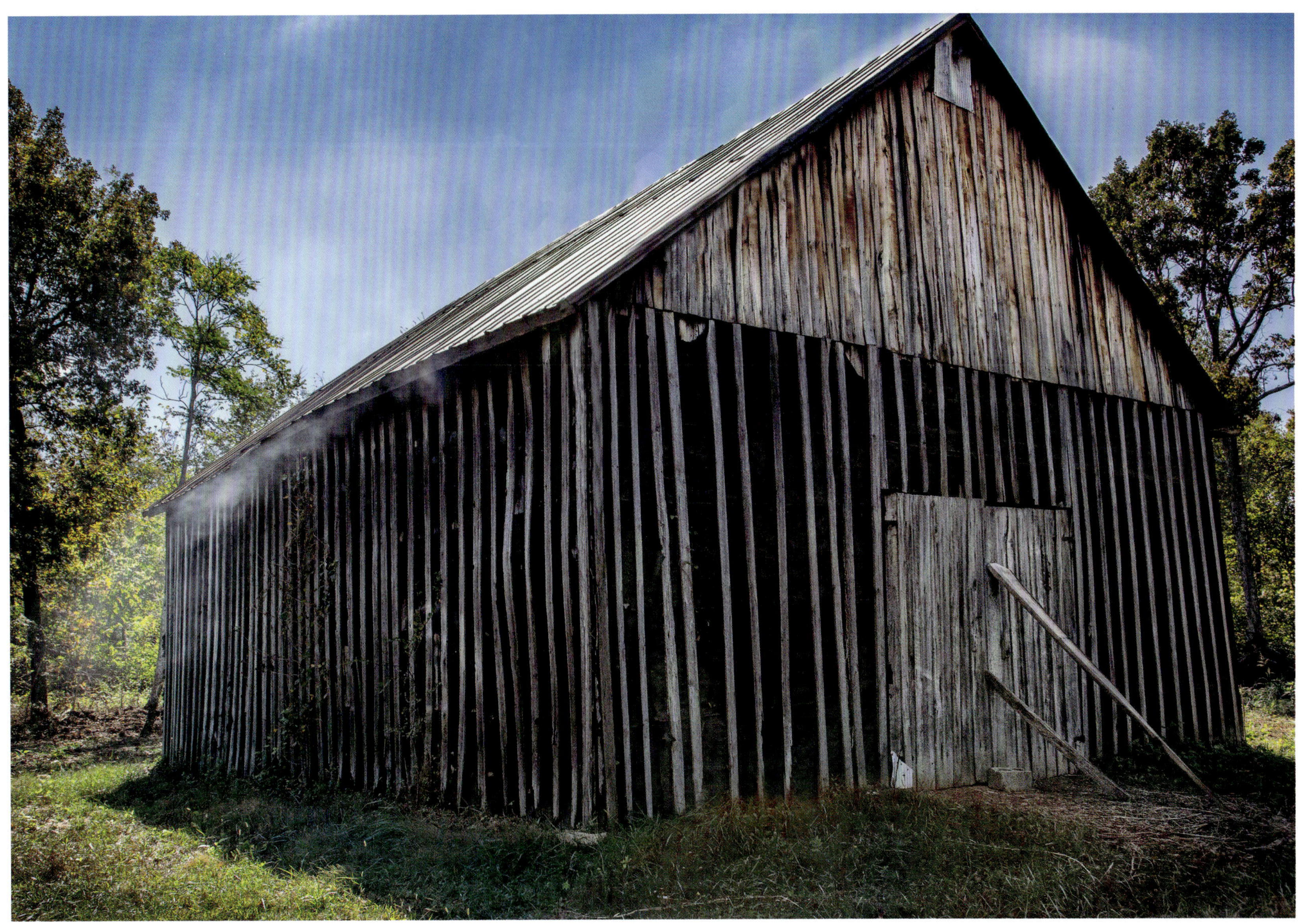

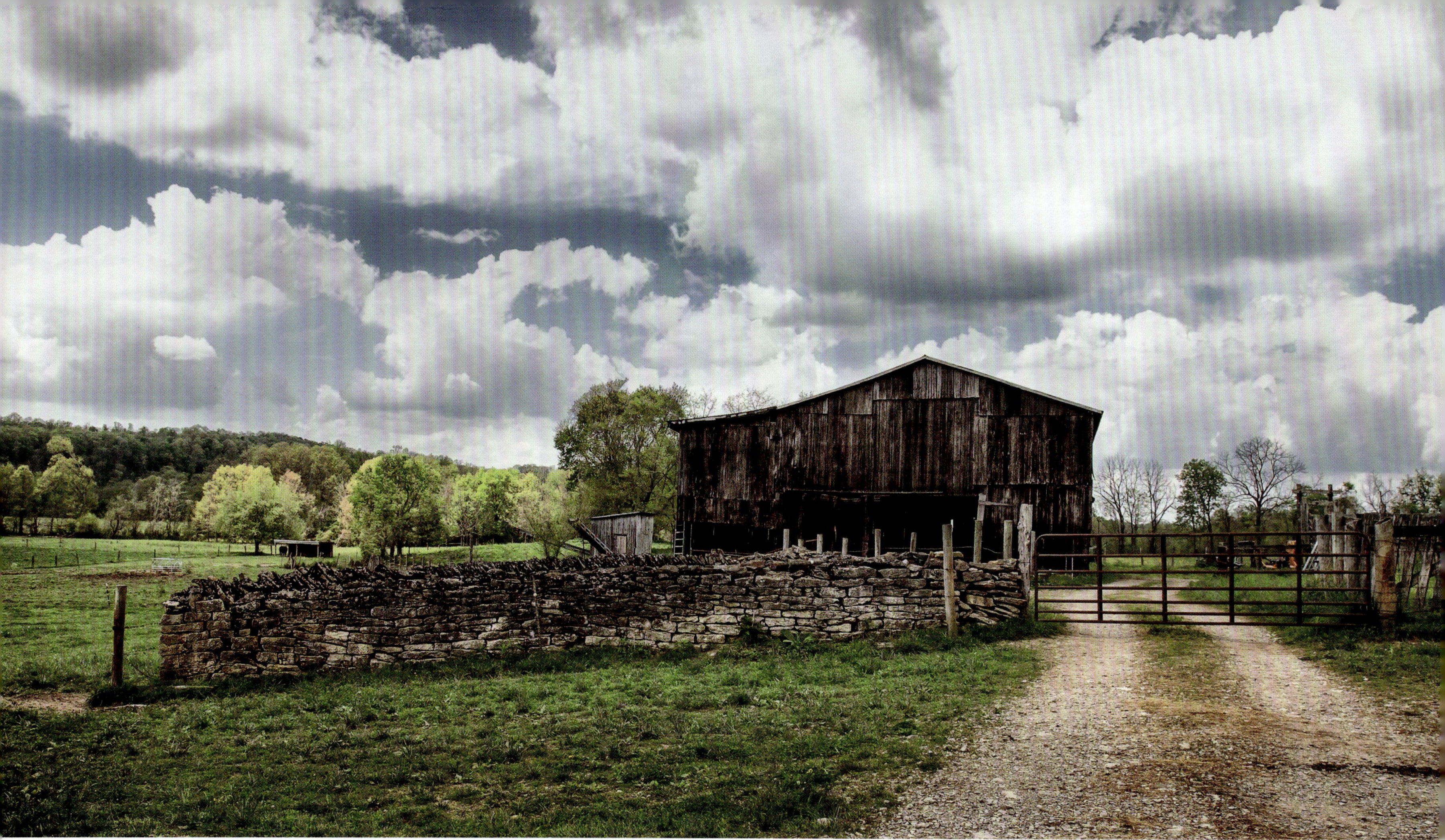

Franklin County.

Quarles Farm includes this nineteenth-century stock
barn with hewn and pegged timbers. The family cemetery
on the left is surrounded by dry-laid stone walls.

Rockcastle County.

Located in a holler, this barn stores hay rather than the
livestock it was originally built for, as younger generations
leave the rural landscape for more urban conveniences.

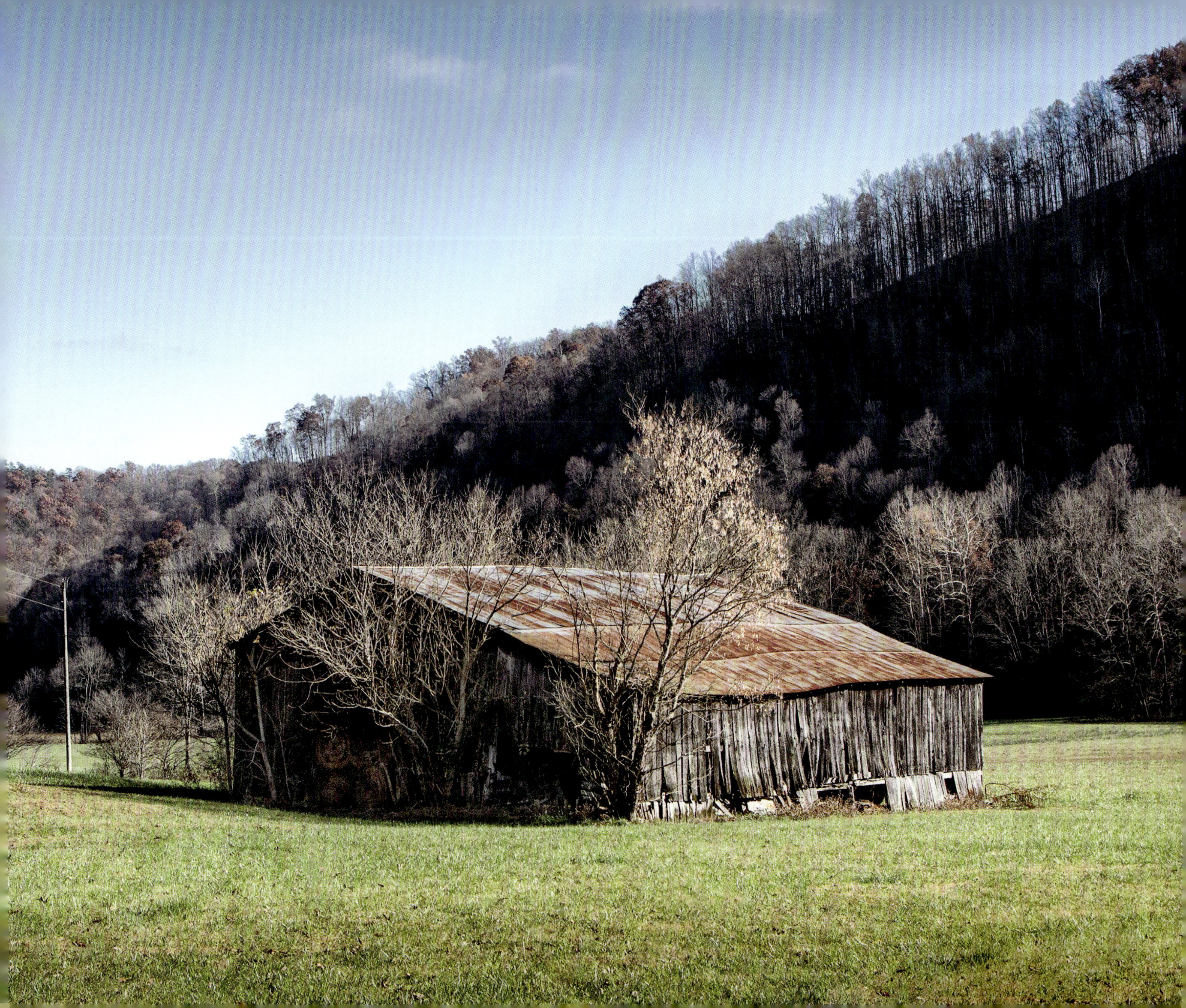

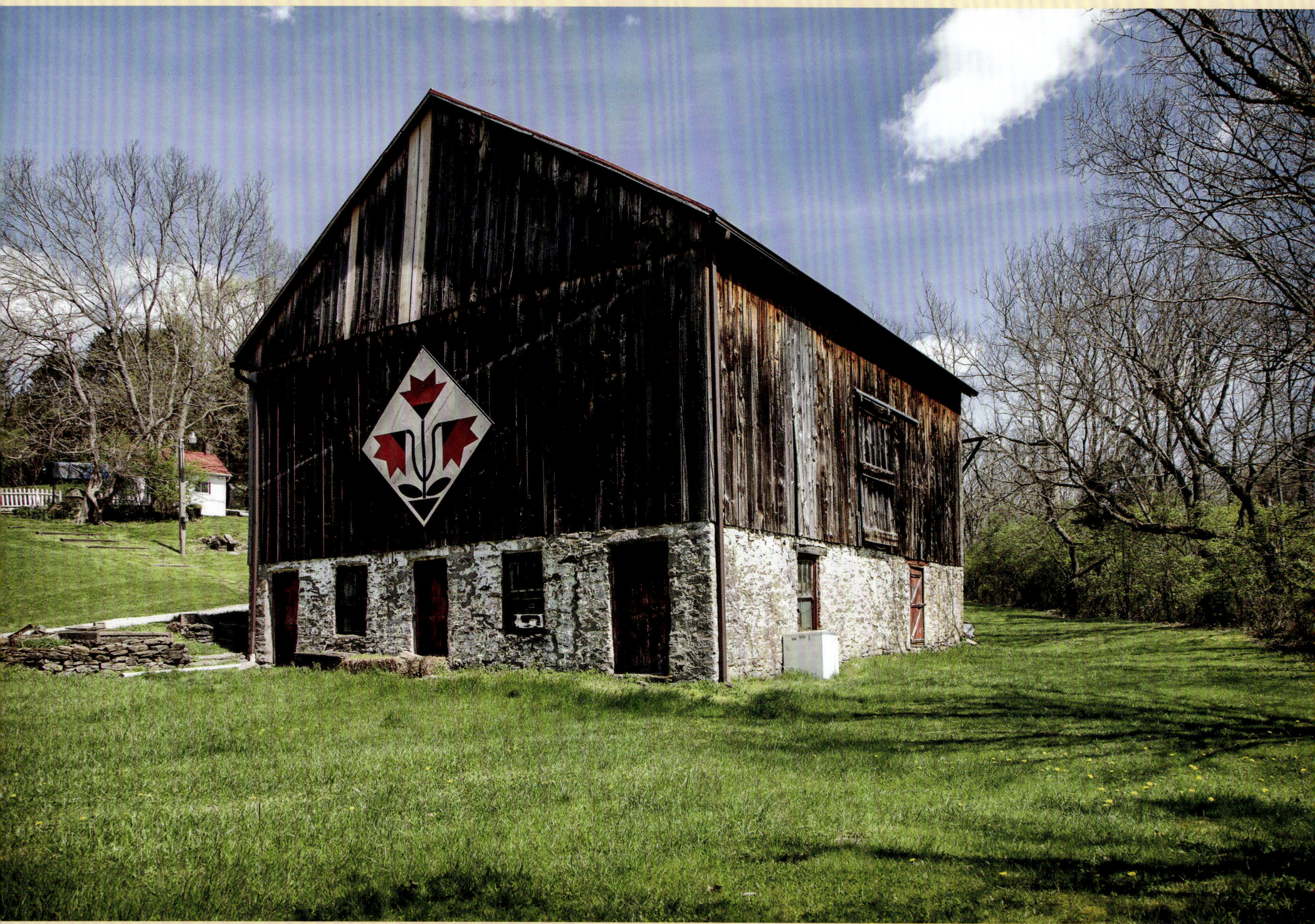

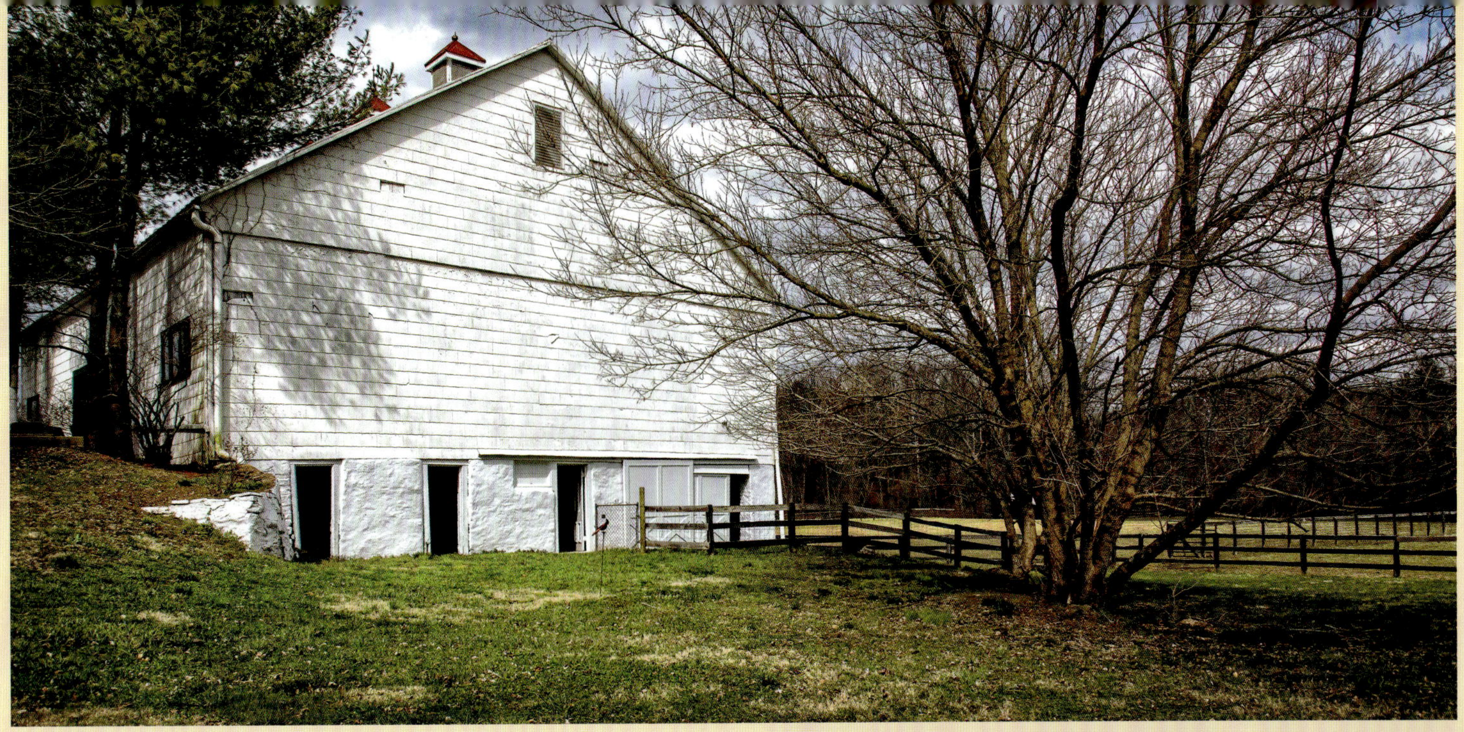

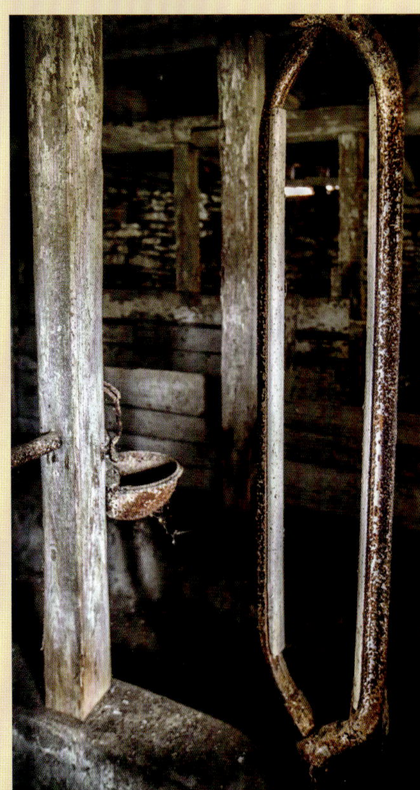

Campbell County.

Built in the German settlement of mid-nineteenth-century Camp Springs, the Nettlers' barn has undergone several metamorphoses. The loft remains for storage, but the underneath has been both a dairy and a stable.

Campbell County.

Members of a German immigration movement who settled in Camp Springs in northern Kentucky built the Kuber-Schuchter barn in the mid-1860s. Later generations have maintained the community's heritage, handing down homes and farmsteads and remaining faithful to the original architecture.

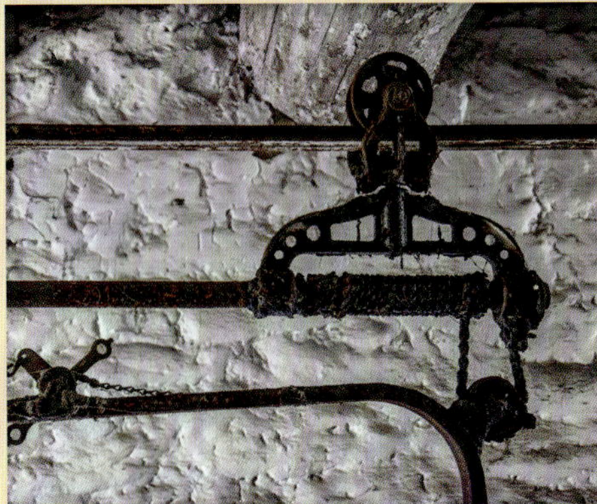

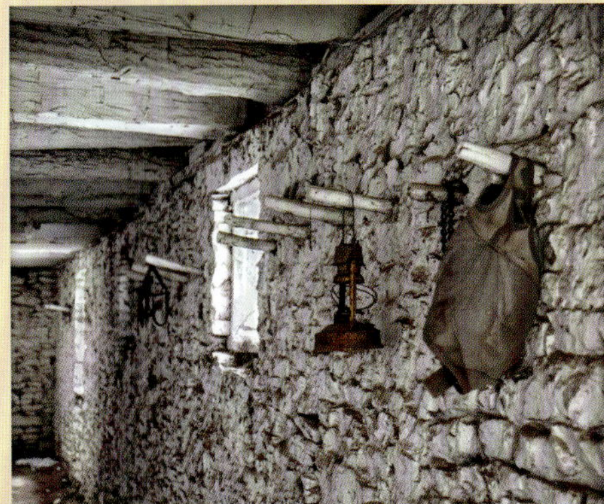

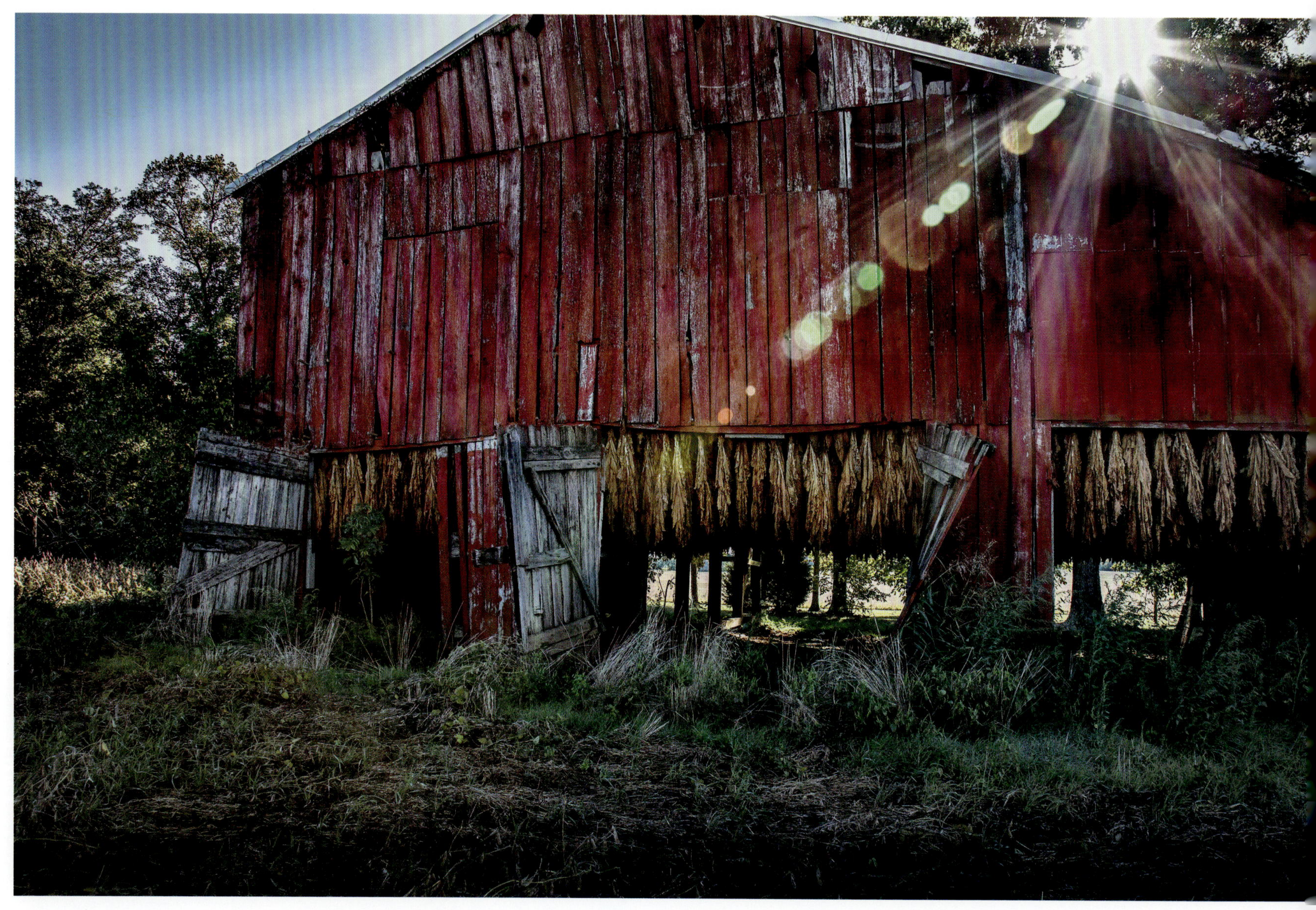

Woodford County.

This silo barn remains, along with one other barn, on a section of the old Chandler/Dunlap property.

Mason County.

Slack Farm is an example of continuous architecture, with connecting barns arranged for passage from one to another without going outside.

Todd County.

Western Kentucky barns cure two types of tobacco: fire-cured dark leaf and air-cured burley. This barn is air-drying burley.

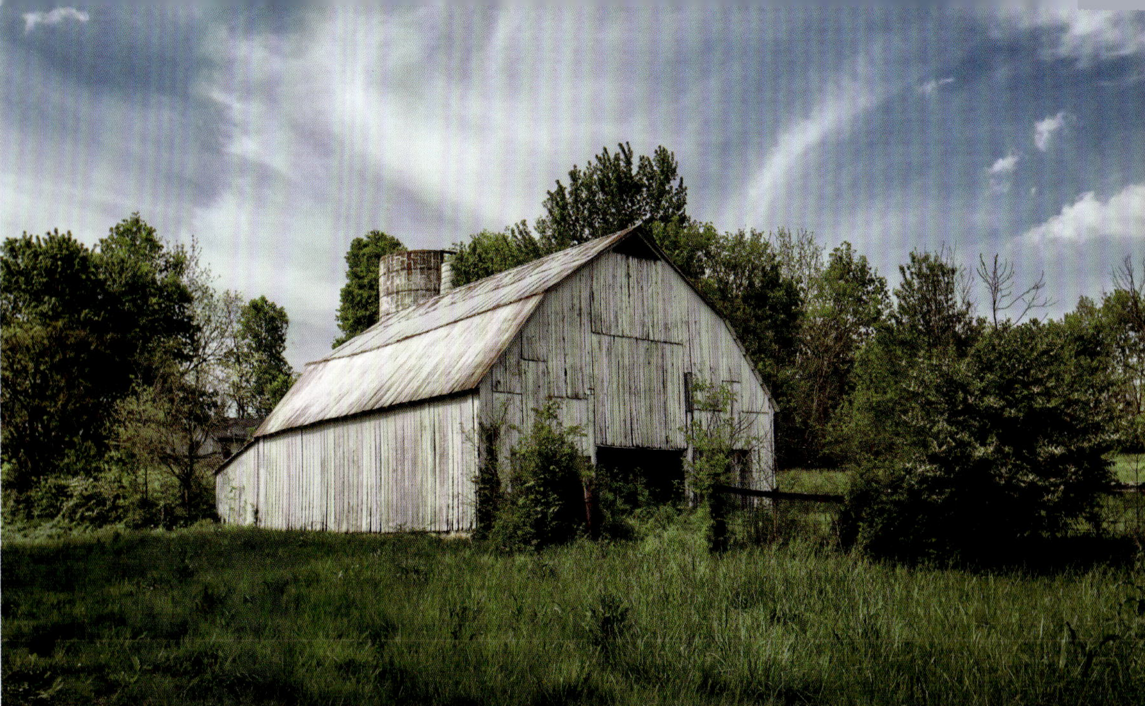

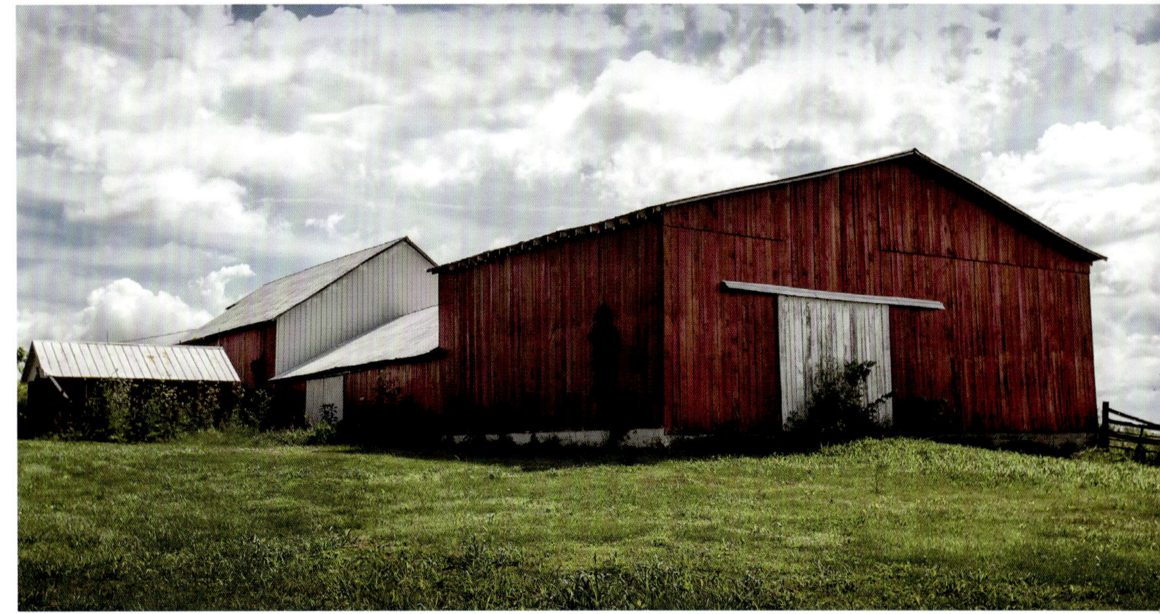

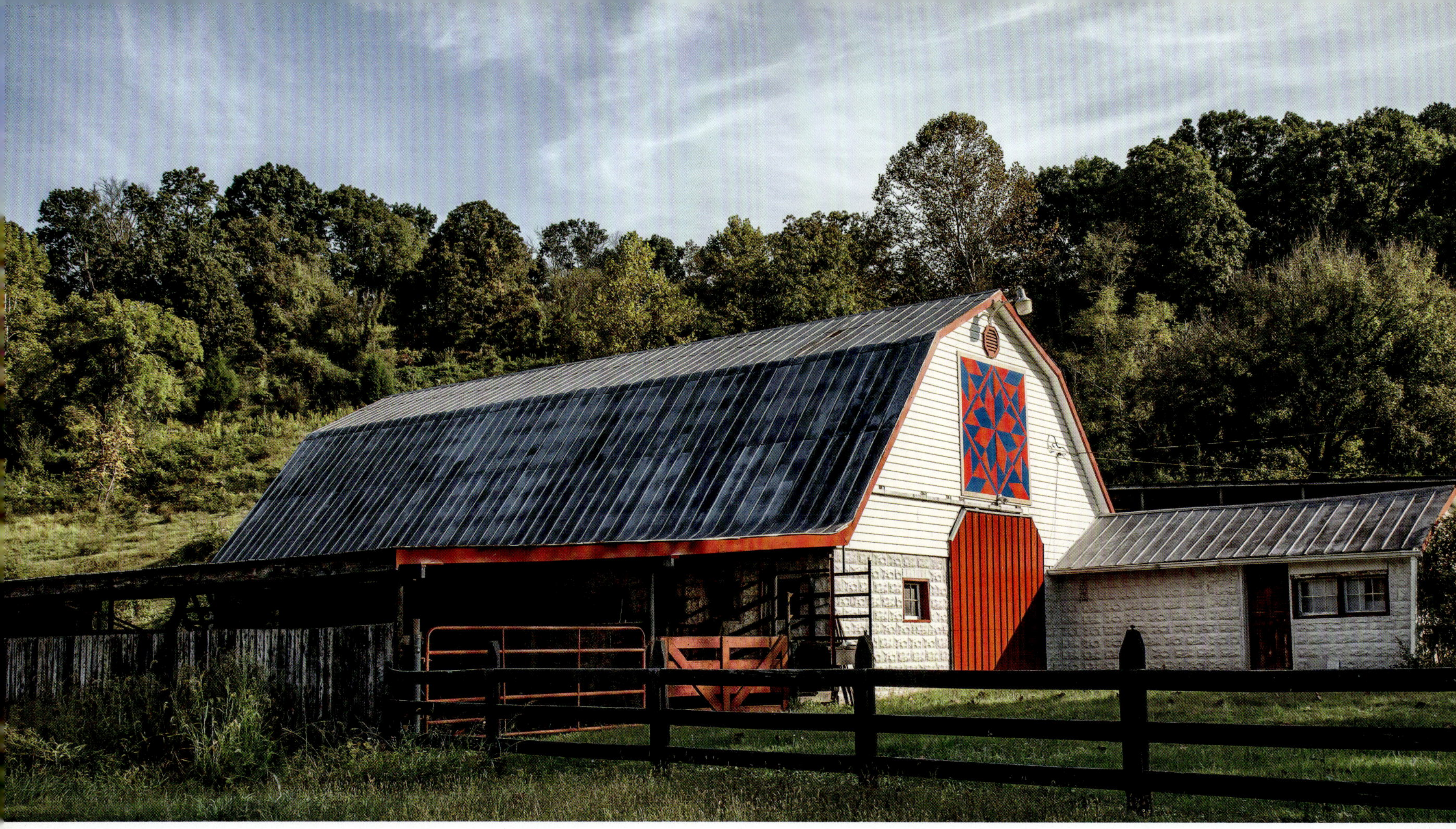

FACING PAGE

Lawrence County.

In dairy operations, the milk shed usually extended from the barn or was its own separate building. The separation of the shed from the barn assured uncontaminated milk.

FACING PAGE

Lawrence County.

Around the state, Chew Mail Pouch Tobacco advertising was painted onto the sides of barns between 1891 and 1992. The advertiser paid the farmer a little extra income and provided a fresh coat of paint for free.

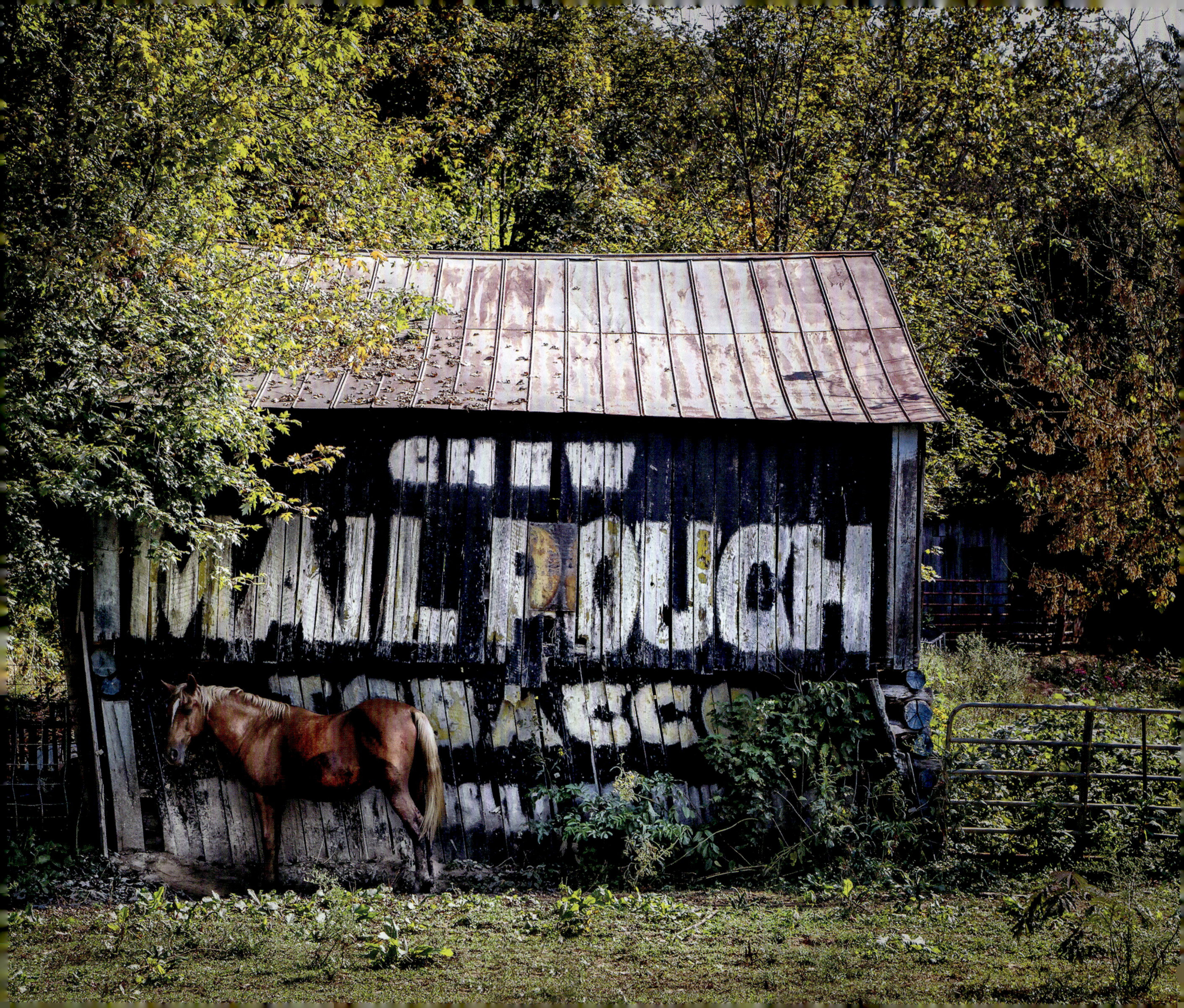
MAIL POUCH

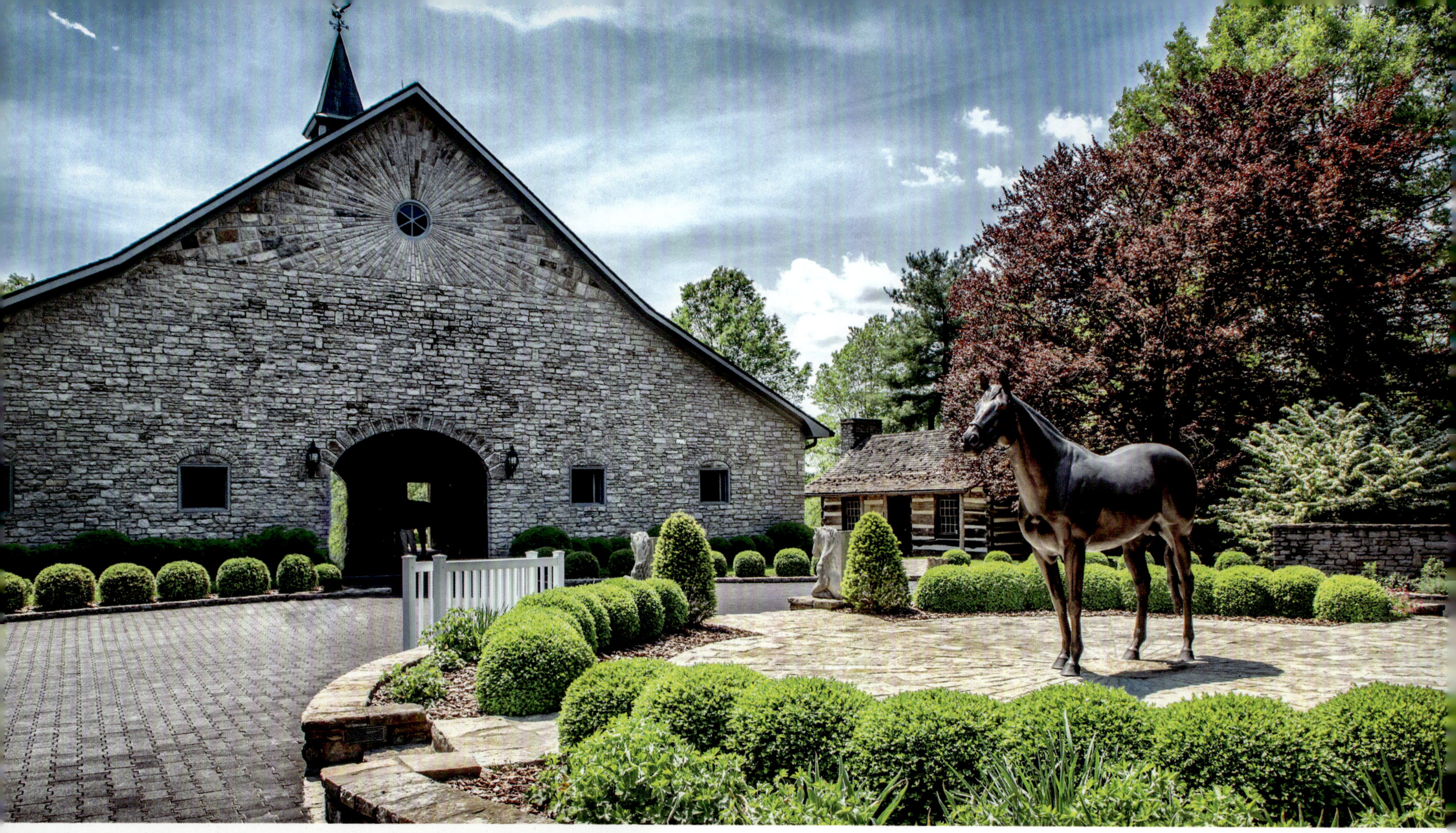

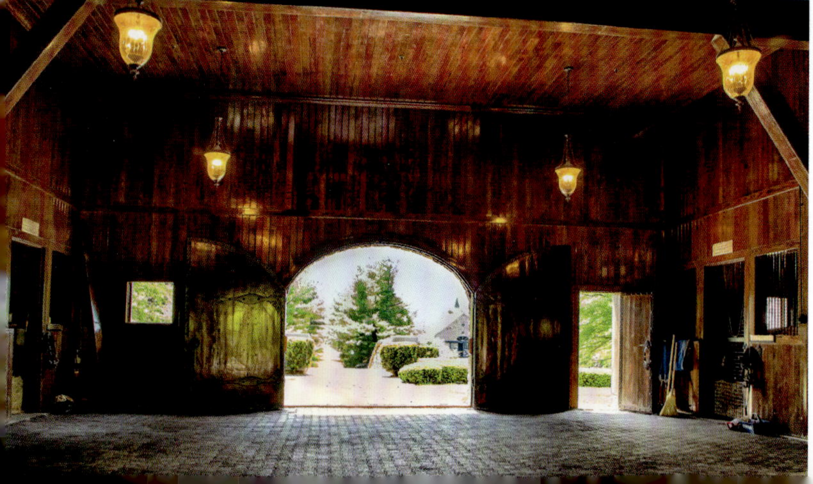

Woodford County.

Ashford Stud, the current home of two Triple Crown winners, Justify and American Pharaoh, was originally a Hereford farm owned by renowned distiller E. H. Taylor. He was known to breed the best Herefords, brought from England. The current stone stallion barn covers an earlier grain barn, and the interior is redone in furnishings fit for the equine celebrity living there.

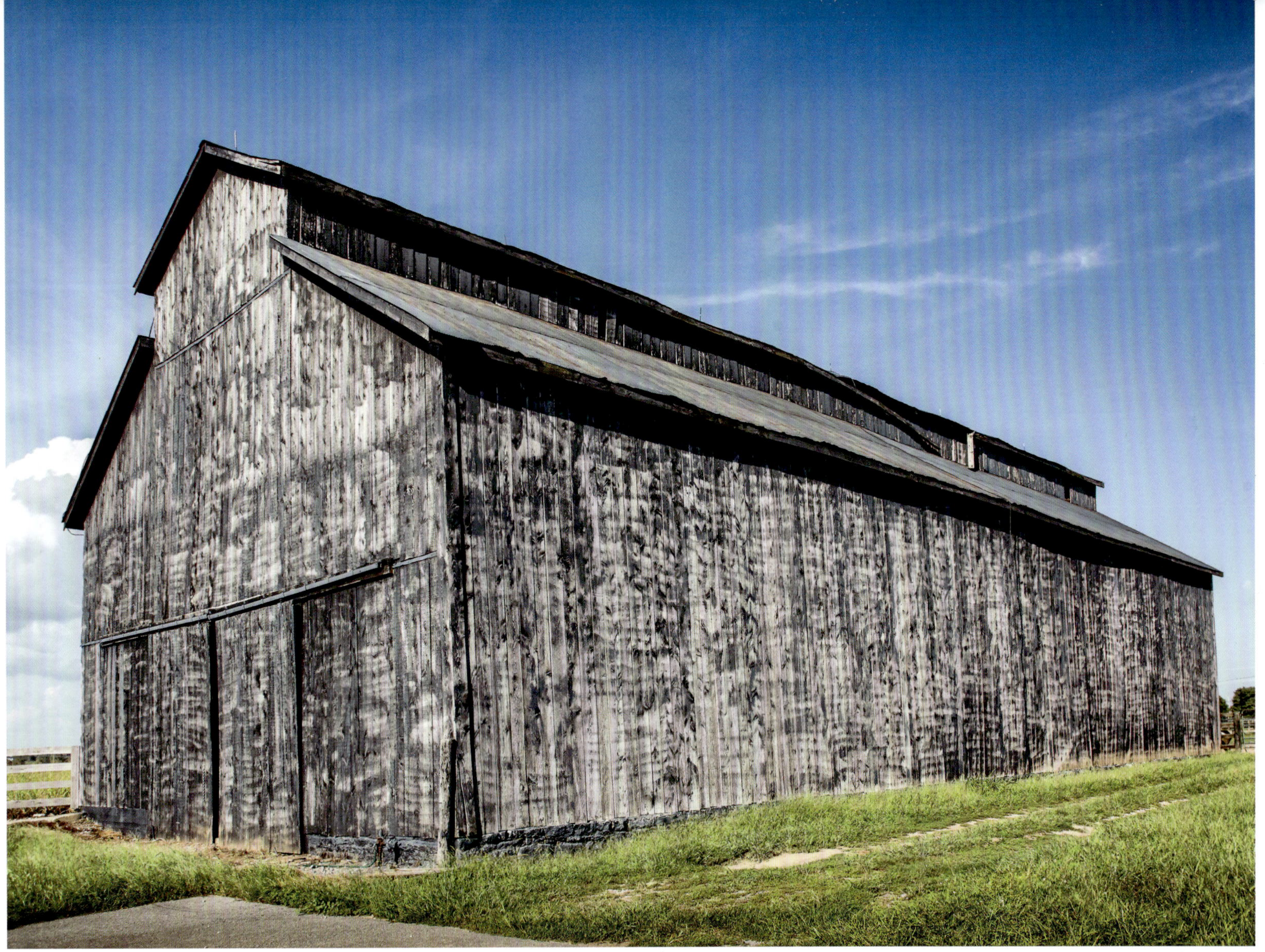

Garrard County.

This unusual stone foundation barn is one of two in
the area and appears to be used for livestock.

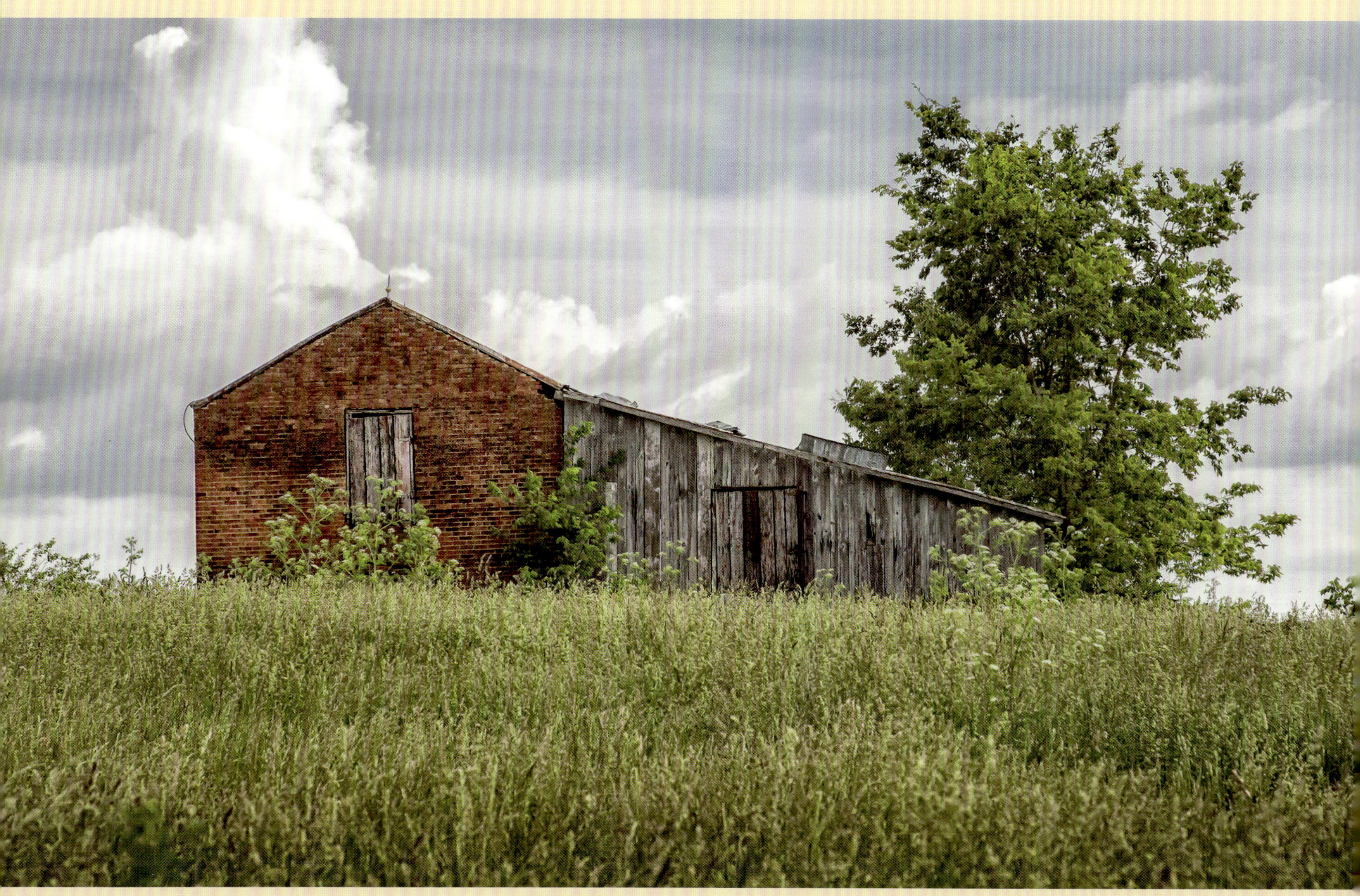

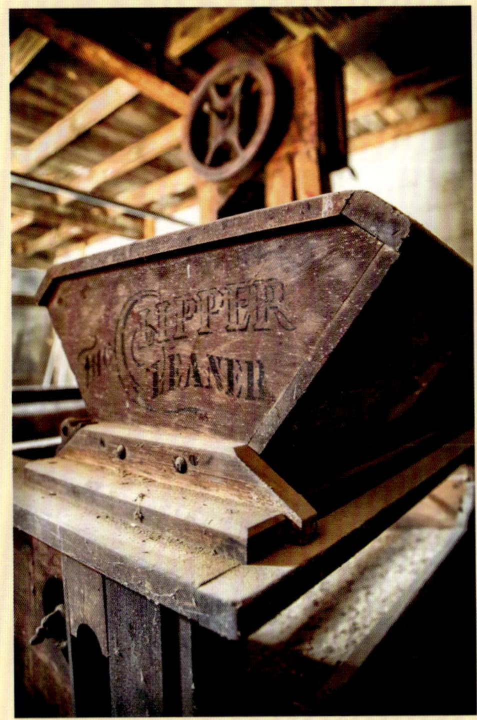 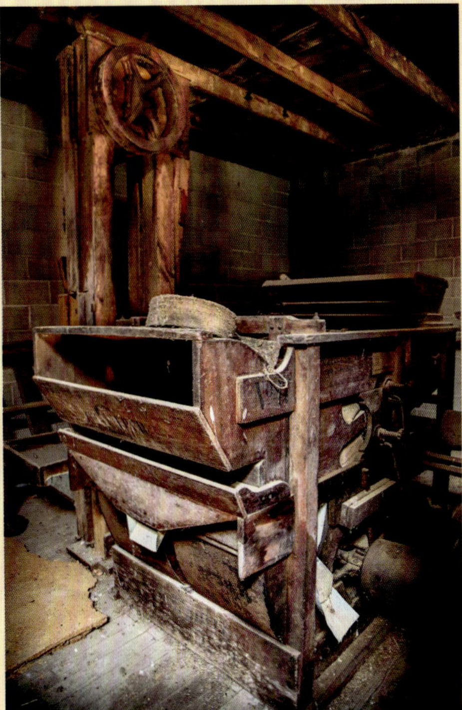 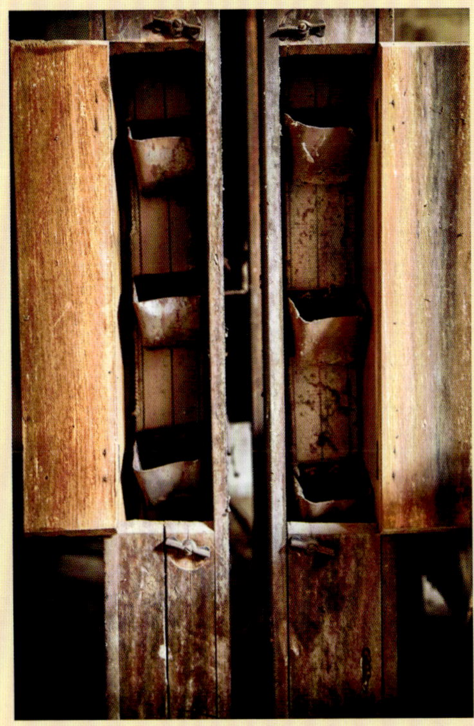

FACING PAGE

Mason County.

Many wooden hemp barns were burned during slave rebellions against slave owners. This mid-nineteenth-century barn on the Slack Farm is typical of the architecture for hemp barns that survived, usually tall and narrow and made of brick. A later side addition was for pigs.

Clark County.

Hemp was part of Kentucky agriculture from 1775 until after World War I, when other sources of fiber were used. This nineteenth-century hempseed cleaner belongs to the Jake Graves hemp barn, which has a drive-in basement as well.

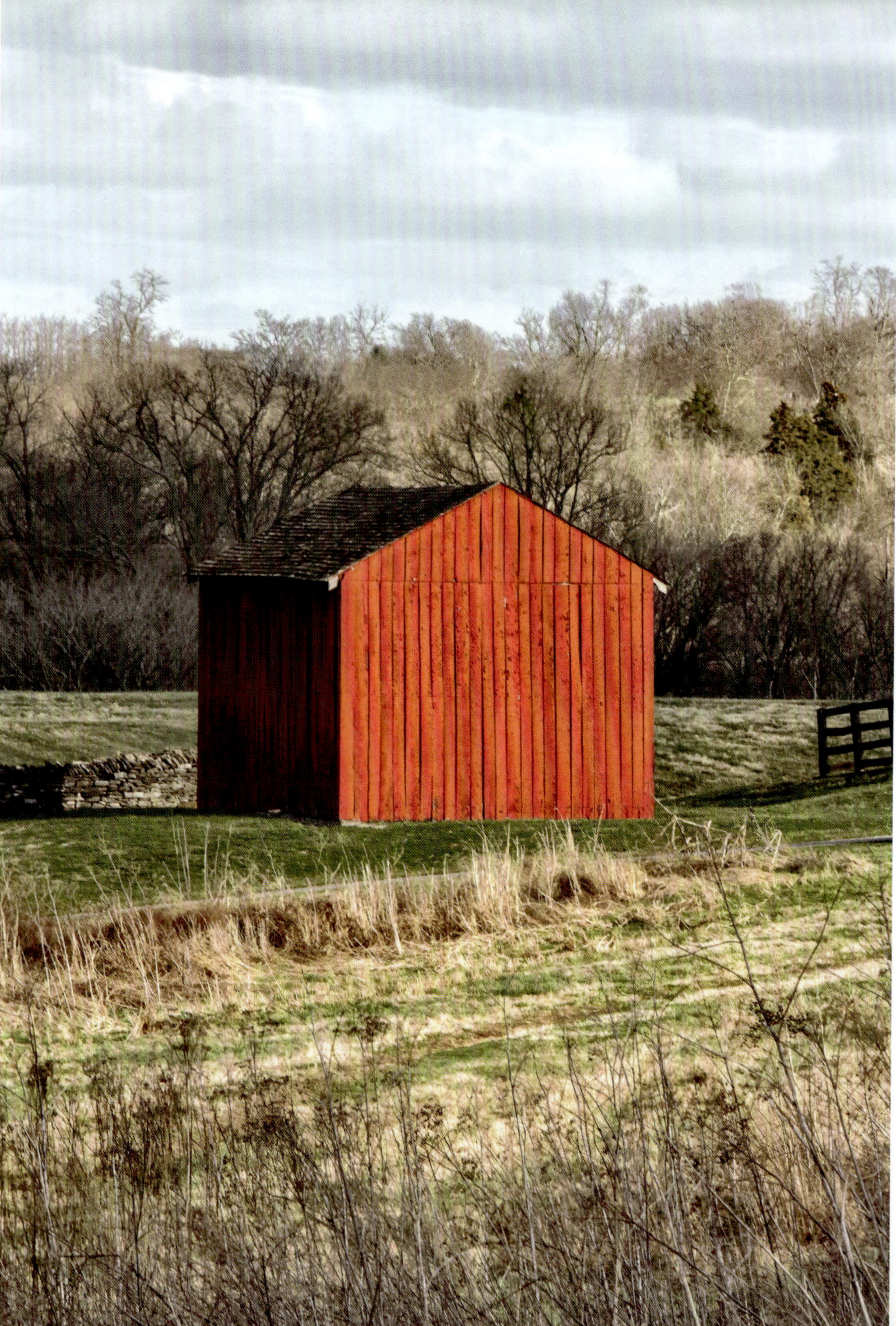

FACING PAGE

Elliott County.

This quilt barn near Isonville is used for storage.

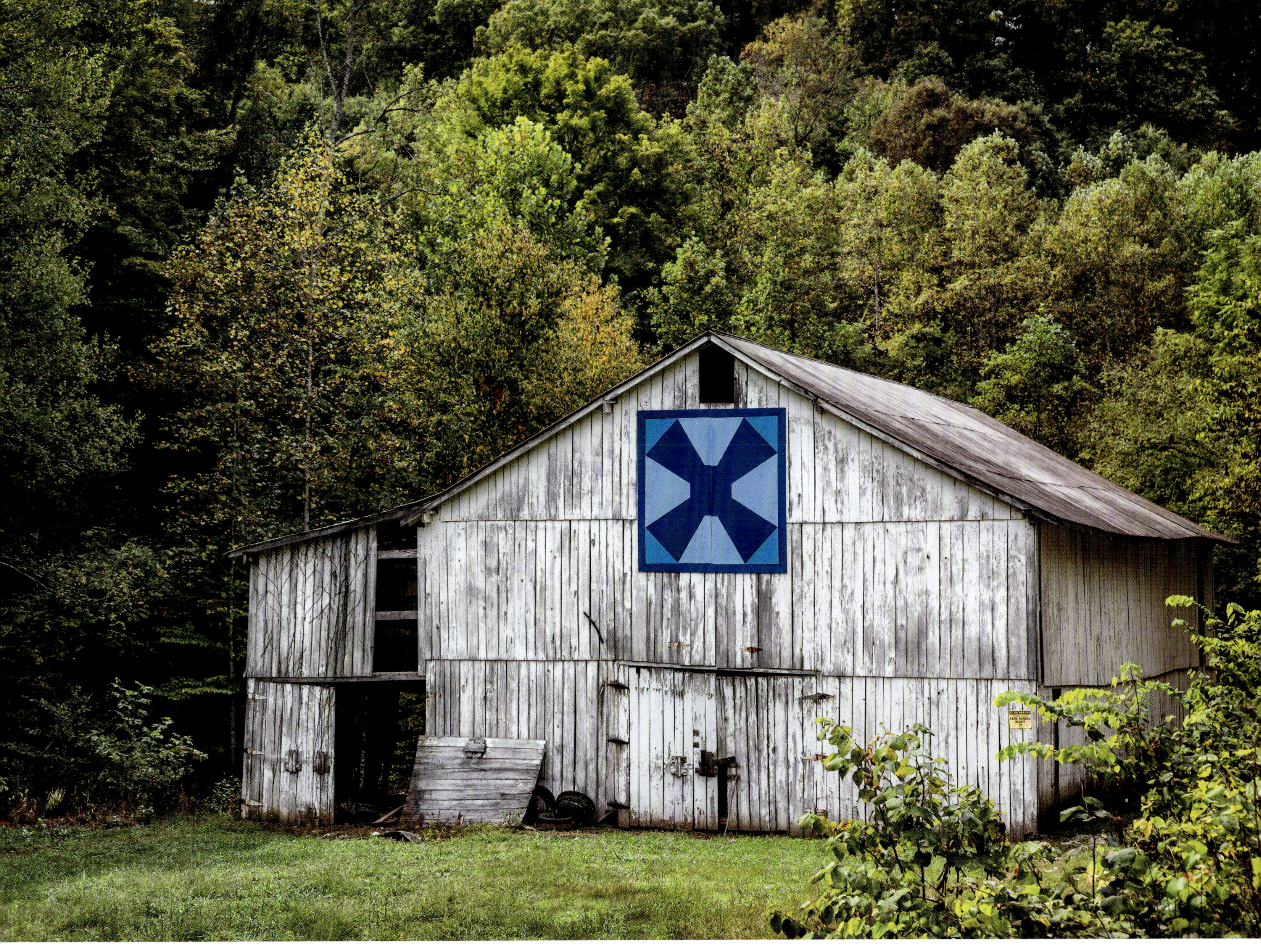

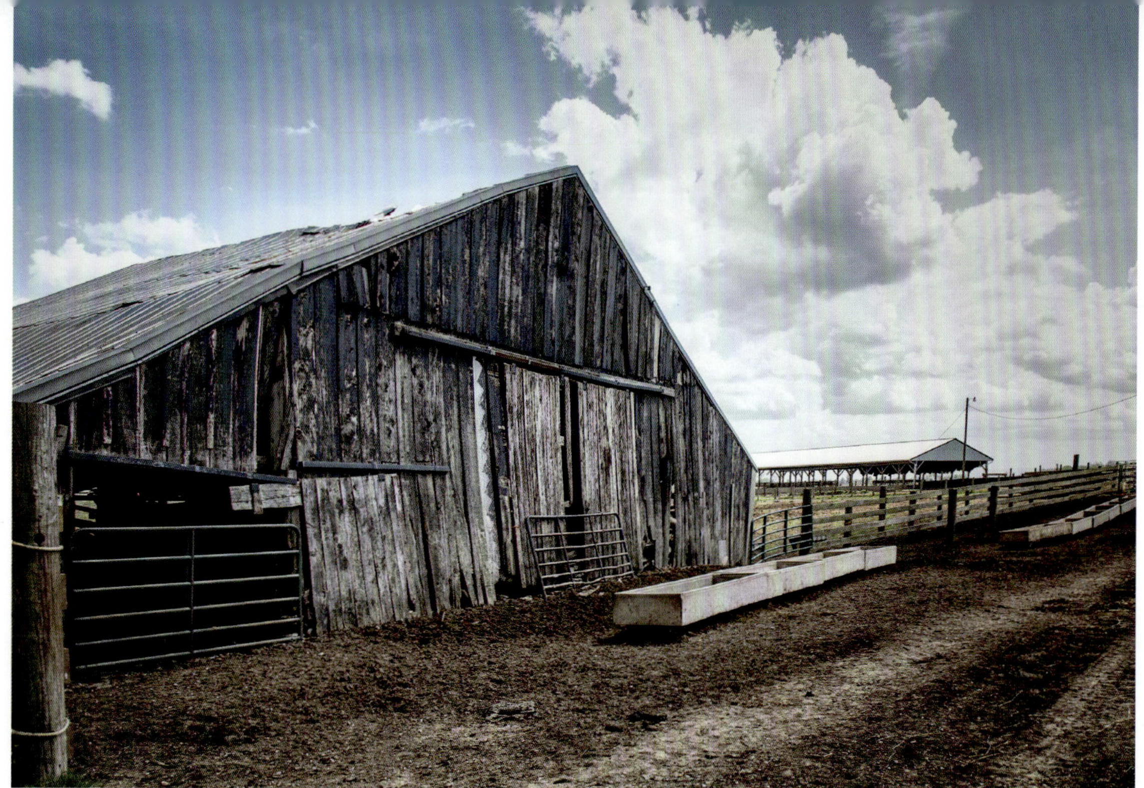

Owen County.

This threshing barn, a rare find these days, boasted a raised threshing floor and massive hand-hewn beams. The length of these beams is a testament to the great size of timber during the nineteenth century.

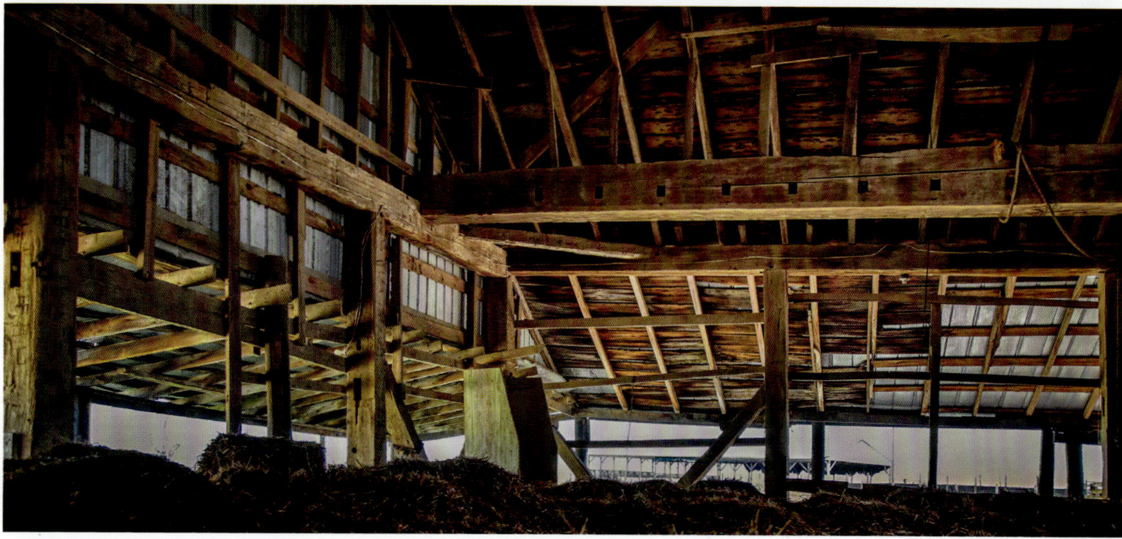

FACING PAGE

Carter County.

Brickley's Mail Pouch barn is one of over twenty thousand barns painted to advertise for West Virginia Mail Pouch Chewing Tobacco Company. It is rare to find these barns in good shape with bright letters.

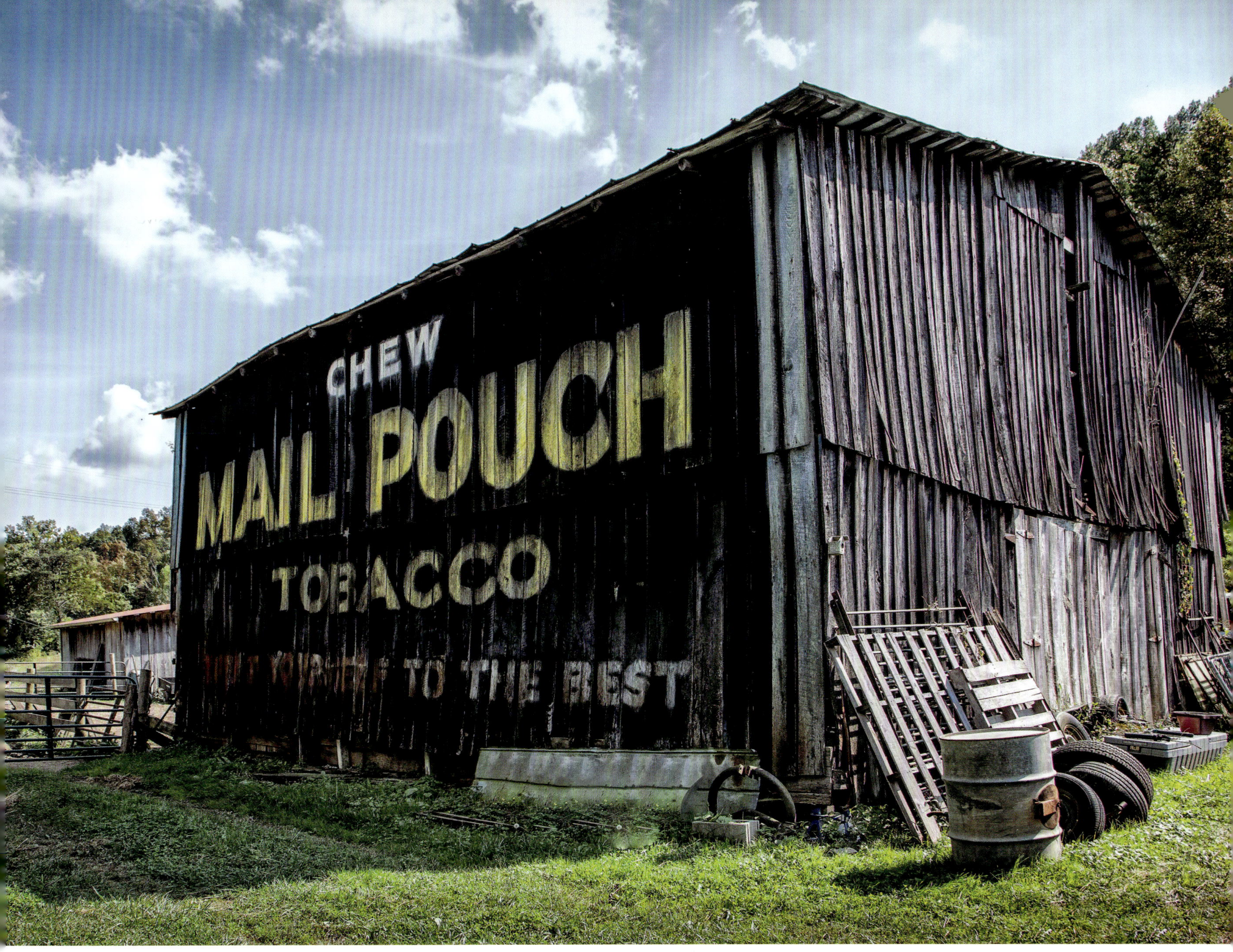
CHEW
MAIL POUCH
TOBACCO
TREAT YOURSELF TO THE BEST

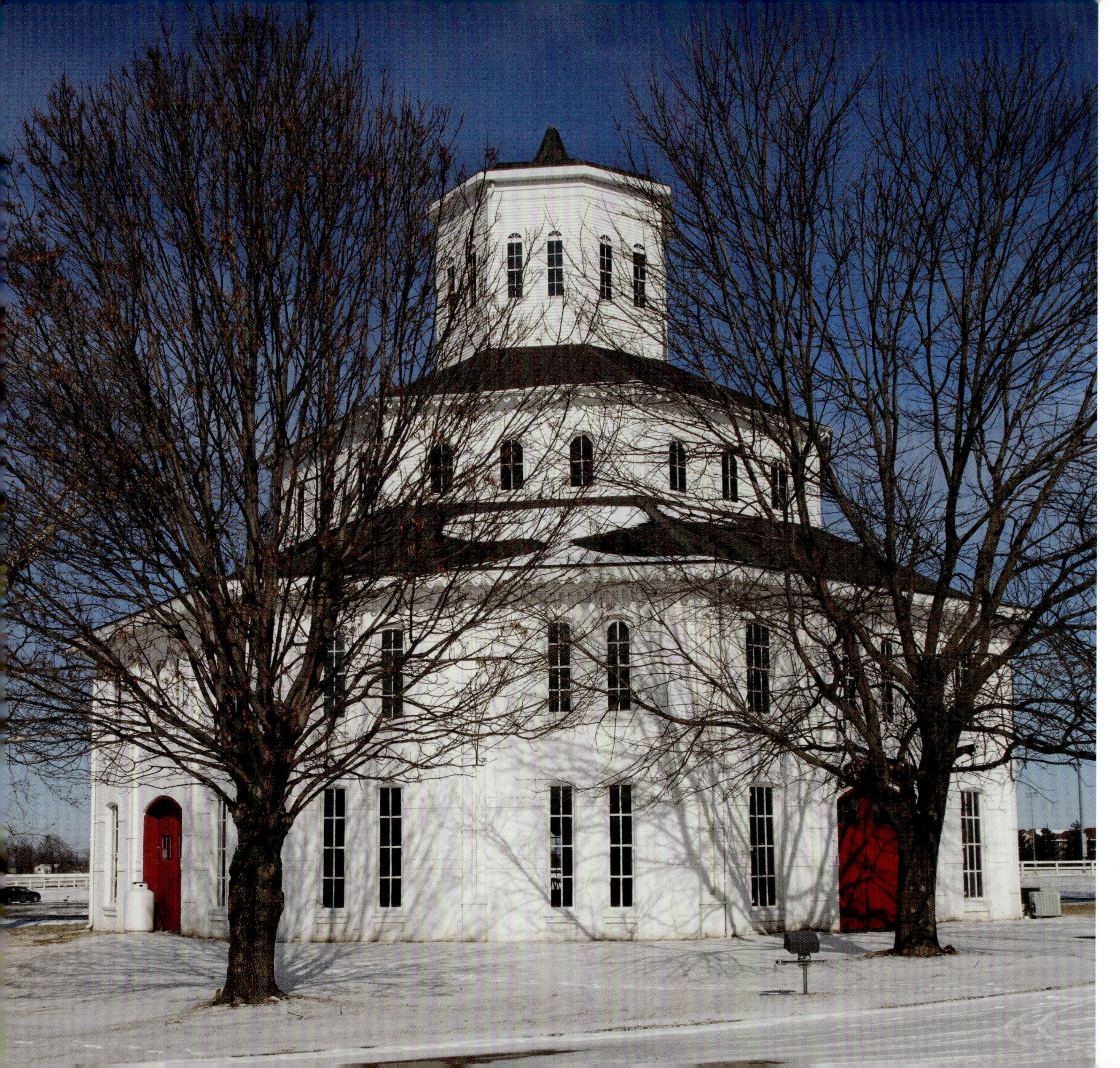

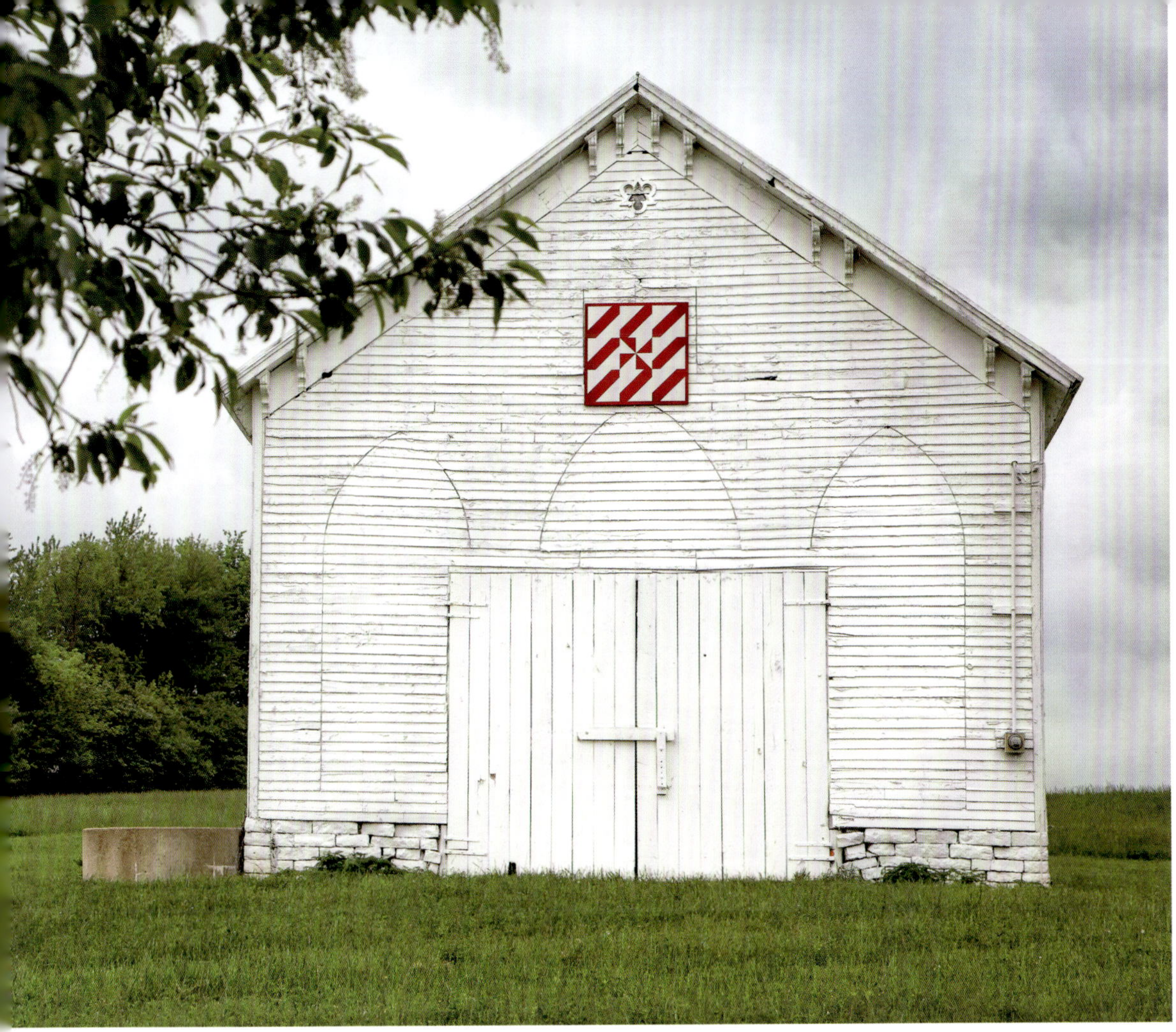

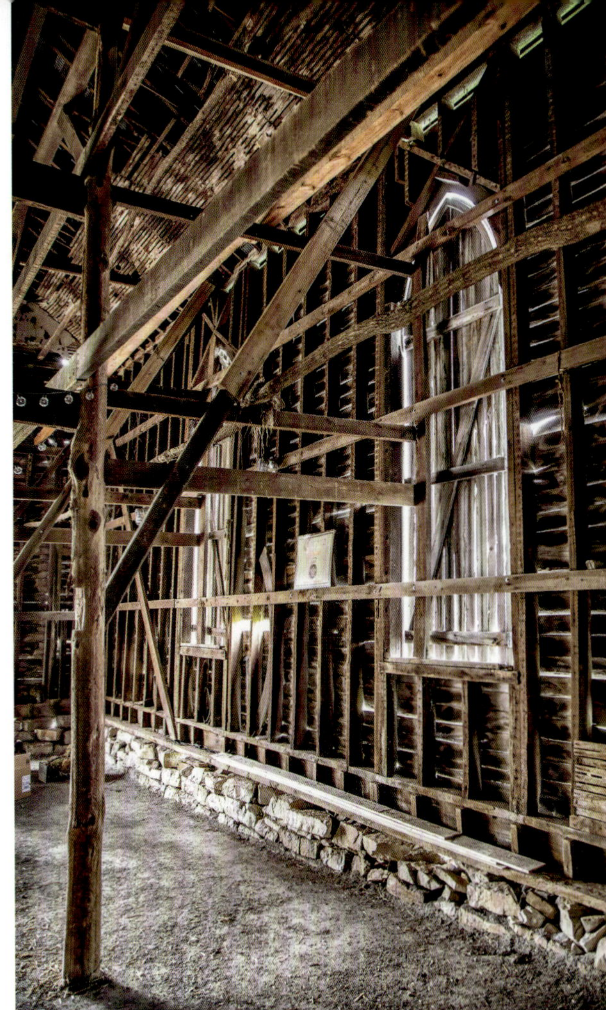

Fayette County.

Floral Hall is a Lexington landmark. It was originally built by John McMurty in 1882 as an exhibition hall for floral displays on the local fairgrounds. The octagonal building is now known as the Standardbred Stable of Memories, located next to the Red Mile harness racing track.

Woodford County.

Known as the church barn, outside of Nonesuch, this rural one-room church was converted to a tobacco barn.

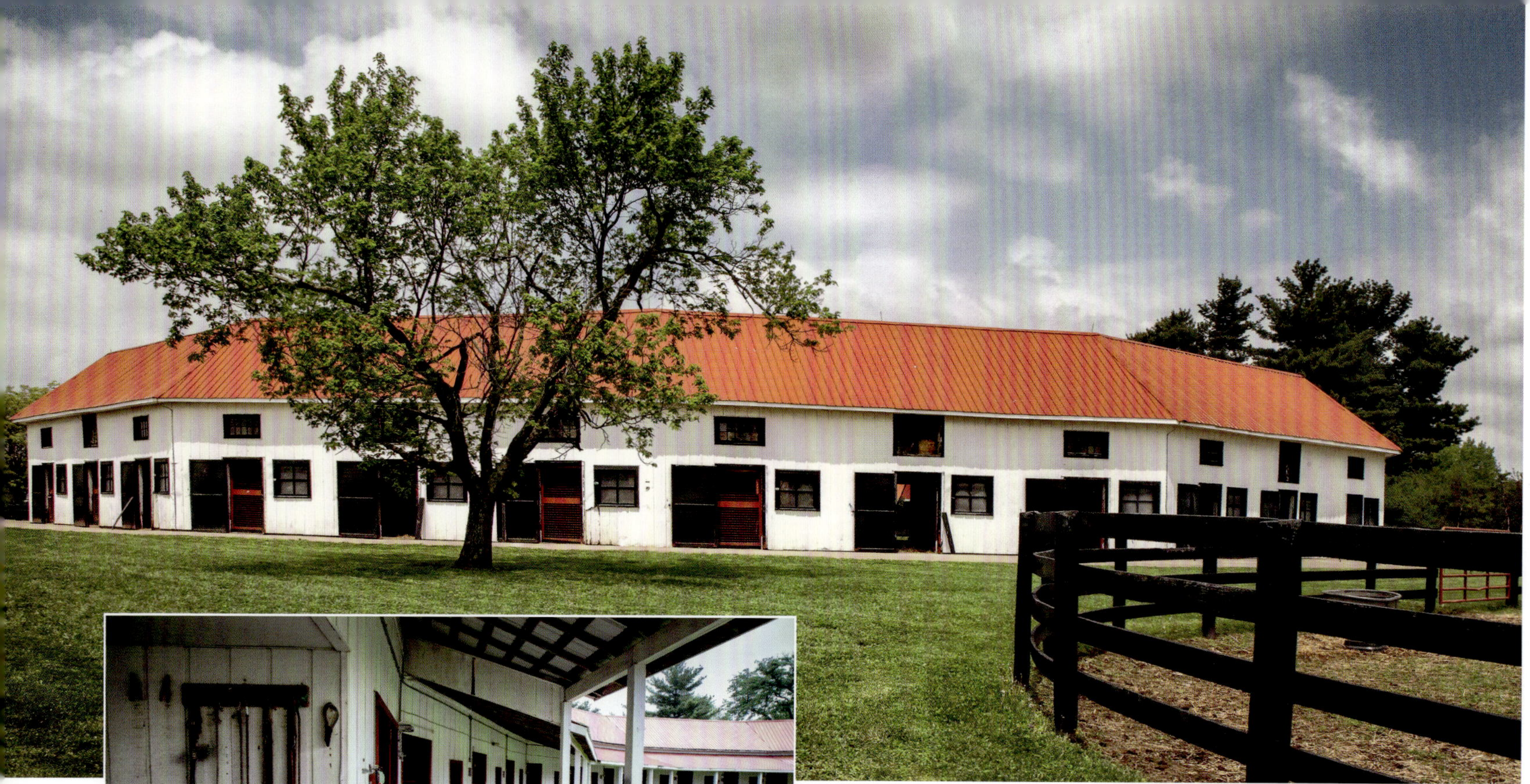

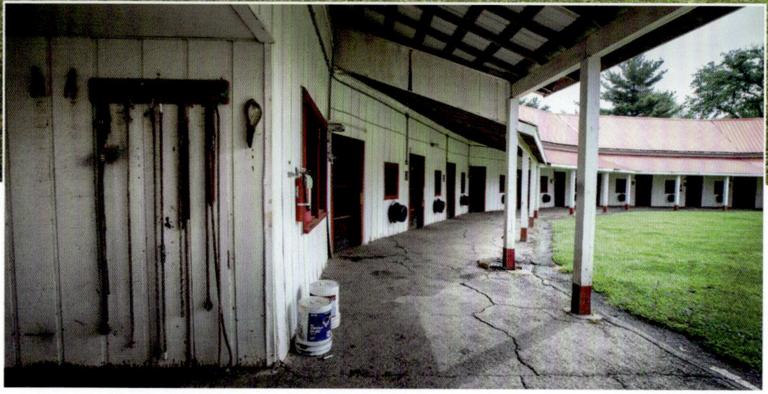

Fayette County.

A tract of land claimed in 1774 by frontiersman John Howard proved to be good ground for raising horses. The property went on to become Greentree Stud and Whitney Farm before becoming the current Gainesway Farm. It is known for its three circle barns, which were built for cattle but later converted for horses. The inside of the circle provides stall doors to an open enclosure, while the outside of the circle provides access to pasture.

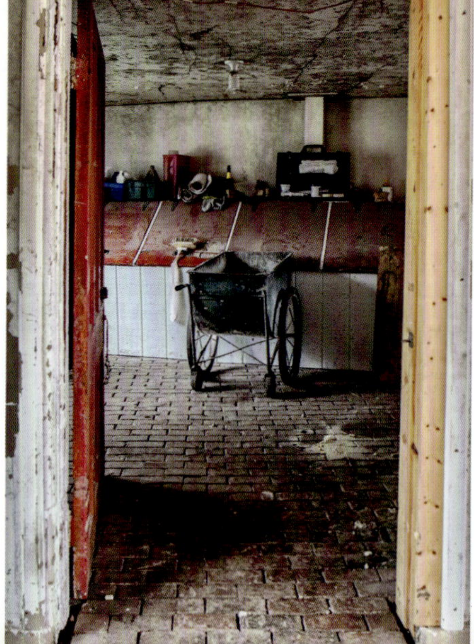

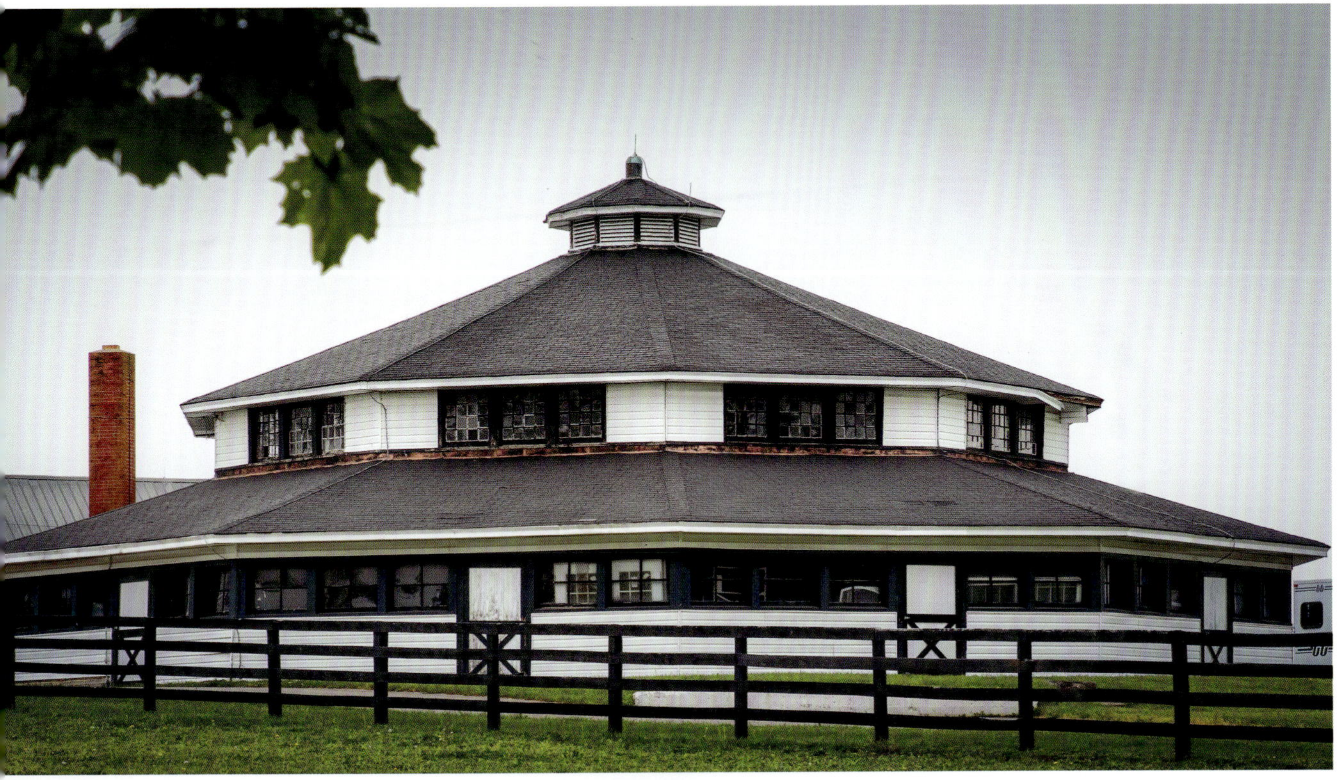

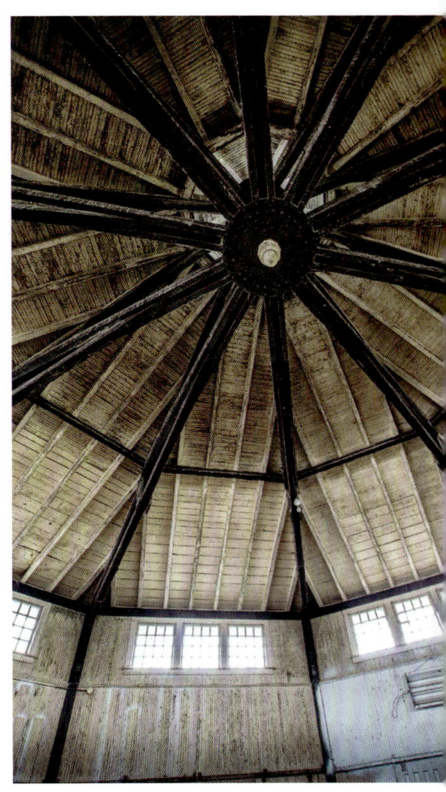

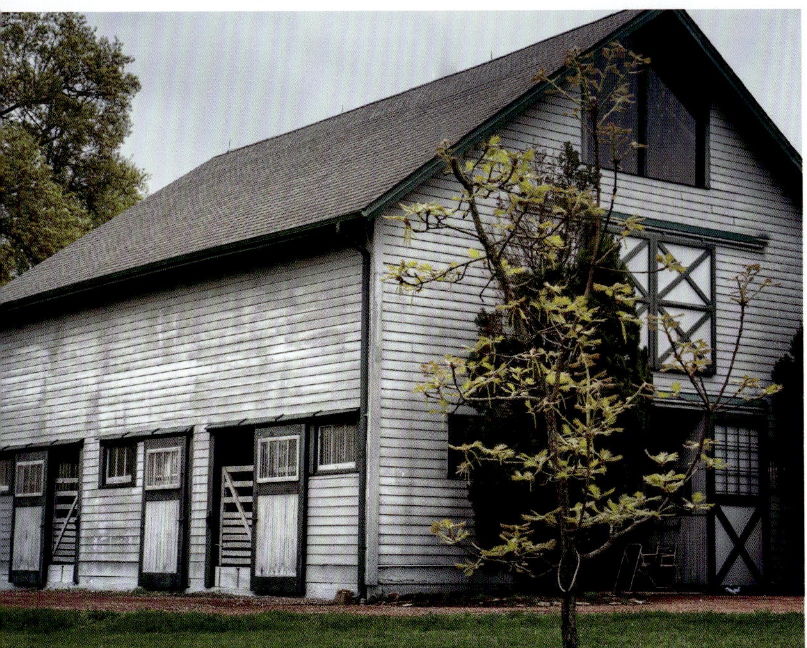

At Gainesway Farm, the two-level octagonal round barn was originally on adjacent property associated with Elmendorf Farm; it was moved to Gainesway in the 1920s. The small six-stall stable is one of the older original Greentree barns.

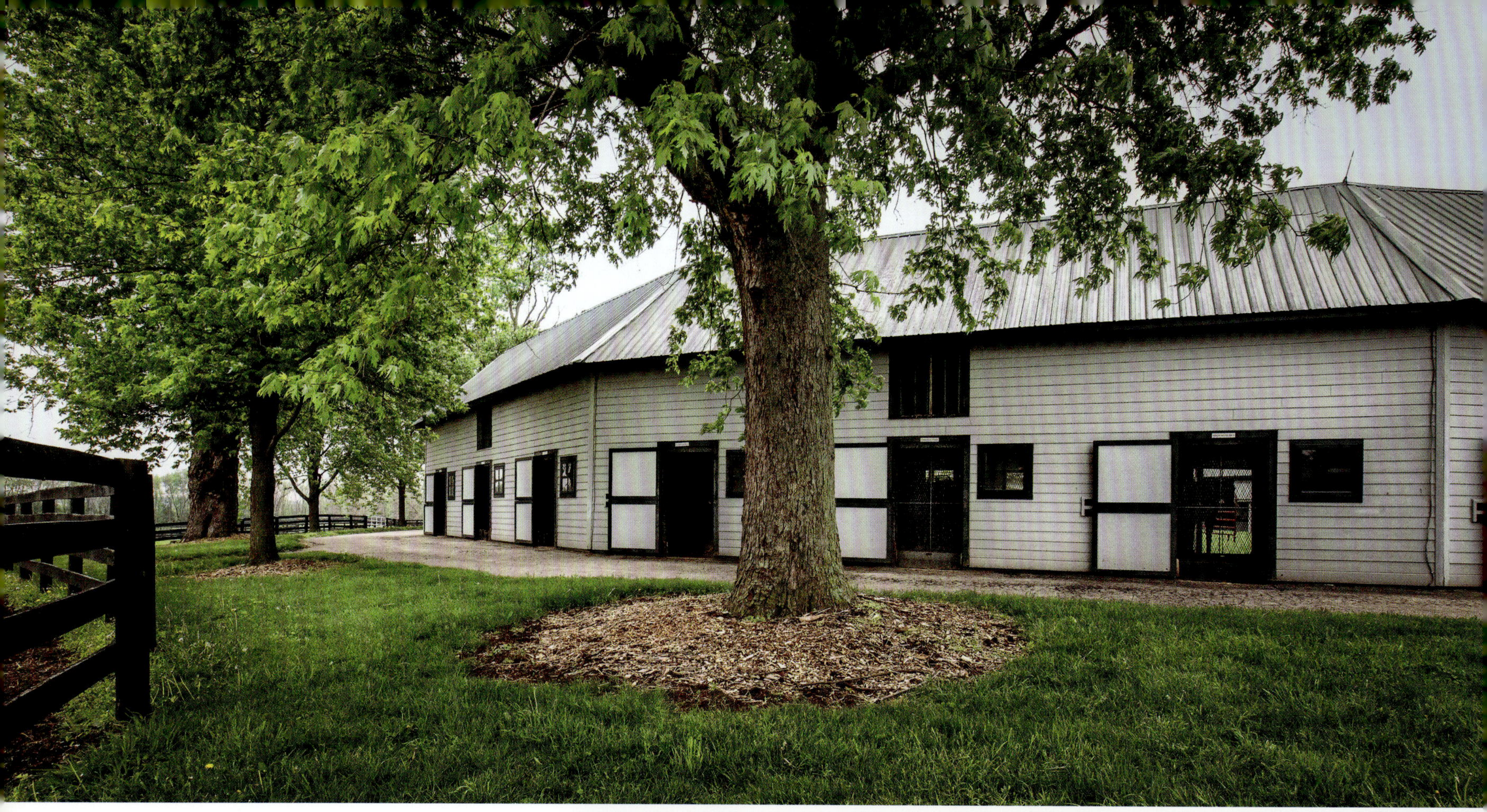

The green-roofed circle barn is located on the old Greentree section of the farm. It is used for broodmares and foals. In the loft are hay and old bridles.

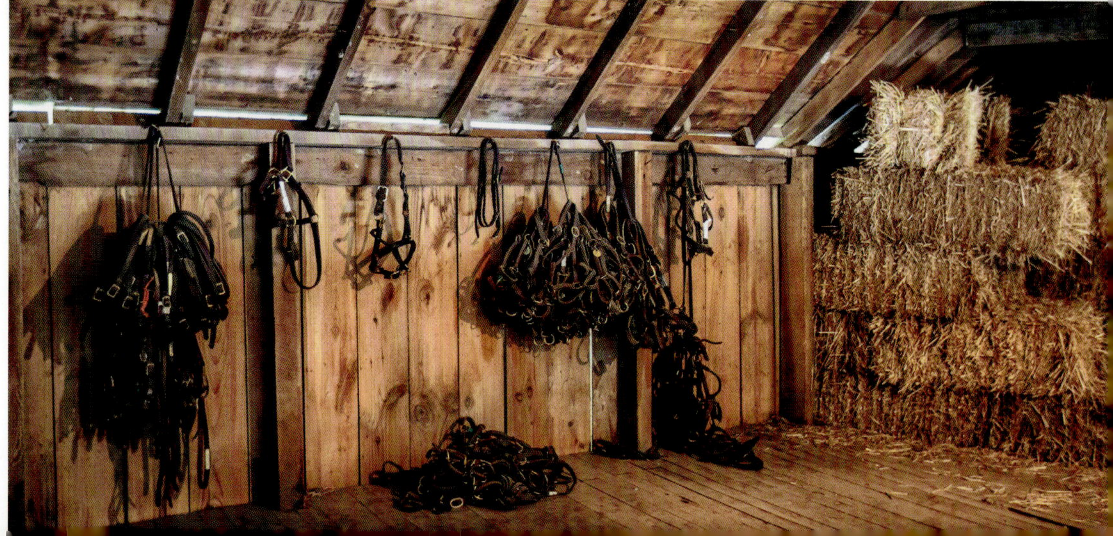

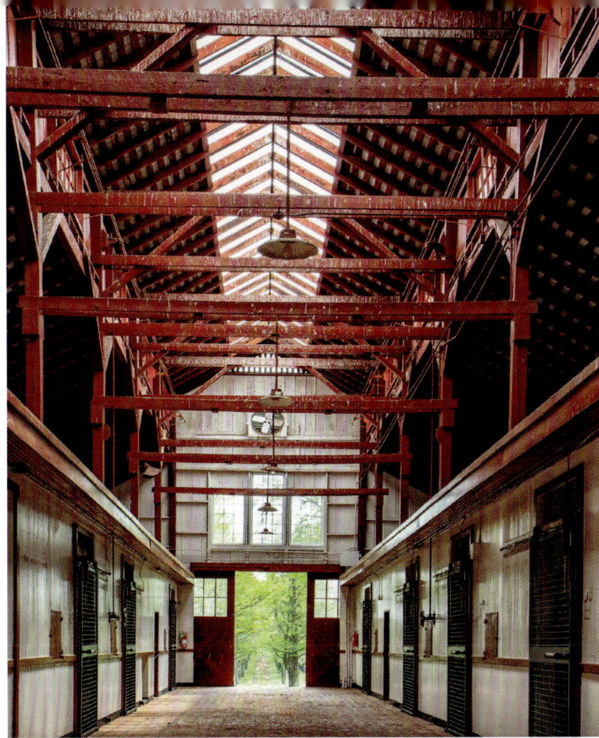

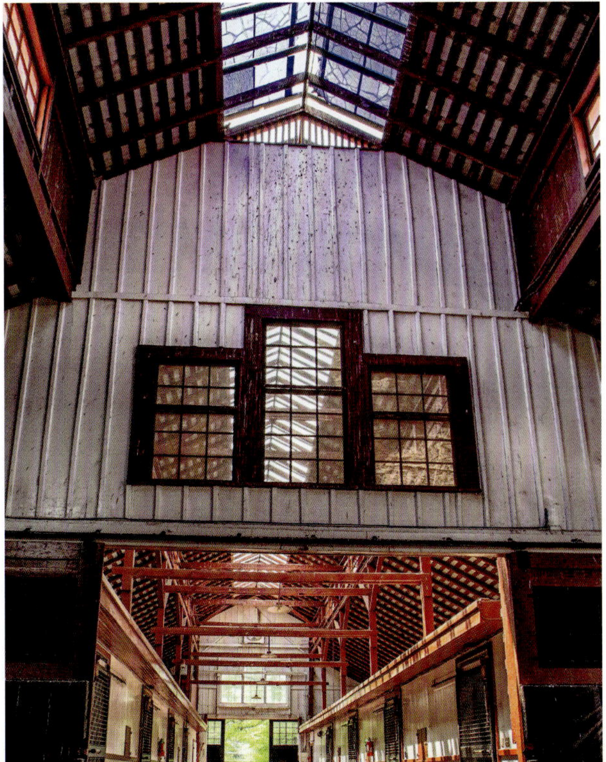

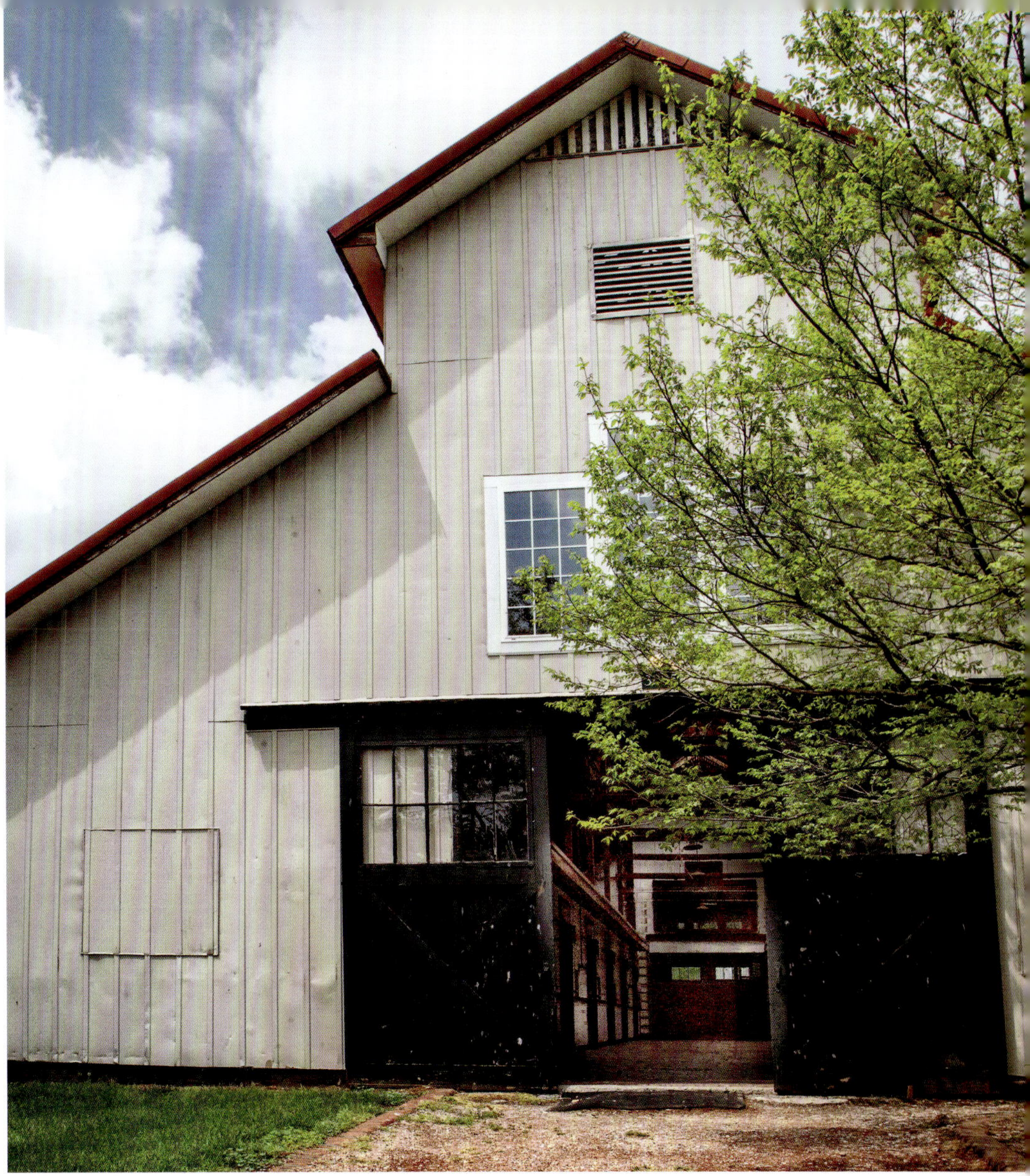

Fayette County.

The stallion barn, with its breeding shed, at Gainesway is over one hundred years old, one of the oldest barns on the property.

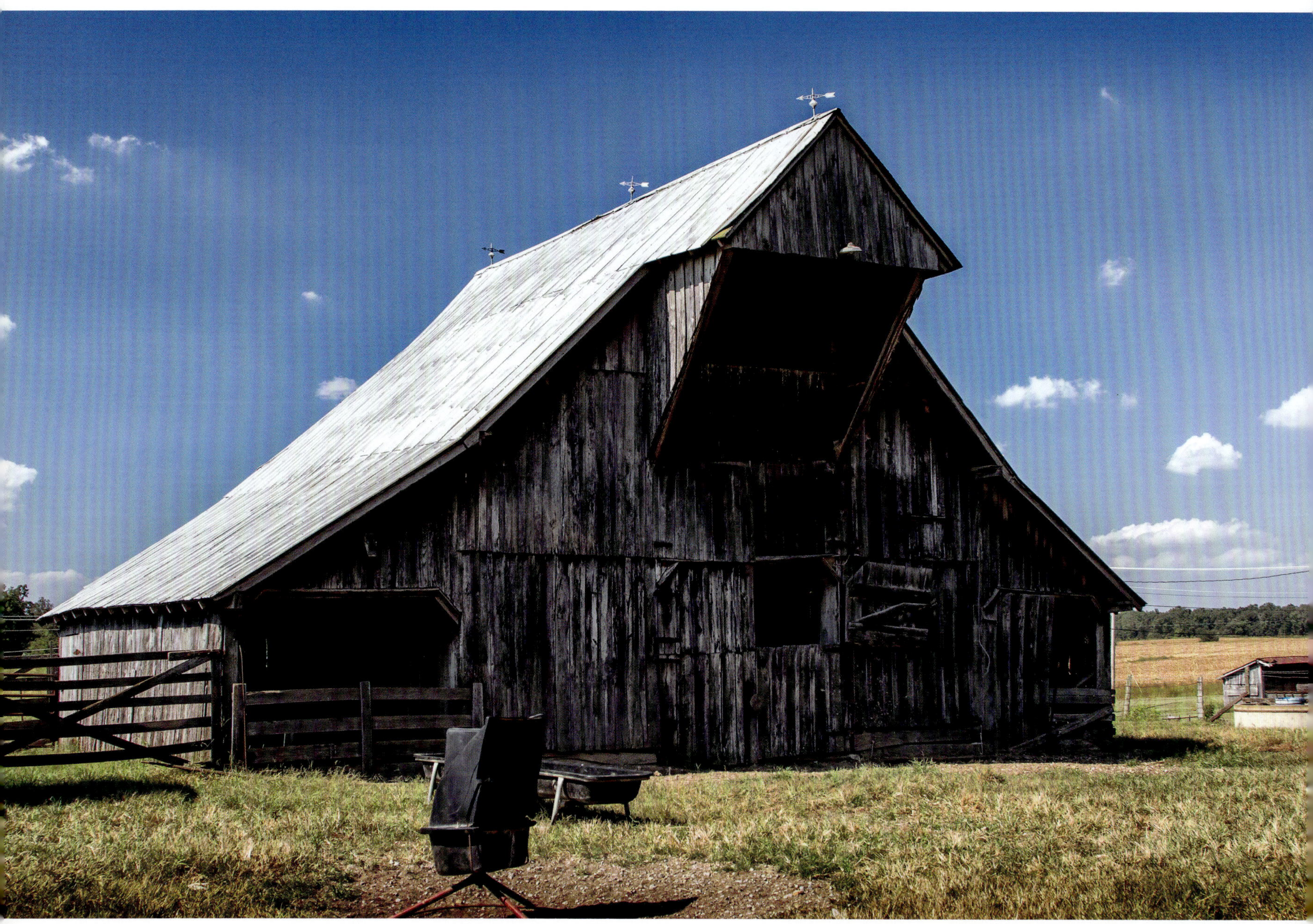

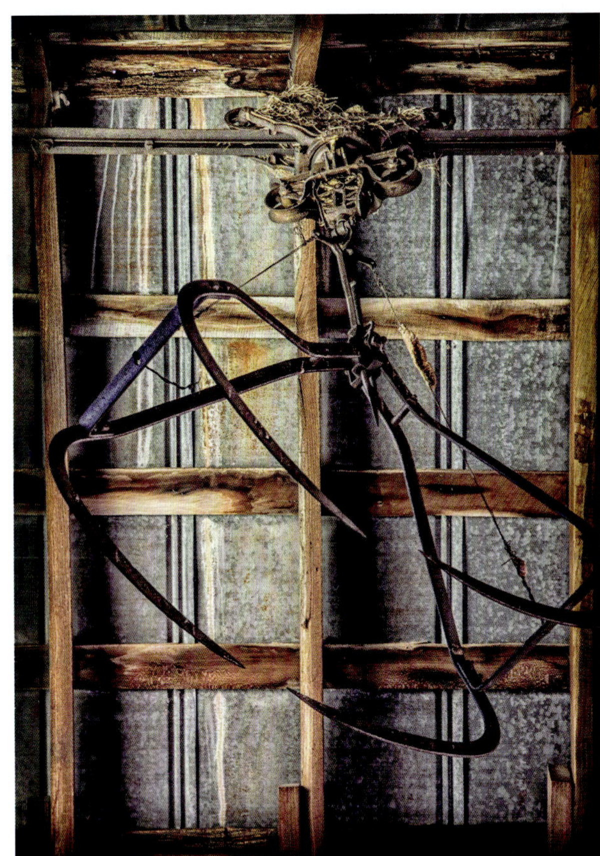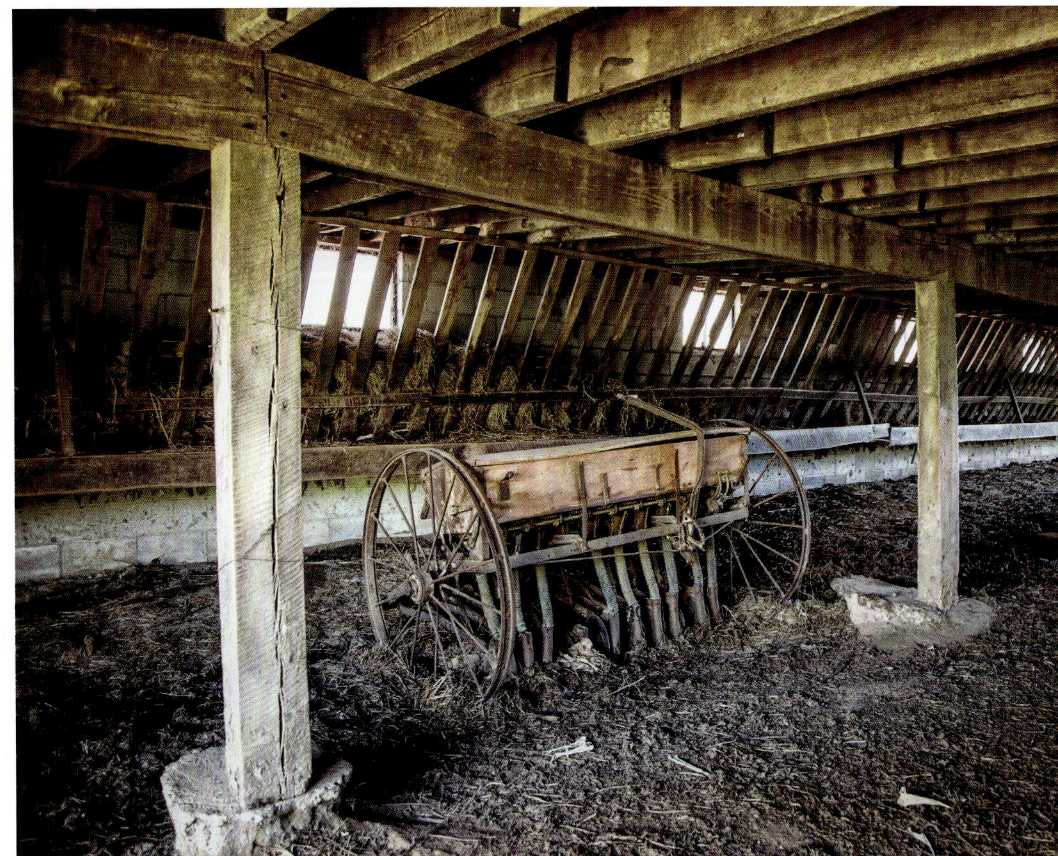

Trigg County.

Taking a back road to Cadiz, I discovered this cattle barn with a hay hood, a very common design in Western Kentucky. Hay in the early twentieth century was cut using a horse-drawn hay reaper, such as the one in the image on the right. Once it was cultivated, the hay would be drawn up into the barn loft through the hay hood by a hayfork pictured on the left.

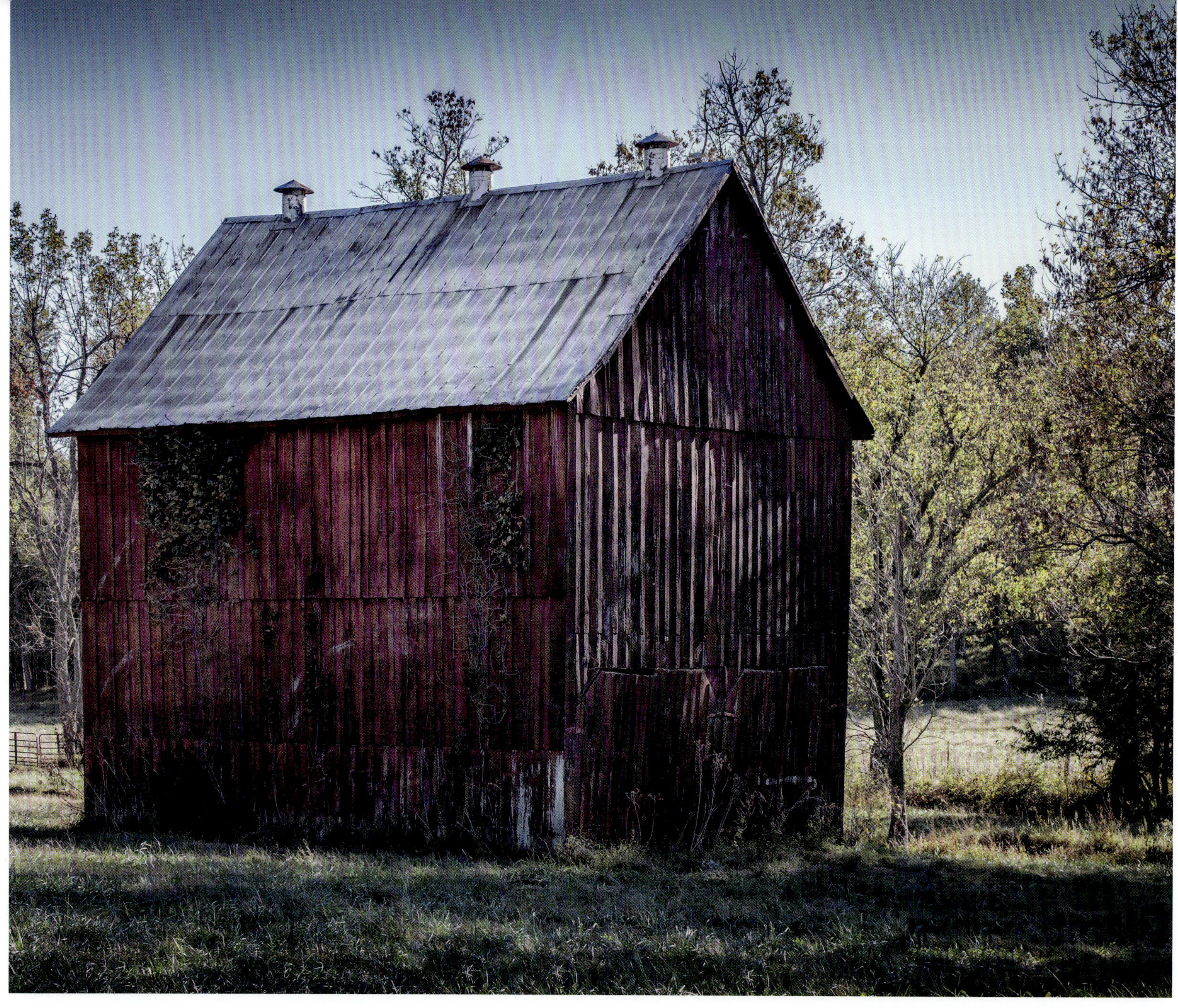

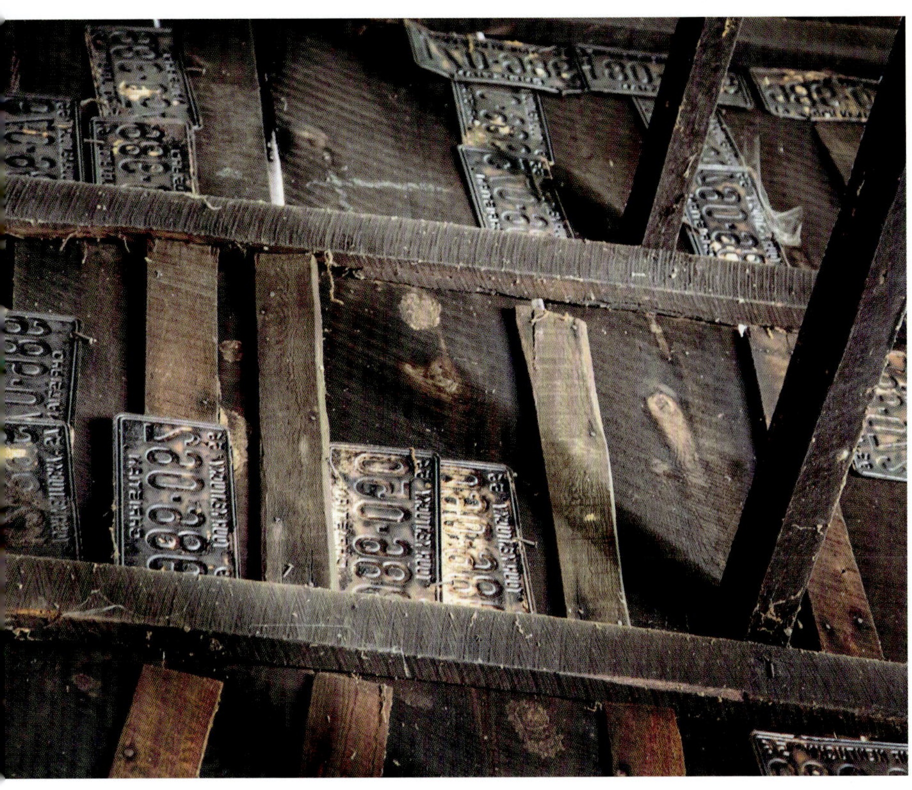

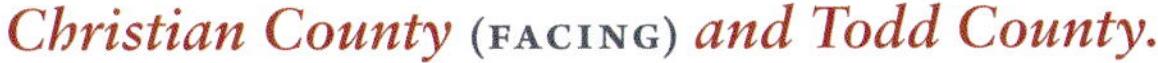

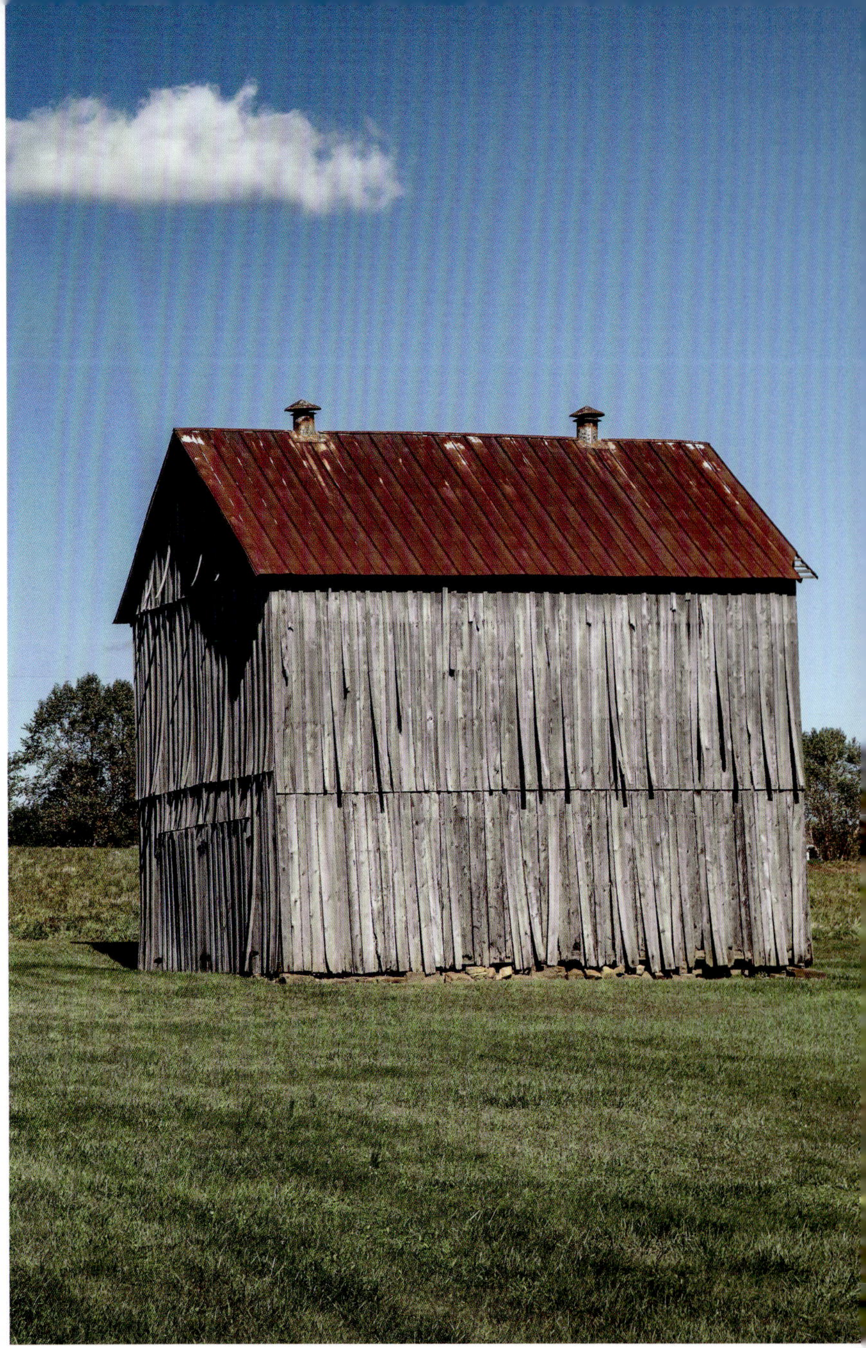

Christian County (FACING) *and Todd County.*

Dark tobacco firing barns in Western Kentucky required a good seal to maintain a dense smoke inside. To cover up holes that appear over time, many farmers used old license plates tacked to the sides. Although picturesque, old wooden dark tobacco firing barns, now abandoned, tell the story of the changing tobacco industry in Kentucky.

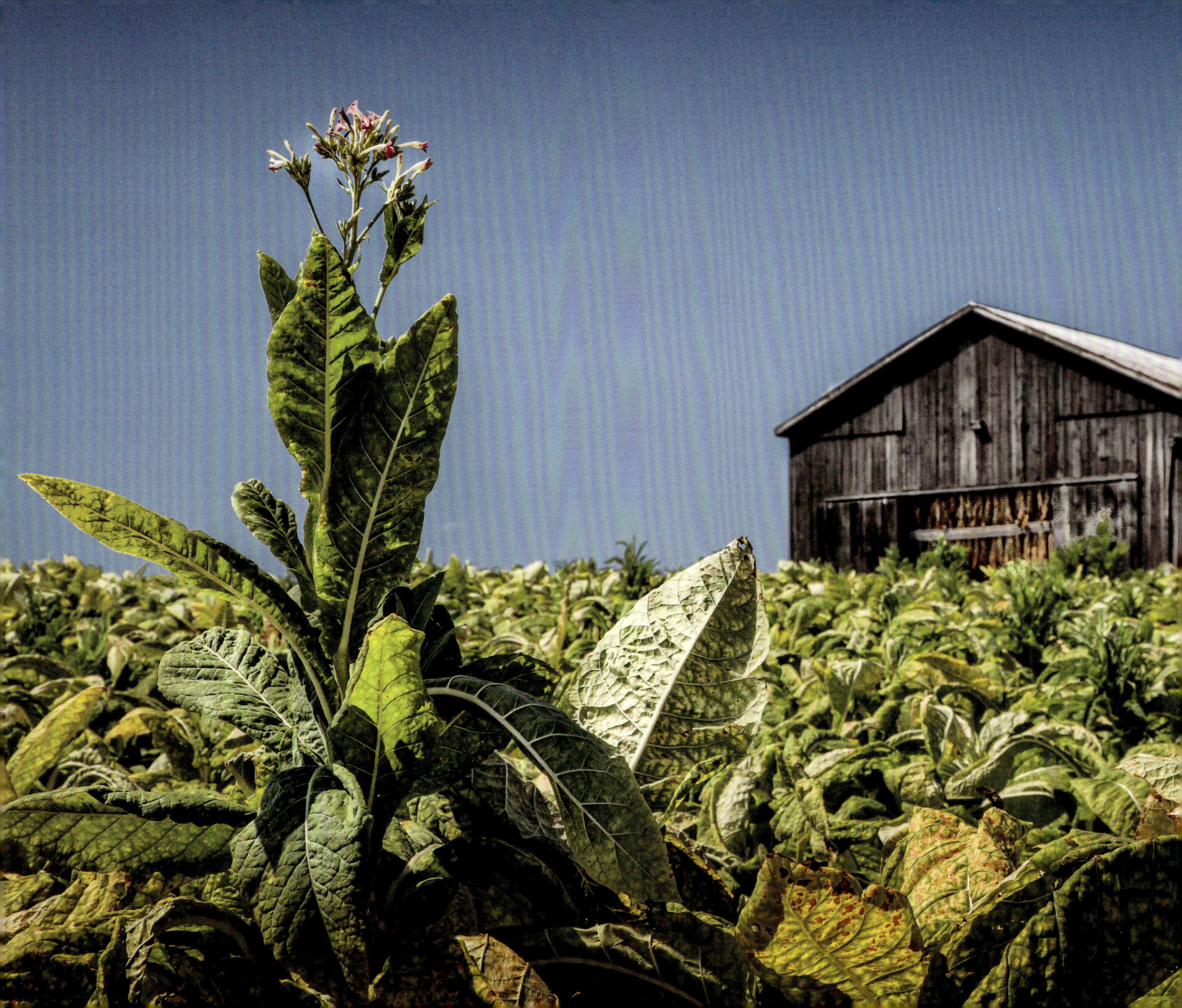

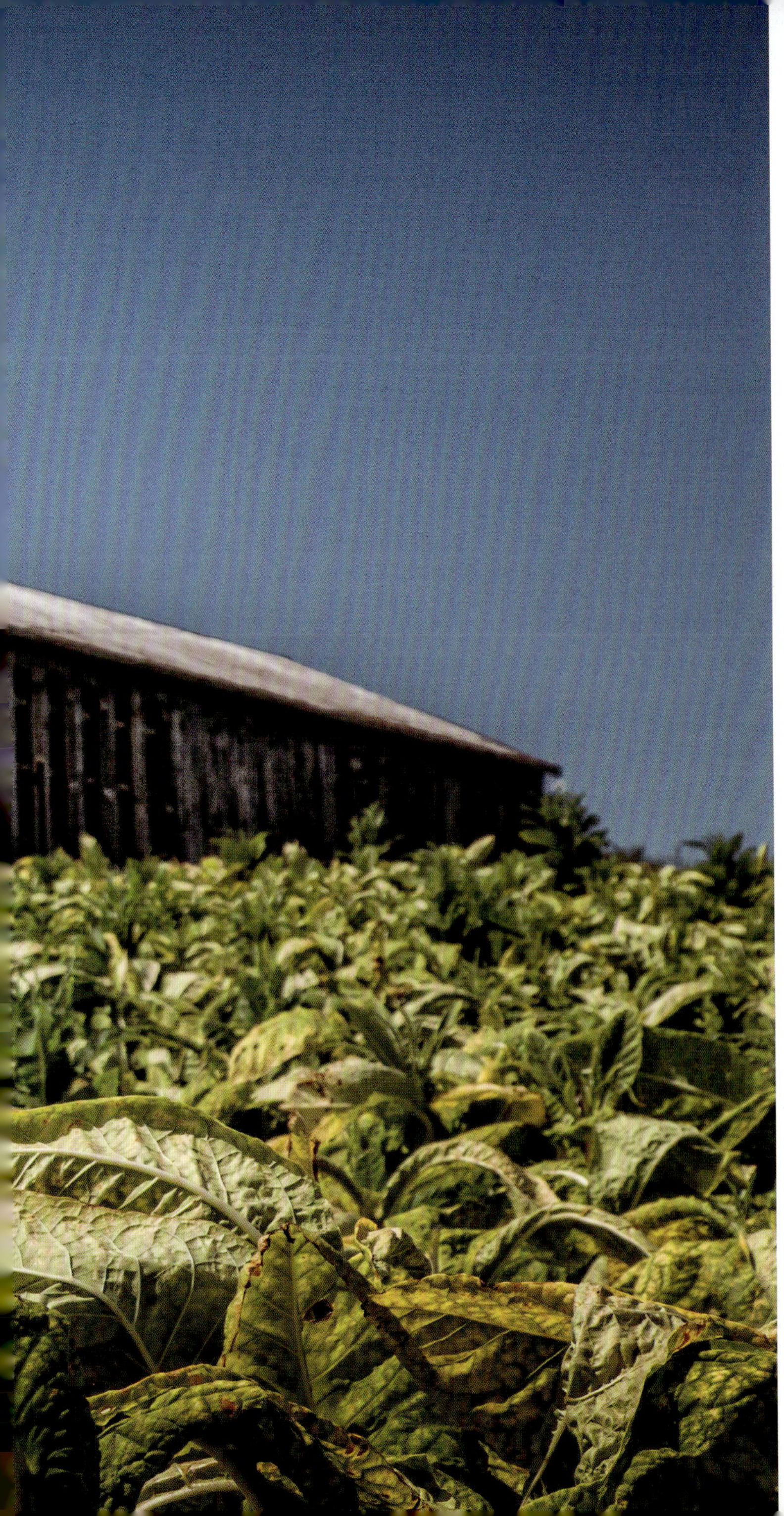

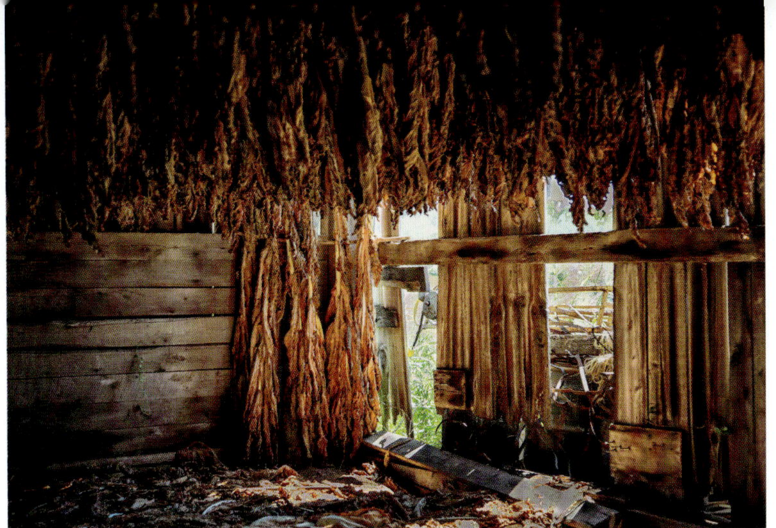

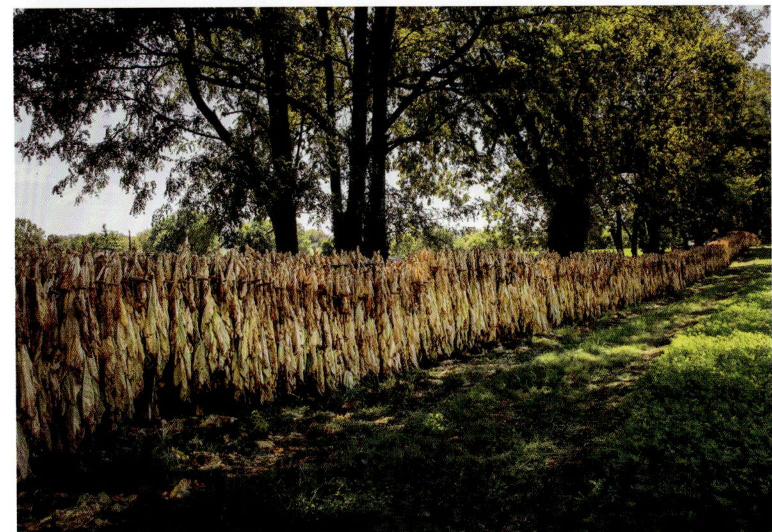

The burley tobacco leaf is an upright, thin-leafed plant compared to dark tobacco. Usually the plant is topped as soon as the flower forms. Roughly three to four weeks later, the plant is harvested in the field and hung by the stalks in tobacco barns with ventilation to air-cure.

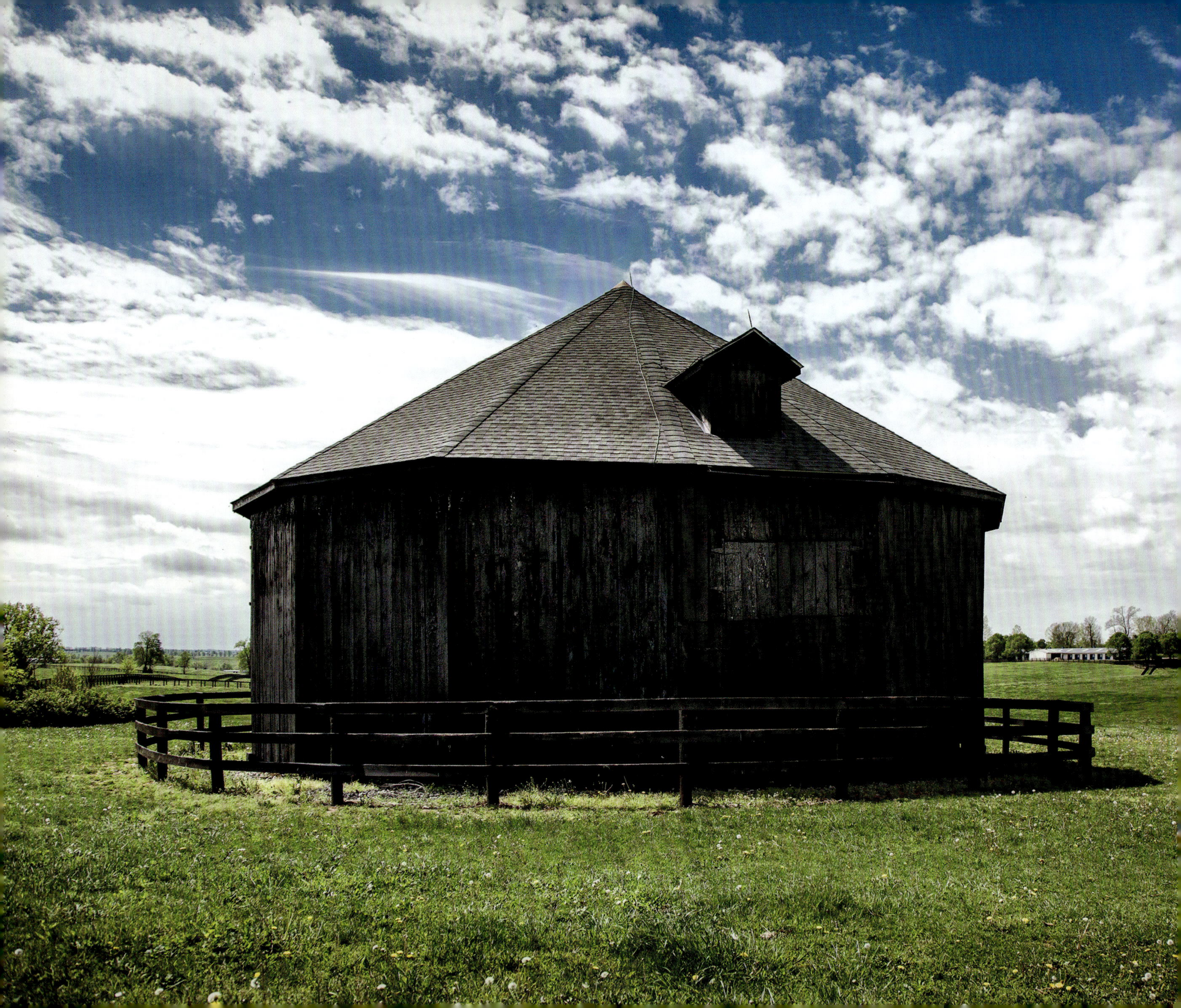

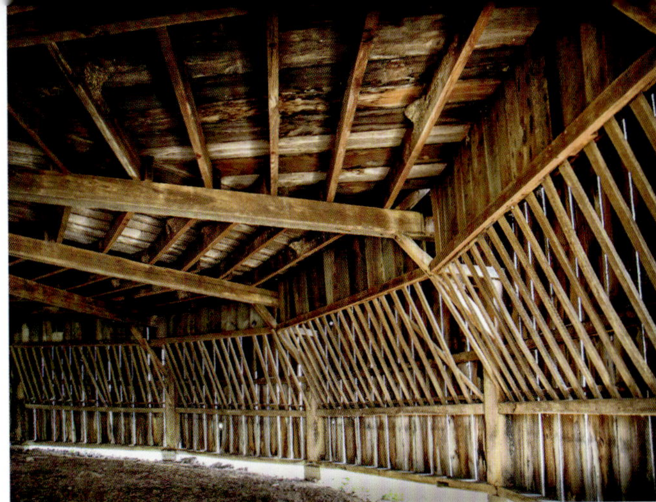

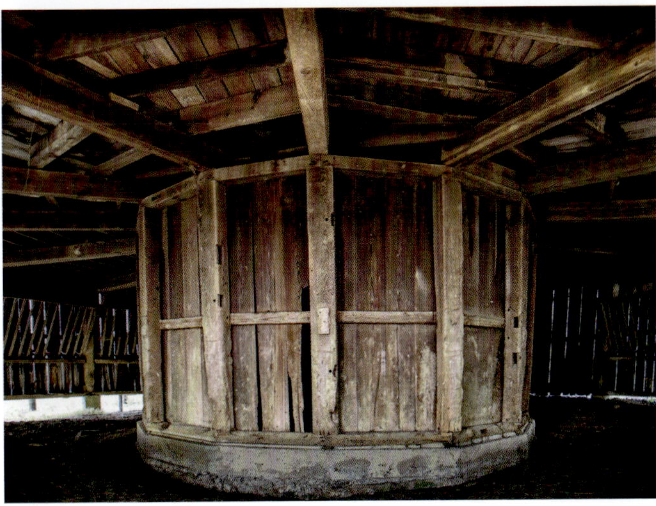

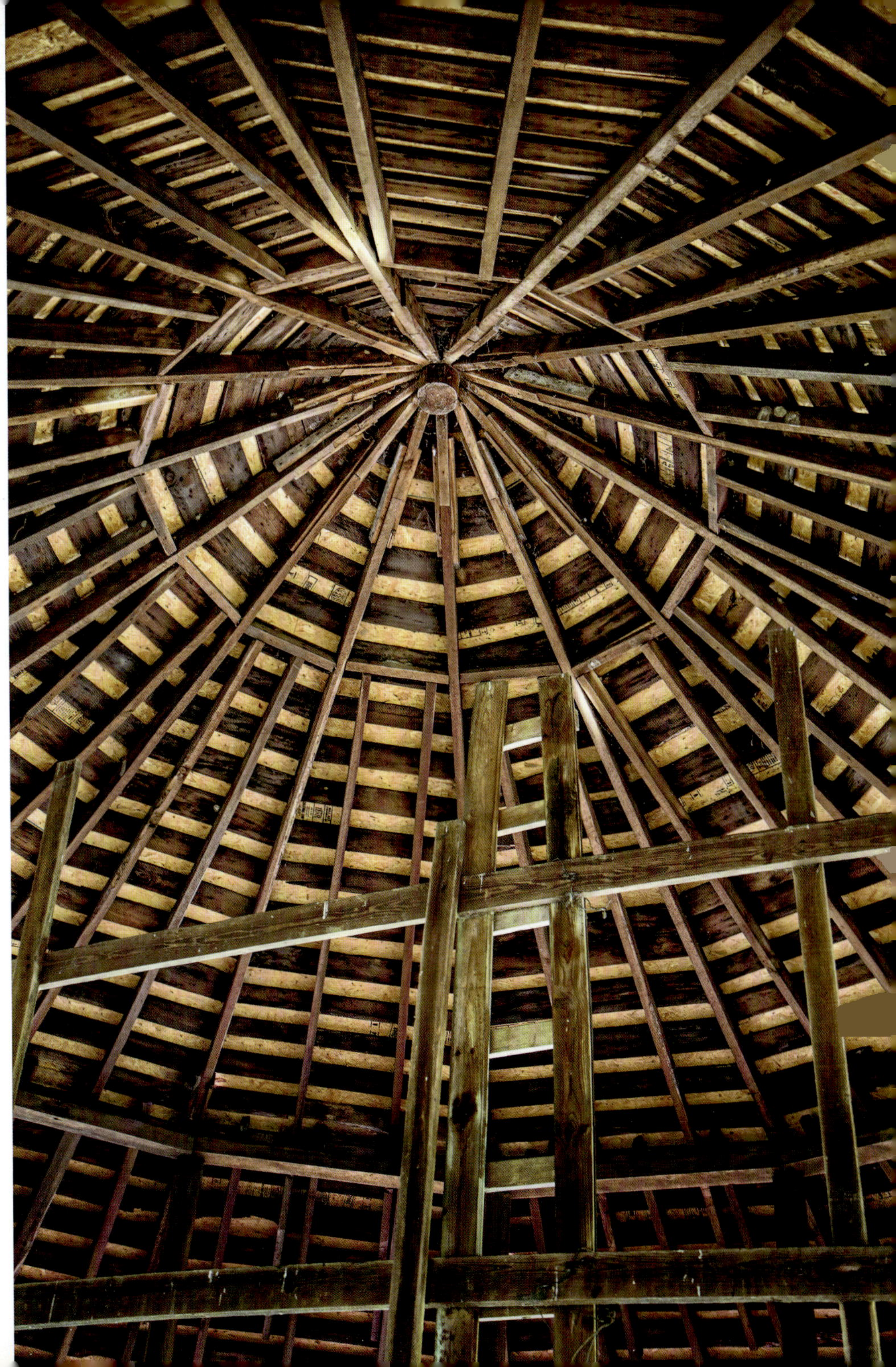

Bourbon County.

Agricultural experiment stations in the late nineteenth and early twentieth century promoted octagonal barns as economical building forms. This octagonal barn, now on the Hunterton Farm at Stoner Creek, has an amazing interior, with a loft designed for hay to be dropped into the racks around the outside wall and an interior room at its core with hand-hewn beams. It was constructed around 1913.

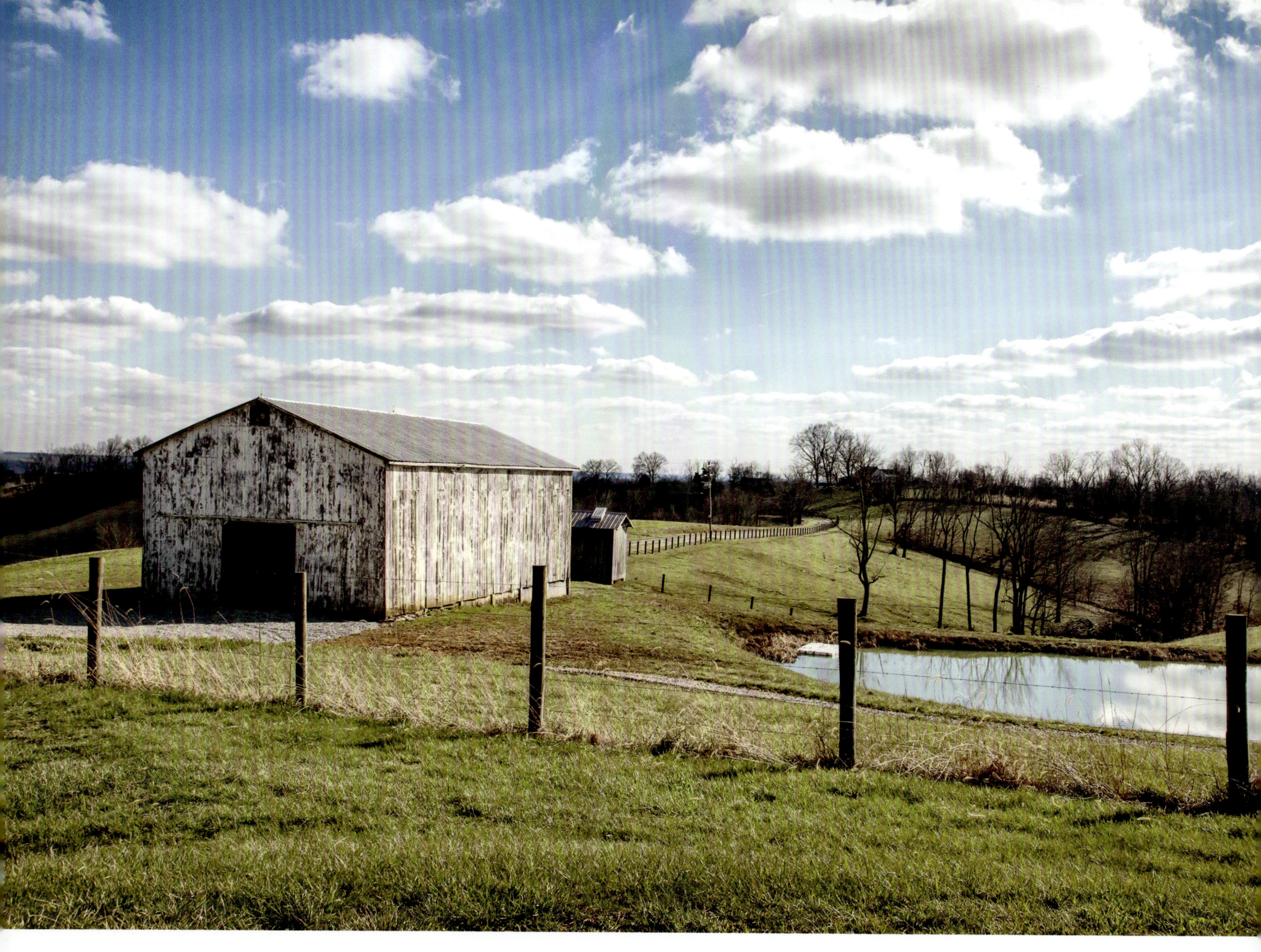

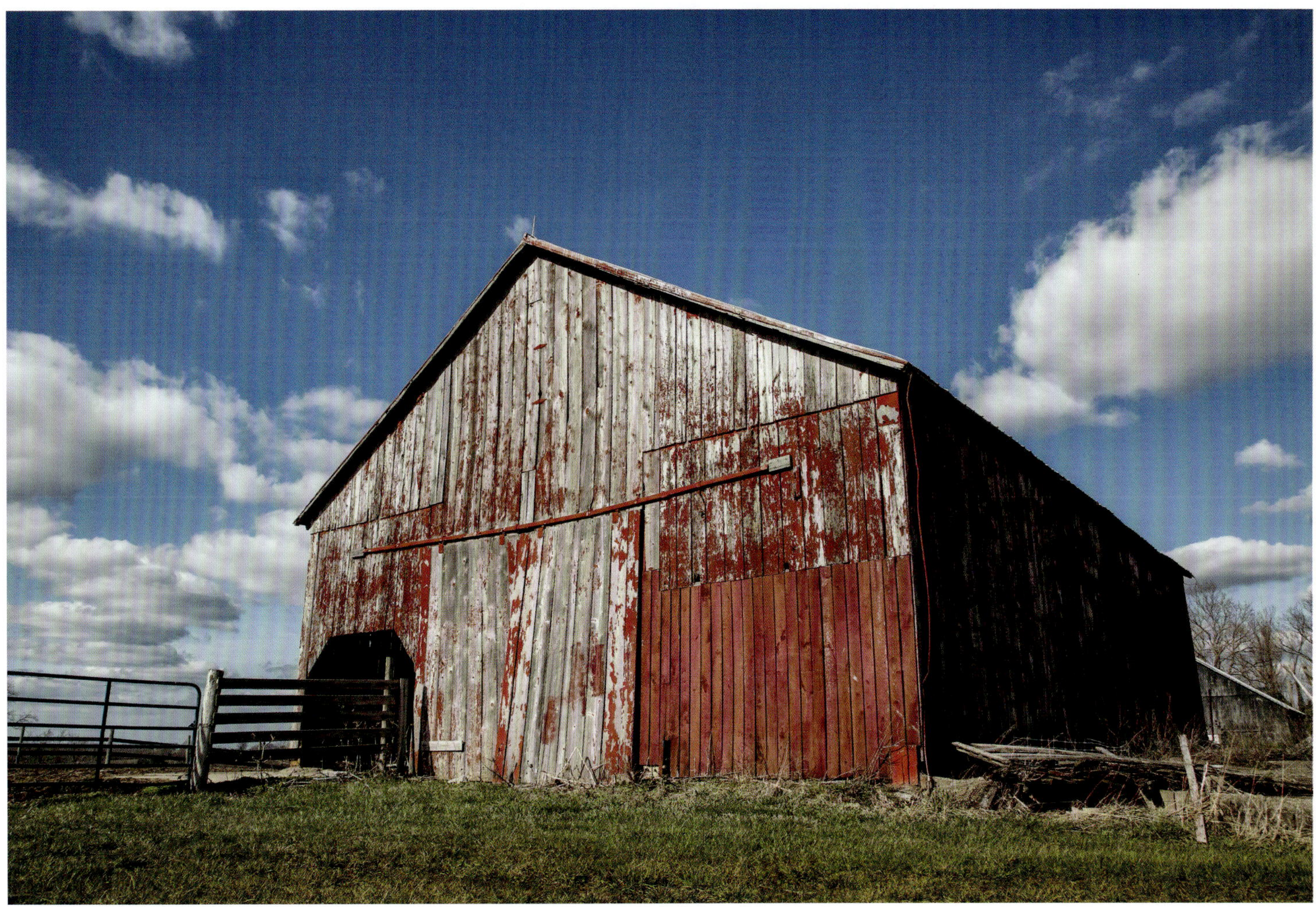

Nicholas County.

Sitting up on a ridge overlooking valleys on either side is this old tobacco barn. Although the area does not look hospitable for fields of tobacco, the crop is grown in a variety of conditions.

Nicholas County.

This peg barn was built by passing a spline, or tenon tongue, through a mortise in a post; it was then pegged rather than nailed to keep it in place.

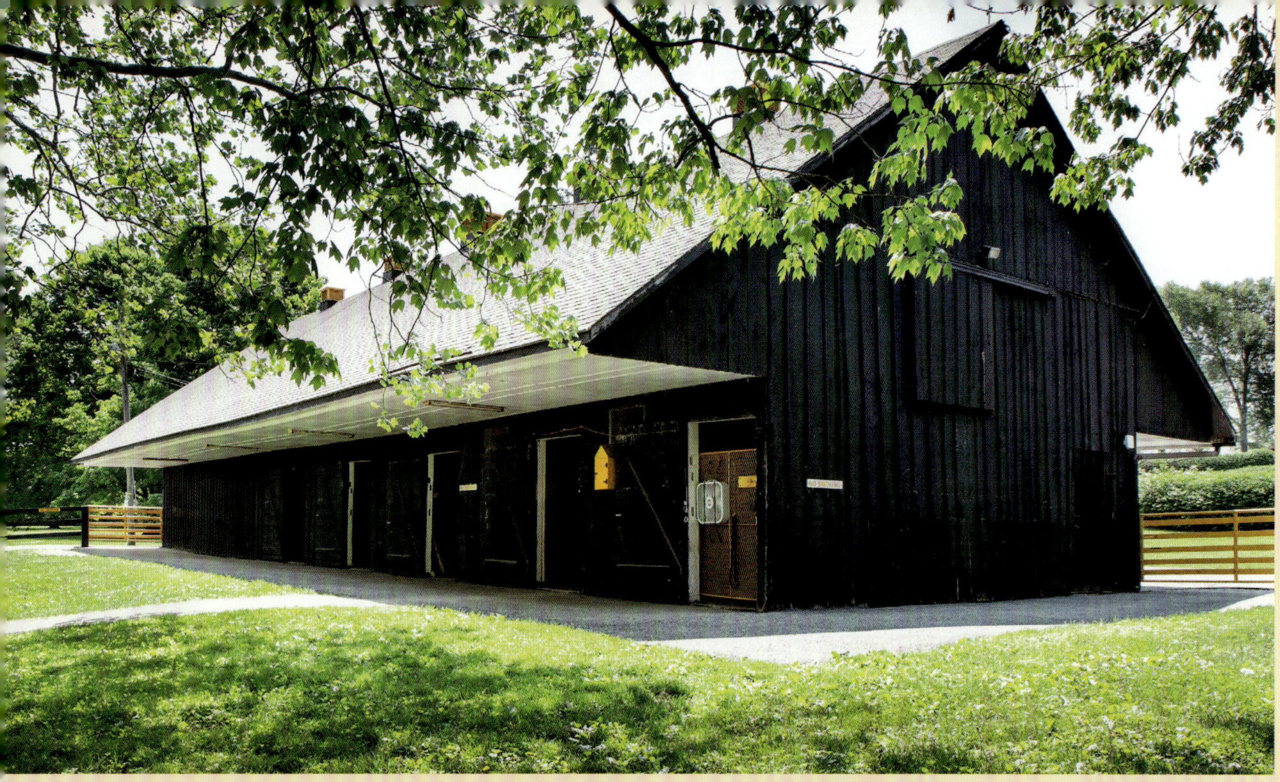

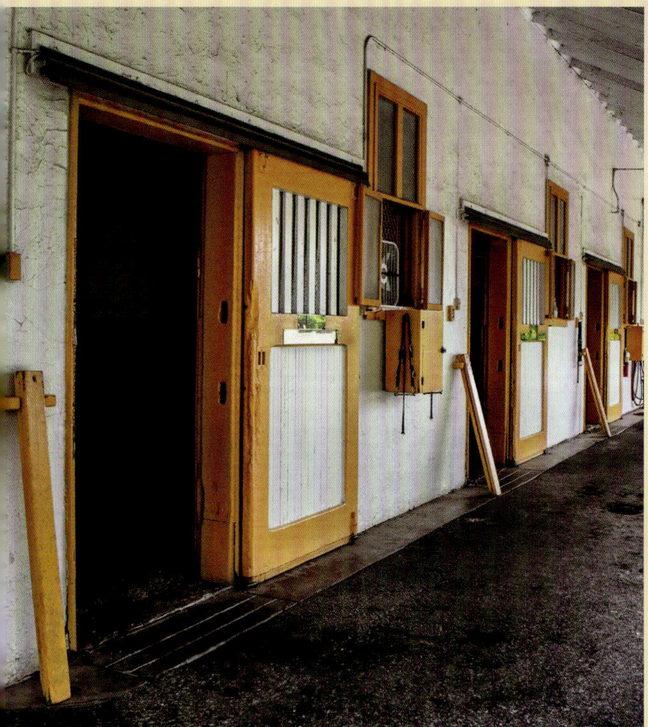
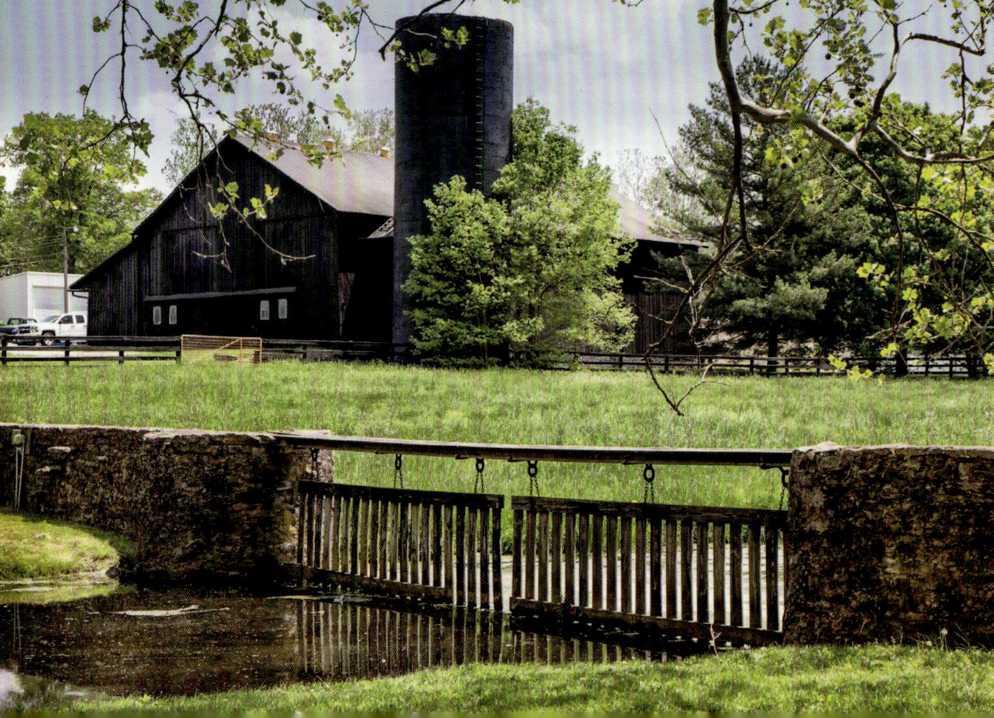

Bourbon County.

Claiborne Farm was established in 1910 by the Hancock family, who continues to run the farm today. On-site is the original wooden stallion barn and breeding shed, built around 1910, and a six-stall stallion barn that was built with block and stucco in the 1920s. Around the farm are other barns resembling the early stallion barn's wooden construction, now used as service barns.

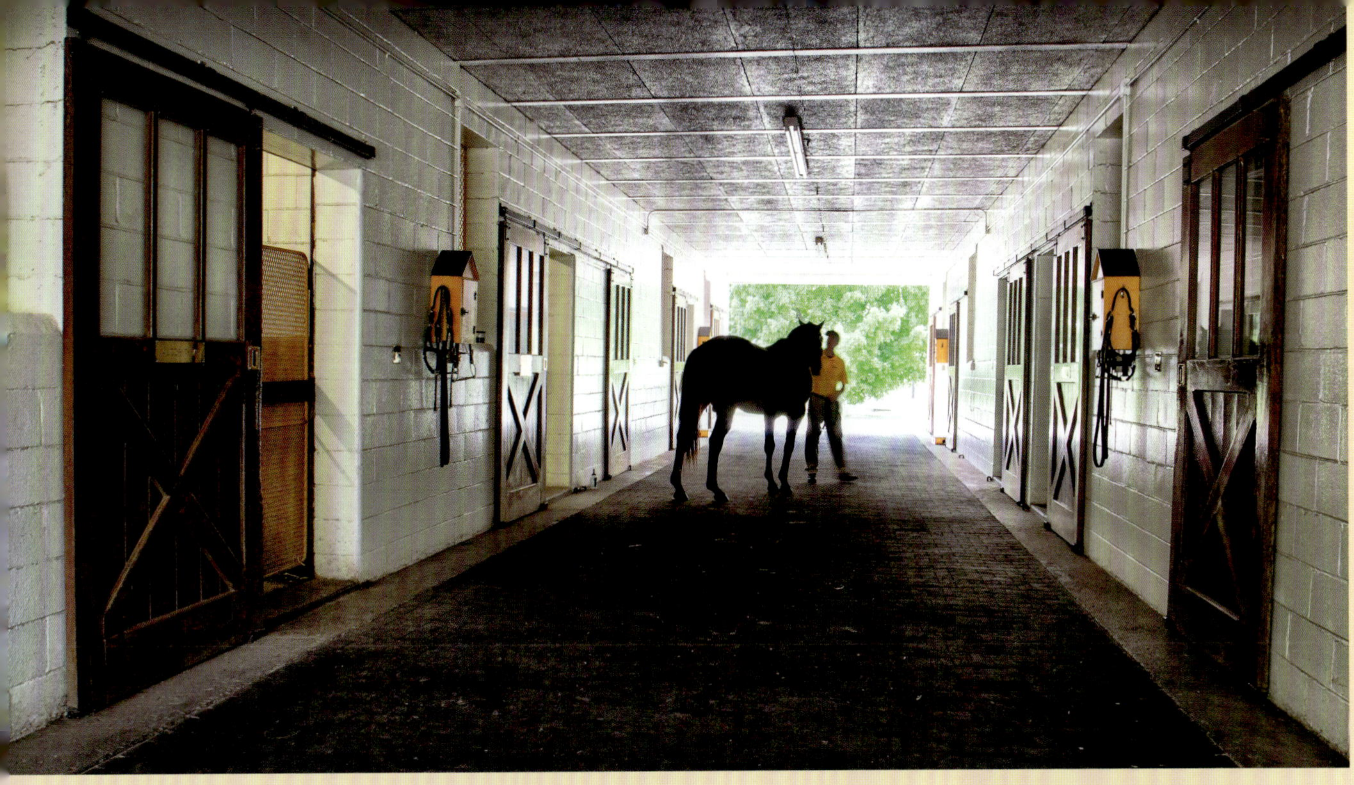

In the 1930s, Claiborne Farm built the block stallion barn where Secretariat stood at stud. Today, his stall is the first on the left as you enter the front of barn with yellow awnings; other successful stallions have used it in the years since.

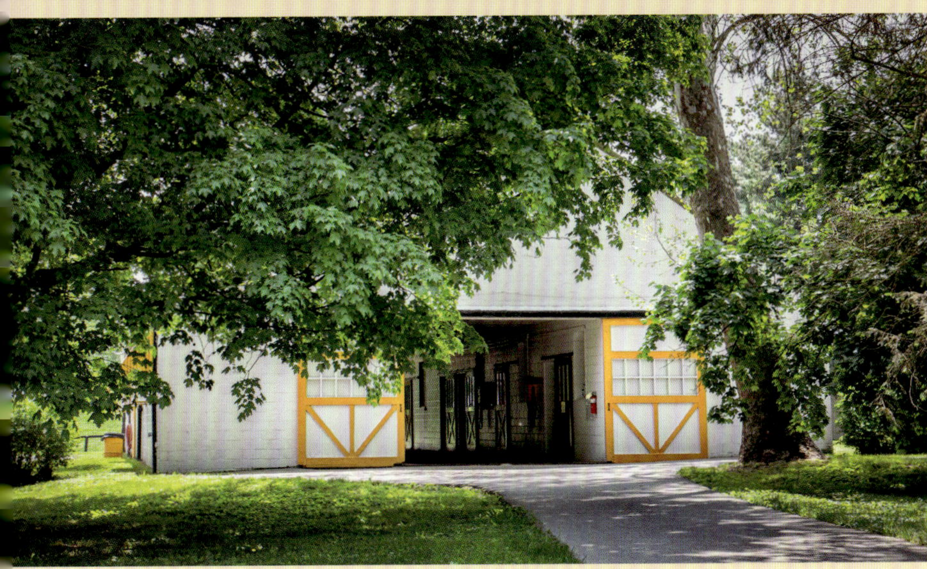

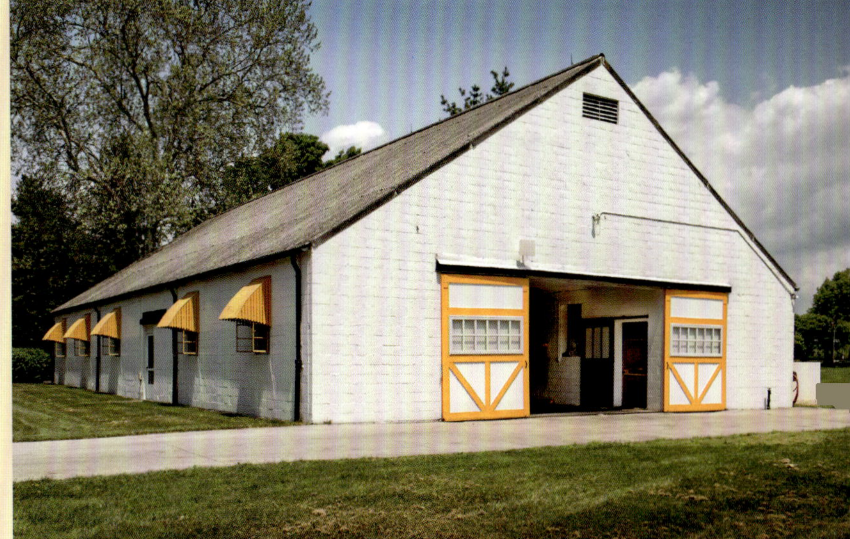

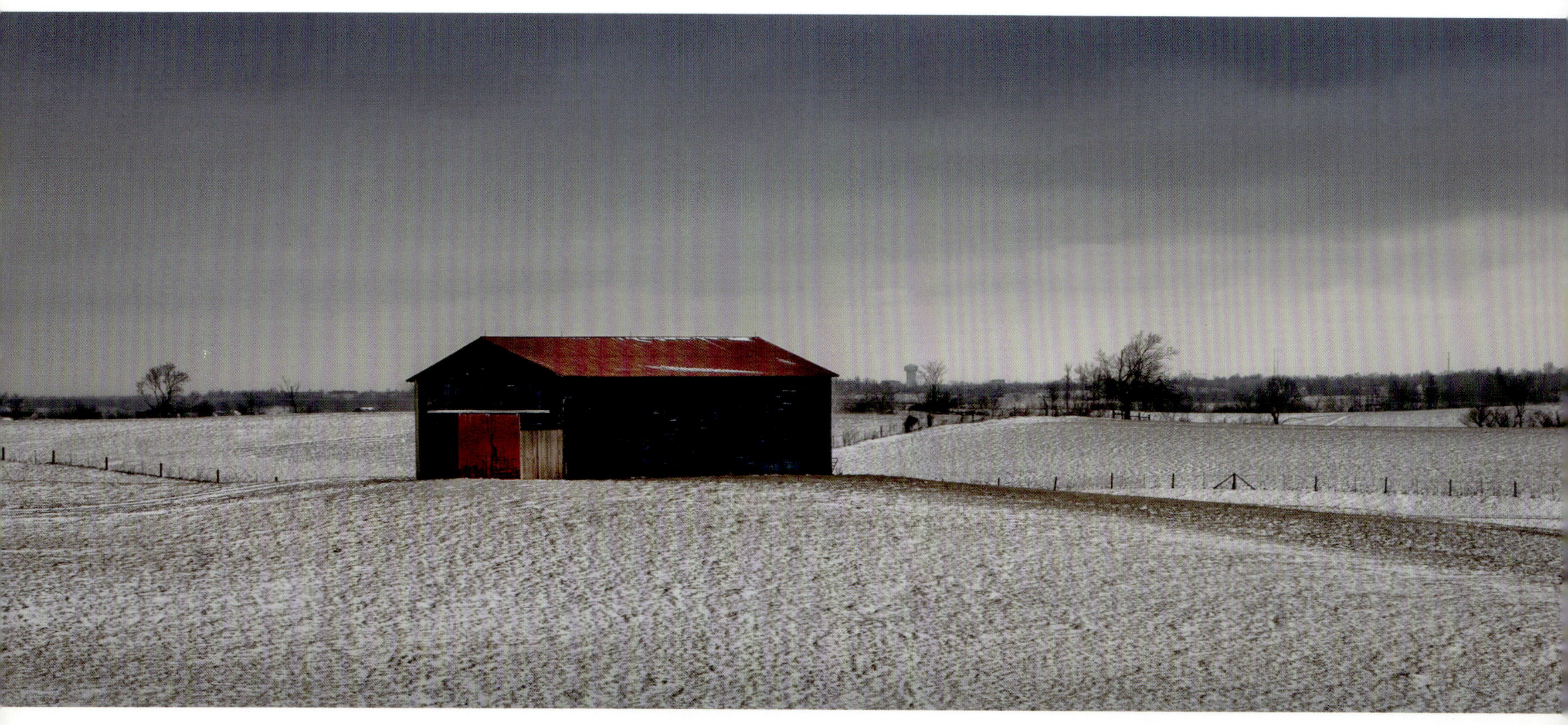

Madison County.

At the time this photo was taken, this barn was sitting out in the middle of a vast rolling field. It is now enclosed by a housing development.

FACING PAGE

Fayette County.

Urbanization has crept in on the horse farms of Fayette County with moments of uneasy cohabitation like this. The barn in this photo was eventually replaced with a big-box store.

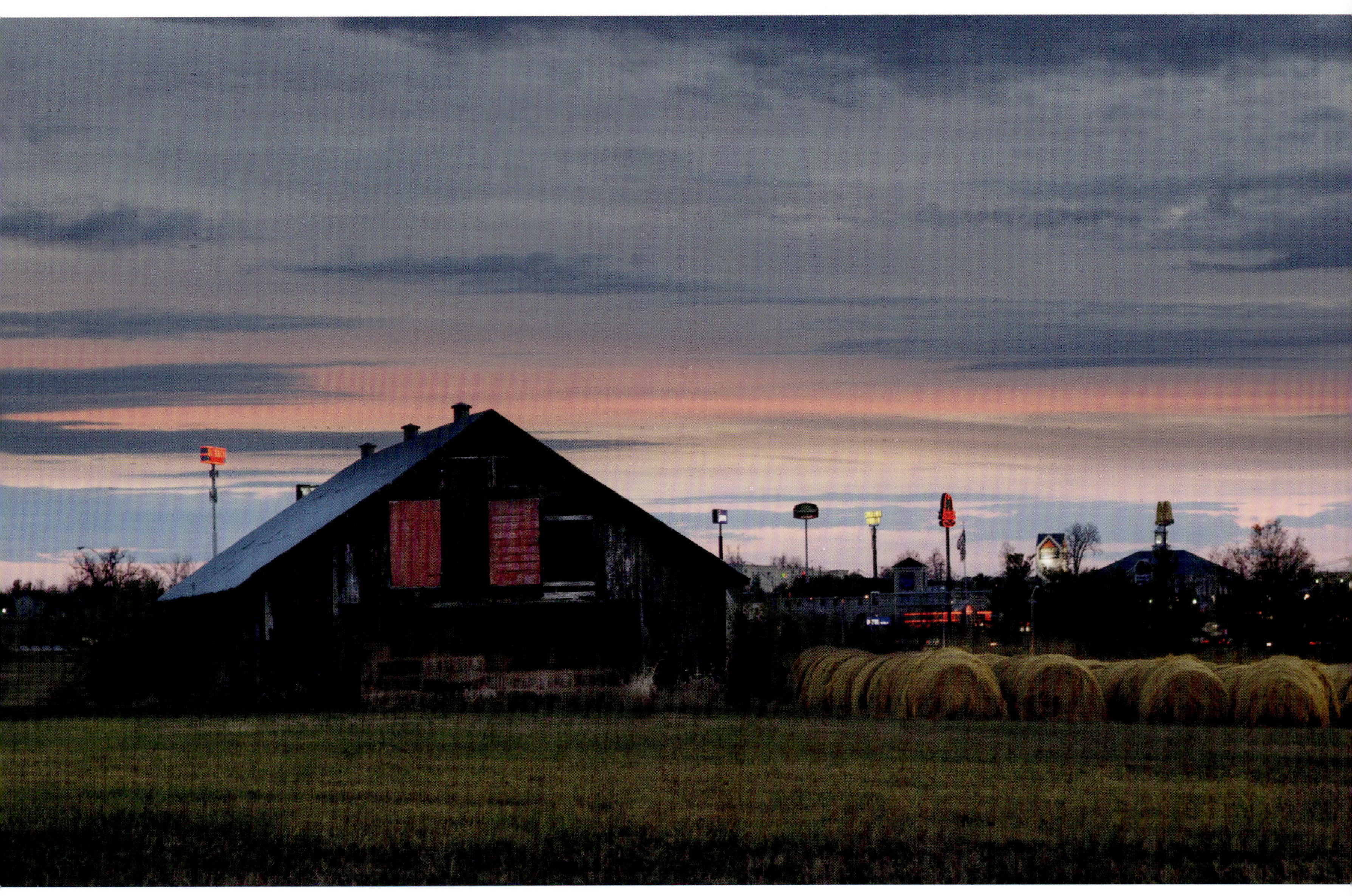

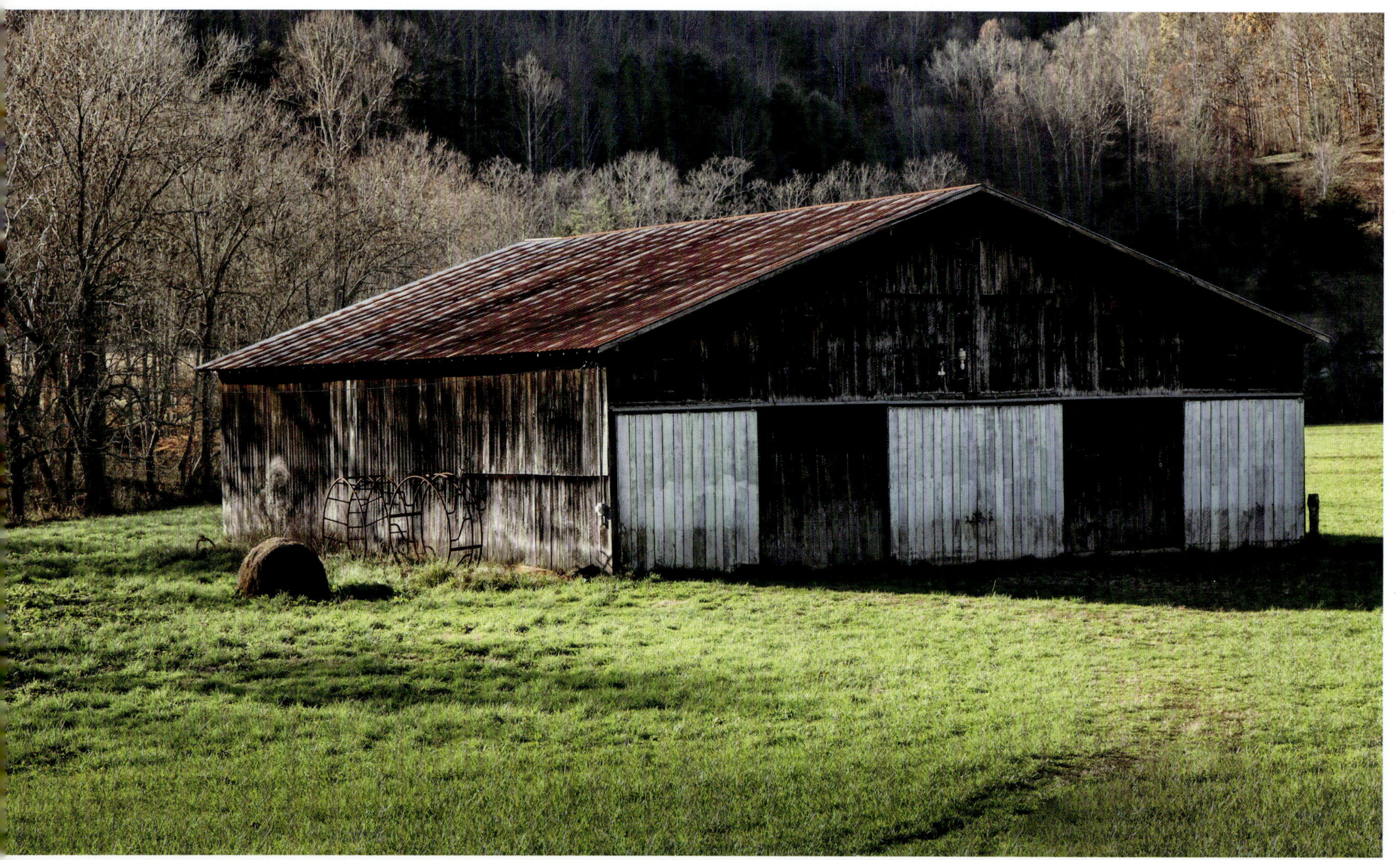

Still used as a stock barn, this wide building
is situated in a holler south of Berea.

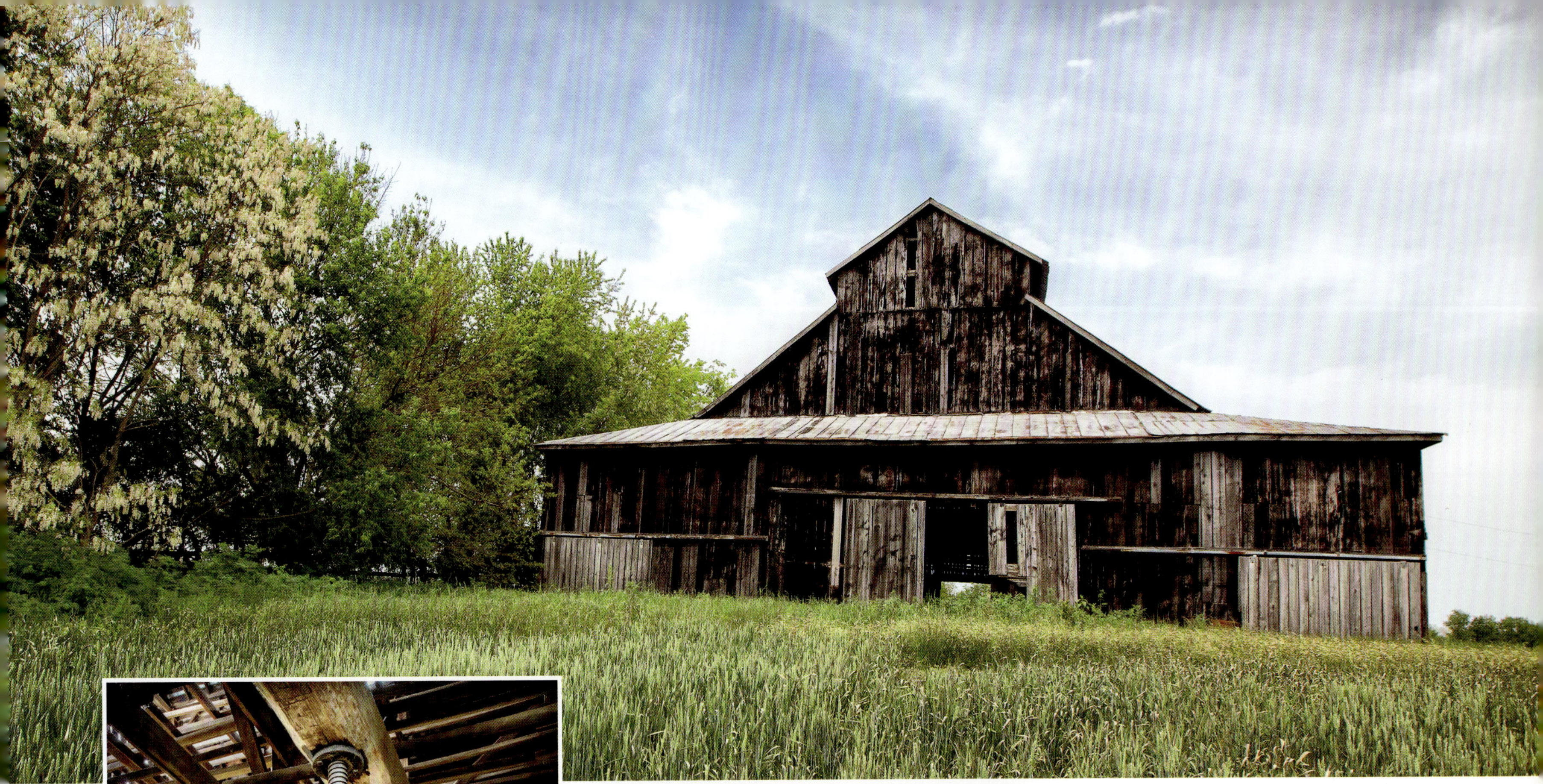

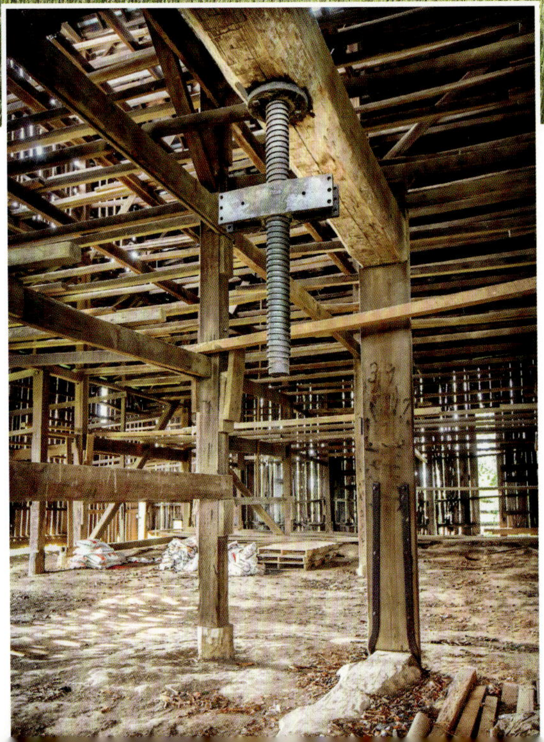

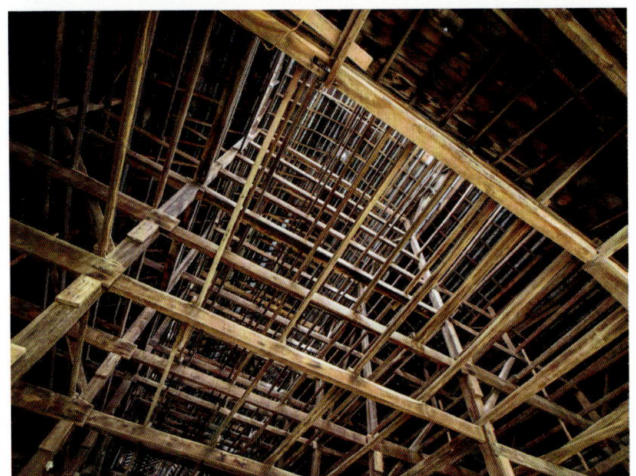

The original part of this barn was built in the nineteenth century without nails and added onto later. Used as a tobacco barn, it still has parts of the press for pressing tobacco into a hogshead inside.

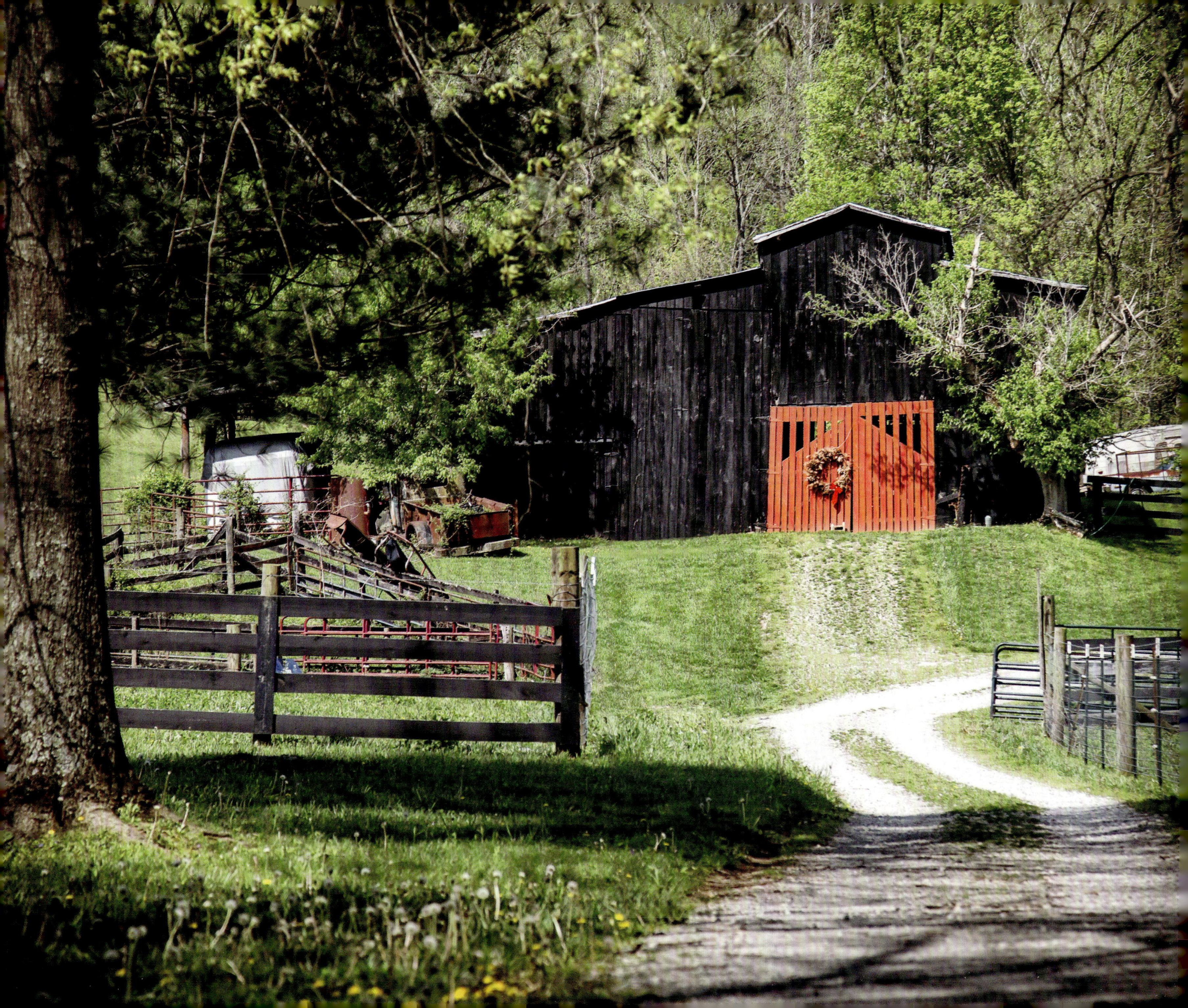

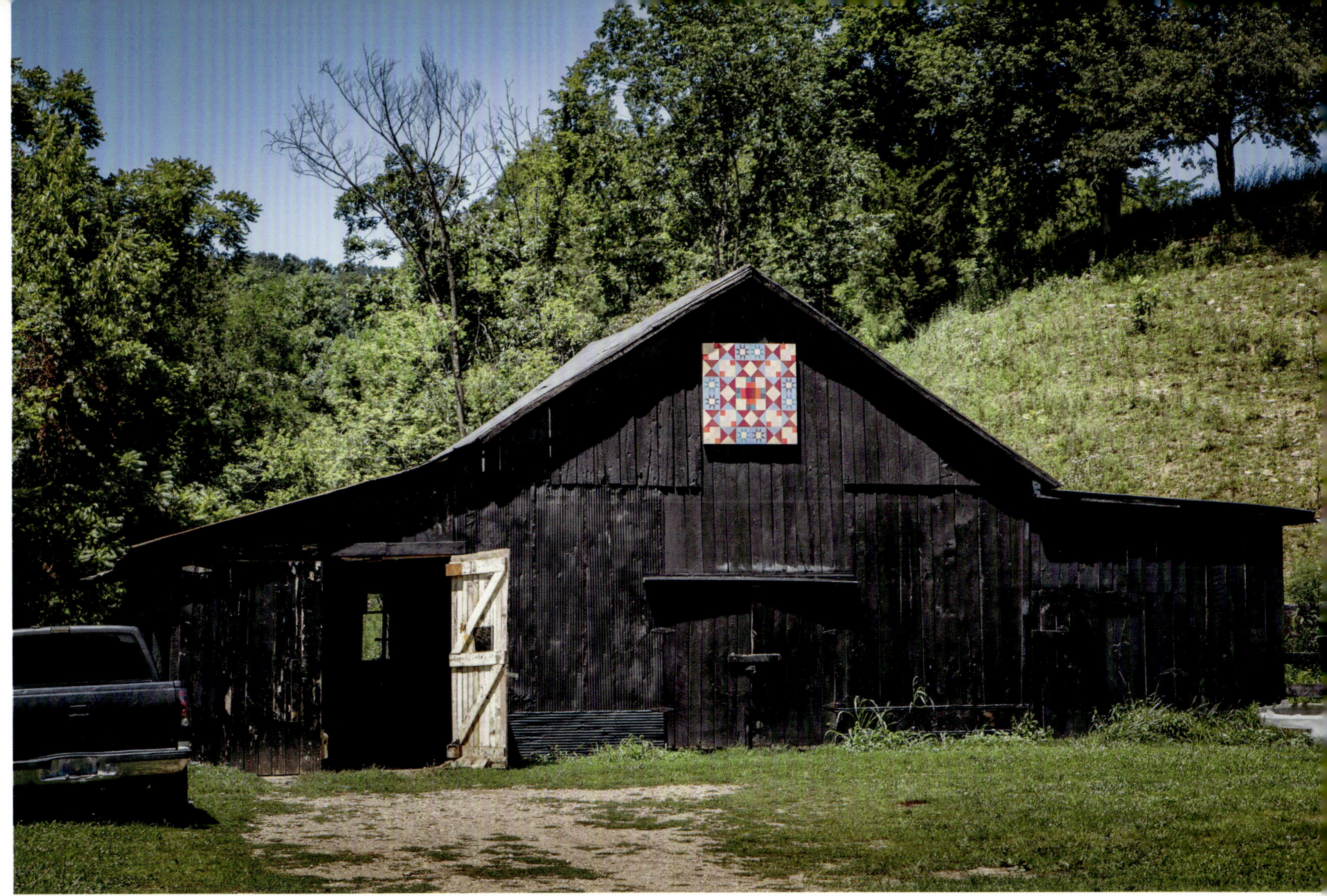

Franklin County.

Vented sliding doors decorate this early stock barn.

Lincoln County.

As barns age, many owners choose to replace the original
wooden siding with sheet metal, as on this barn.

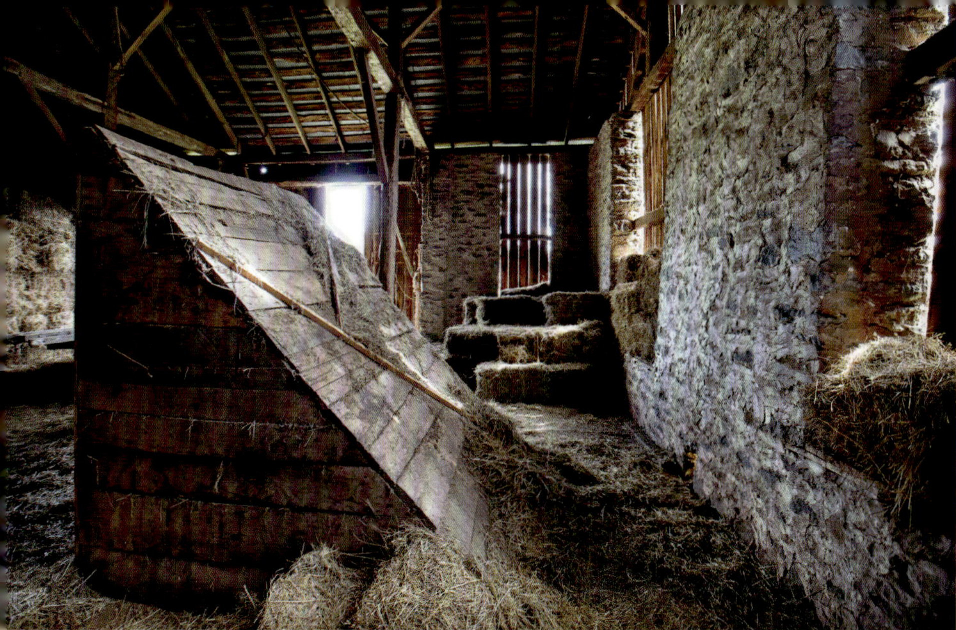

Campbell County.

Andrew Ritter helped build many of the stone barns in the German settlement of Camp Springs during the mid-nineteenth century. His own farm is a fine example of his handiwork, with timber loft and stonemasonry for the livestock area underneath. Upstairs in the hayloft, an enclosed stairwell leads downstairs, and hand-hewn ladders, short and tall, aid in access to the hayfork track. The carved cross with two hearts is a trademark of his.

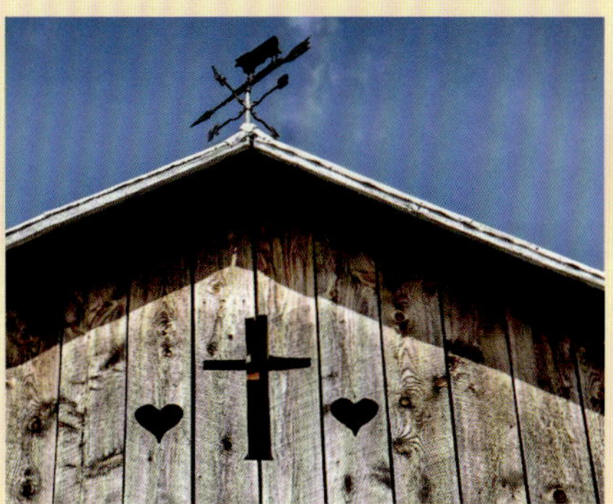

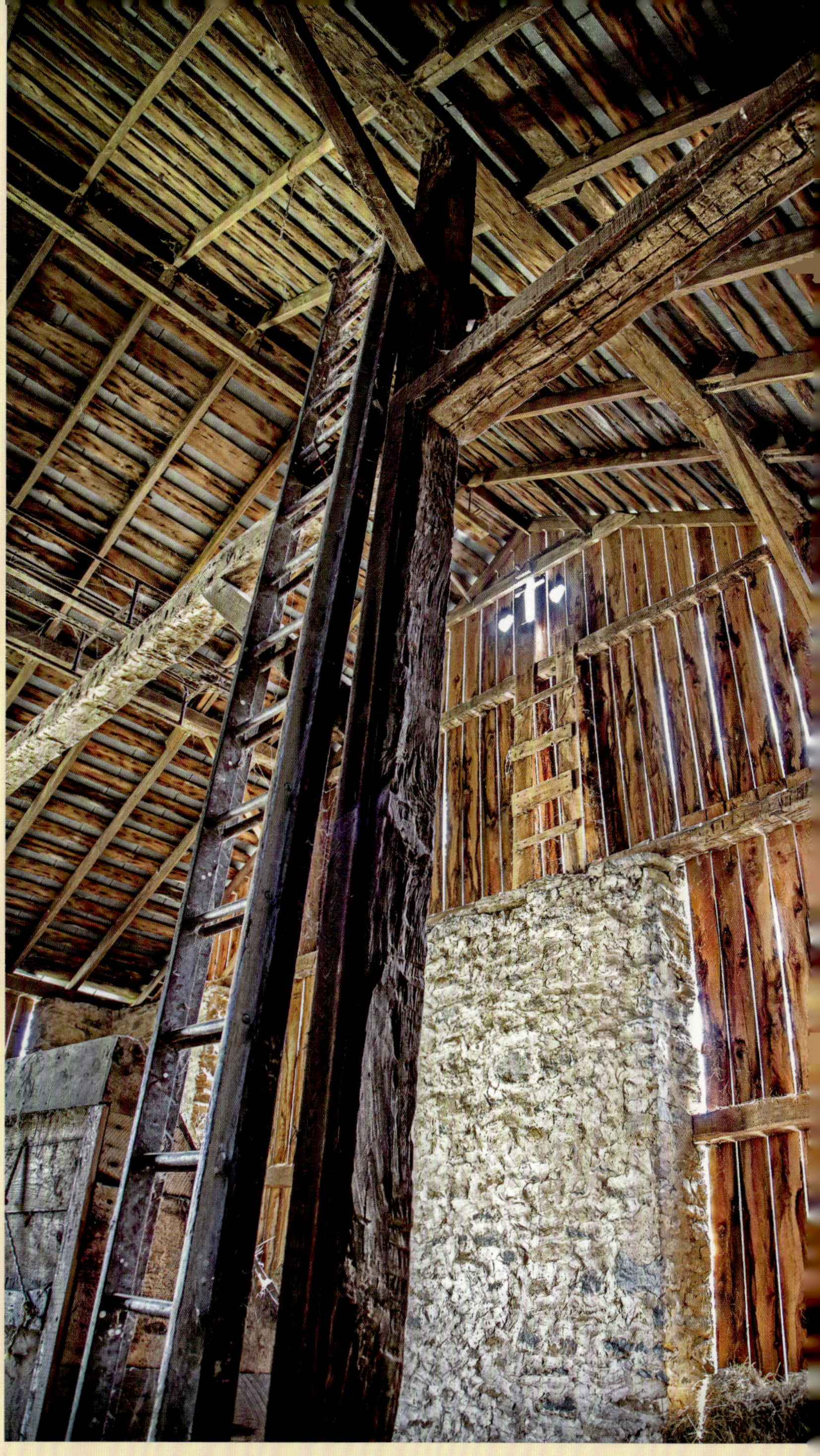

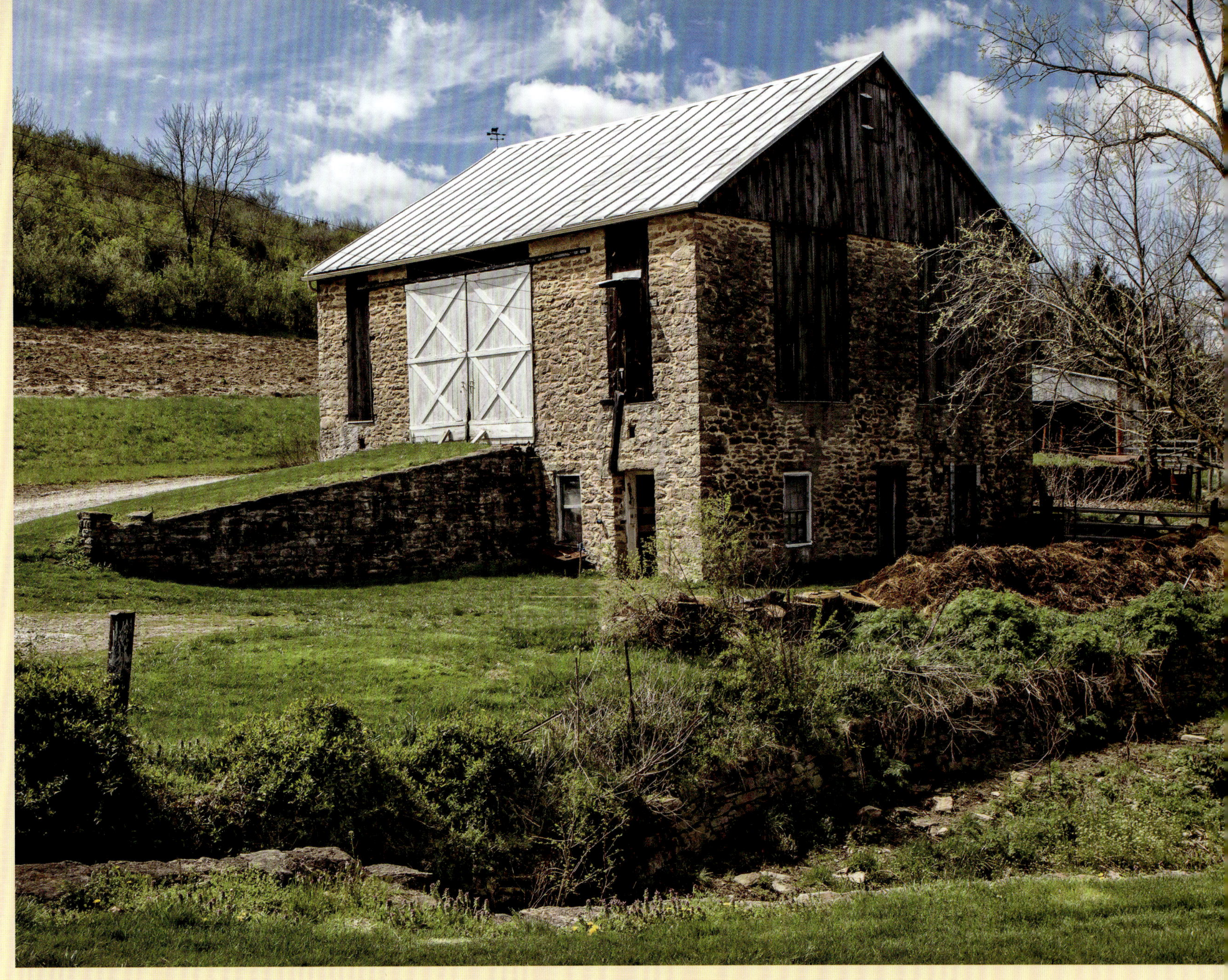

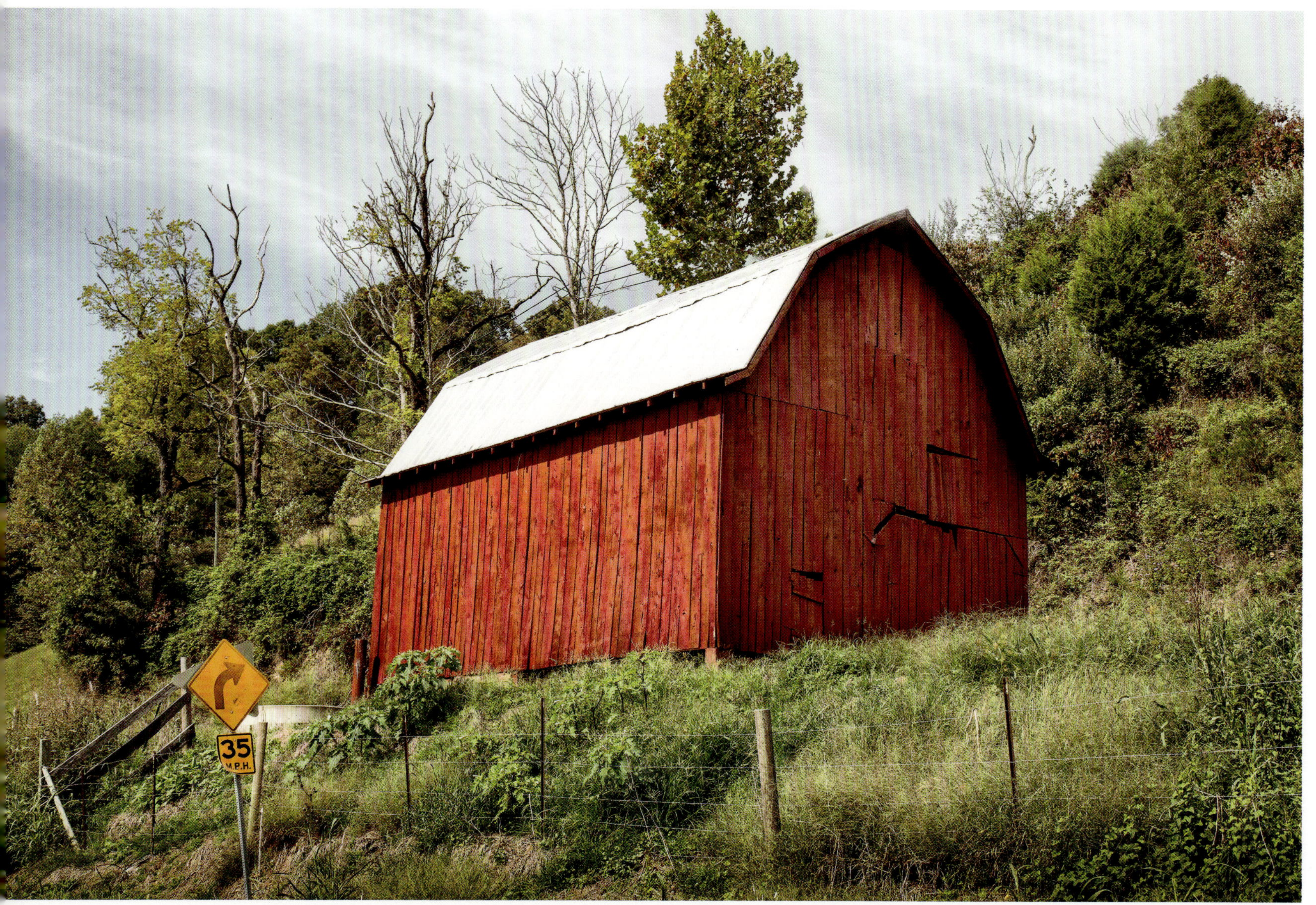

35
M.P.H.

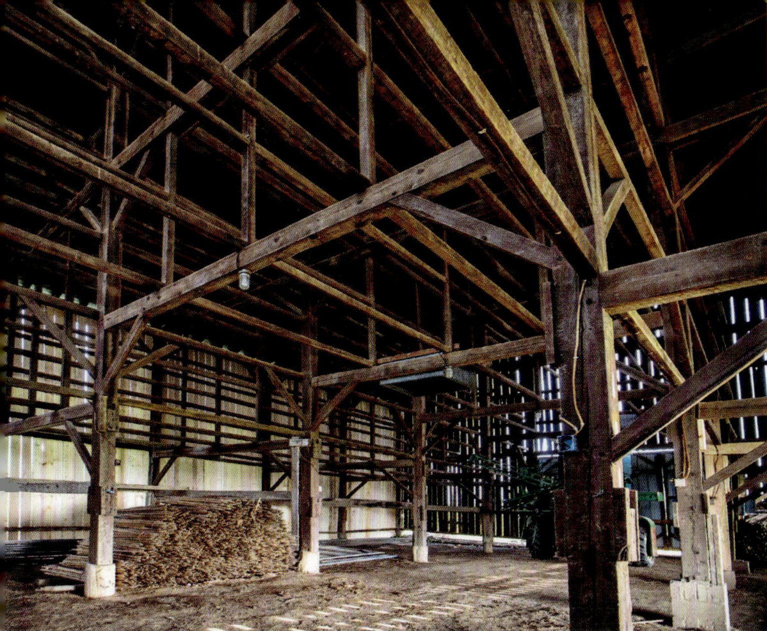

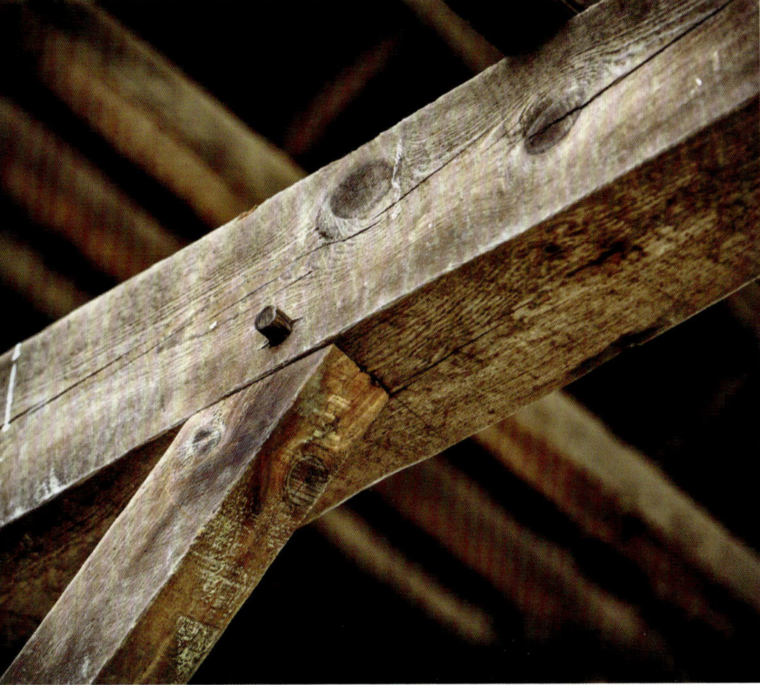

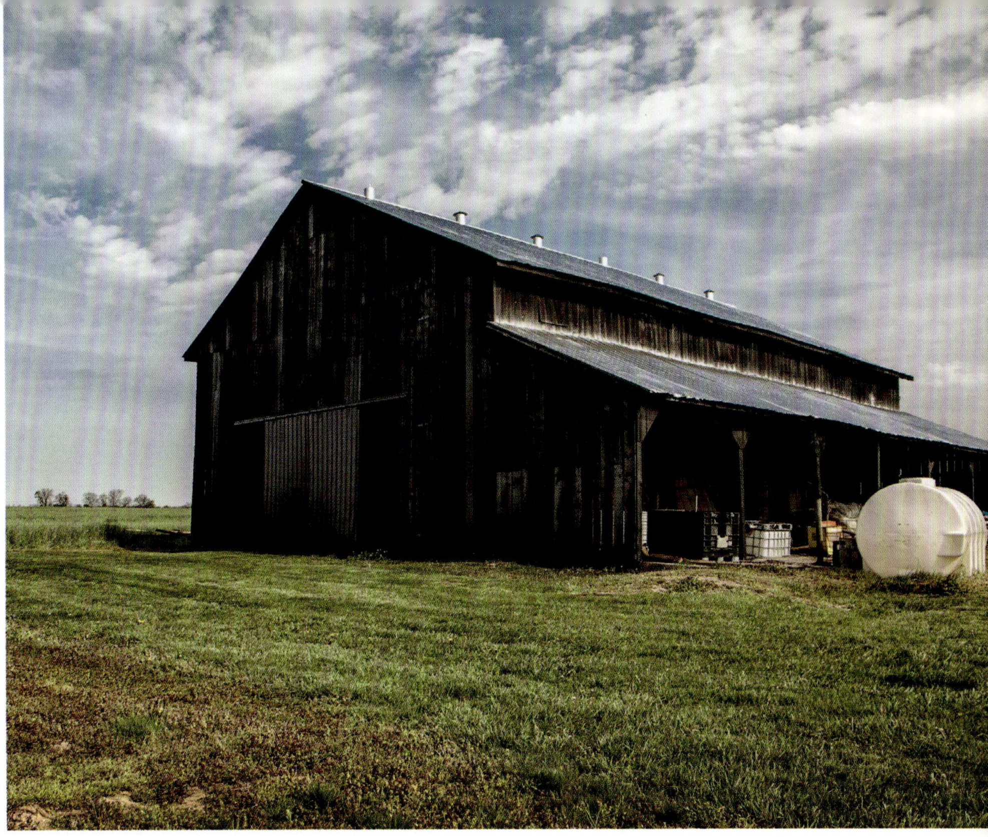

Clark County.

Peg barns were put together using tenon and mortise joints, secured with a peg. The barn pictured here was used as a tobacco barn.

FACING PAGE

Lawrence County.

In the hilly regions of eastern Kentucky, barns can be found built on improbable patches of ground, such as this barn sidled up to the hillside and right above the road.

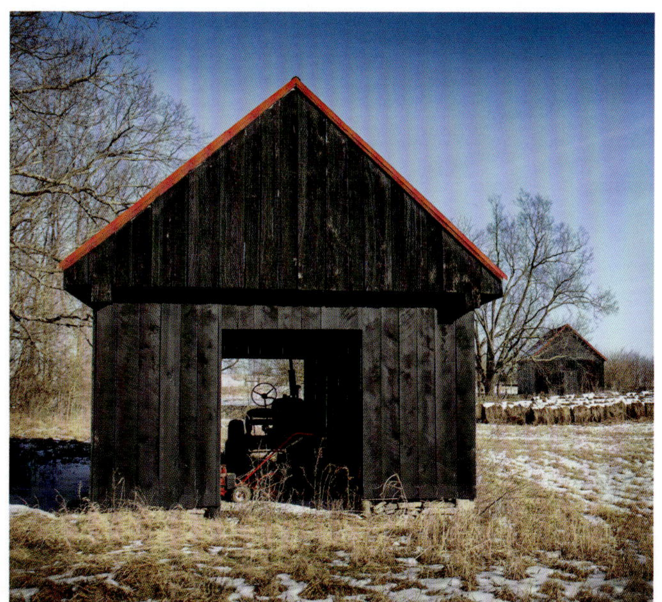

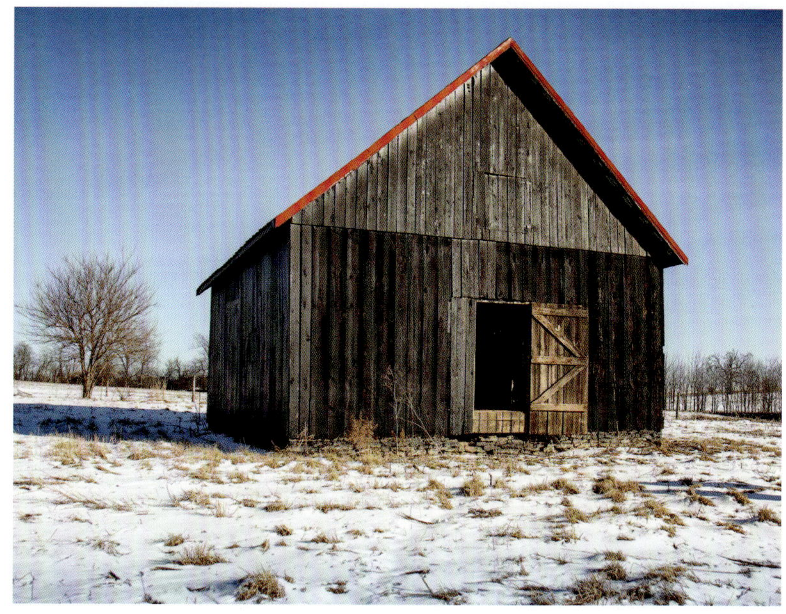

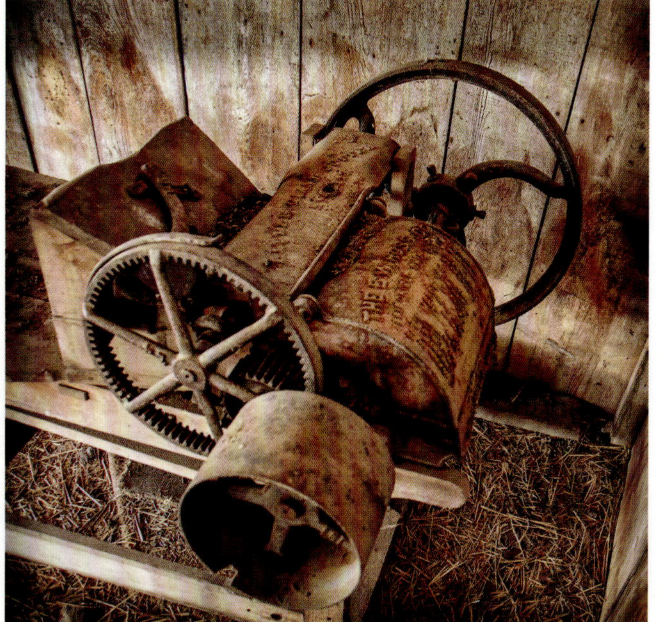

Bourbon County.

Auvergne is the historic homestead of Brutus Clay, maintained as a living historical capsule by his descendent Berle Clay. Most of the barns are timber framed, built circa 1840–1842 by a Mr. Layton. The smaller barns are built with a simple plan that can be adapted according to functional needs. One barn is a threshing barn or cutting-up barn with a raised floor; the other is a general-purpose barn. Inside the threshing barn is a machine for chopping fodder, from around 1894. The brick building was originally built as a hemp barn and later used as a lumber barn. The tobacco barn was built with a very wide central aisle, which, it is speculated, was intended to make turning carts around easier. The buggy resides inside a carriage house, one among many other single-purpose buildings on the place.

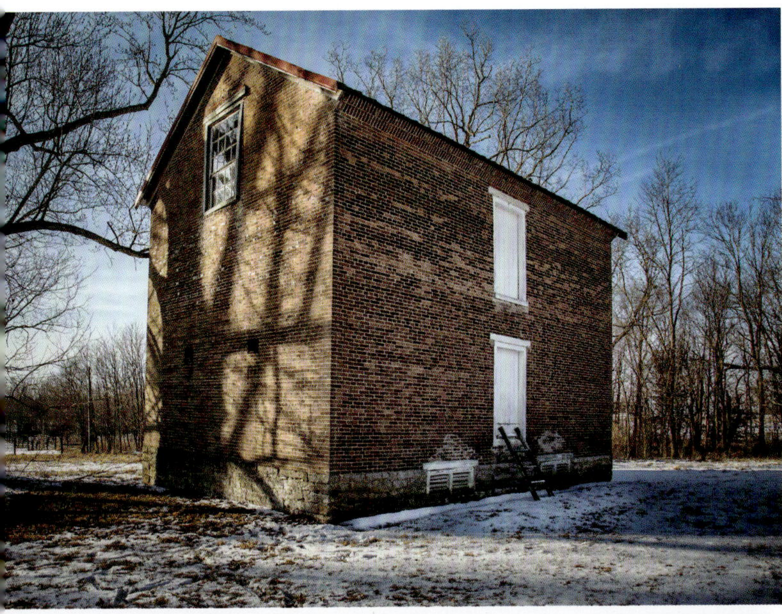

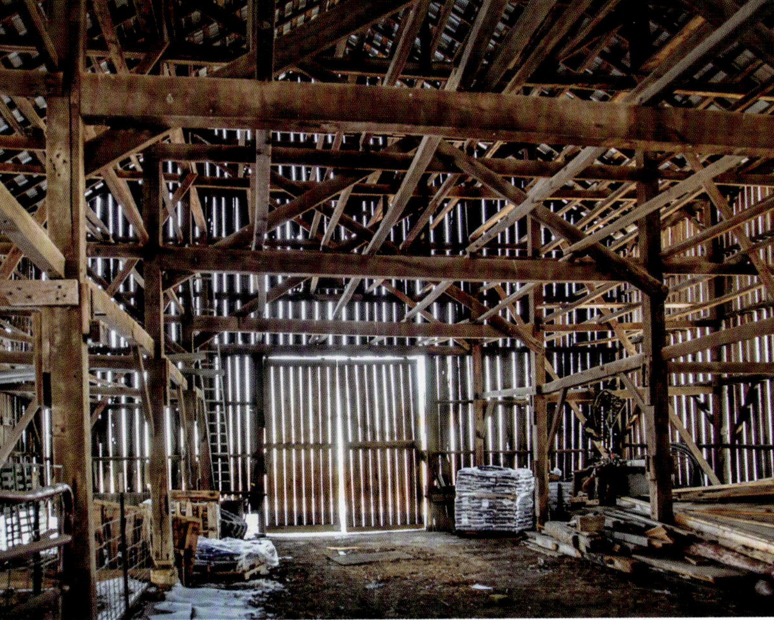

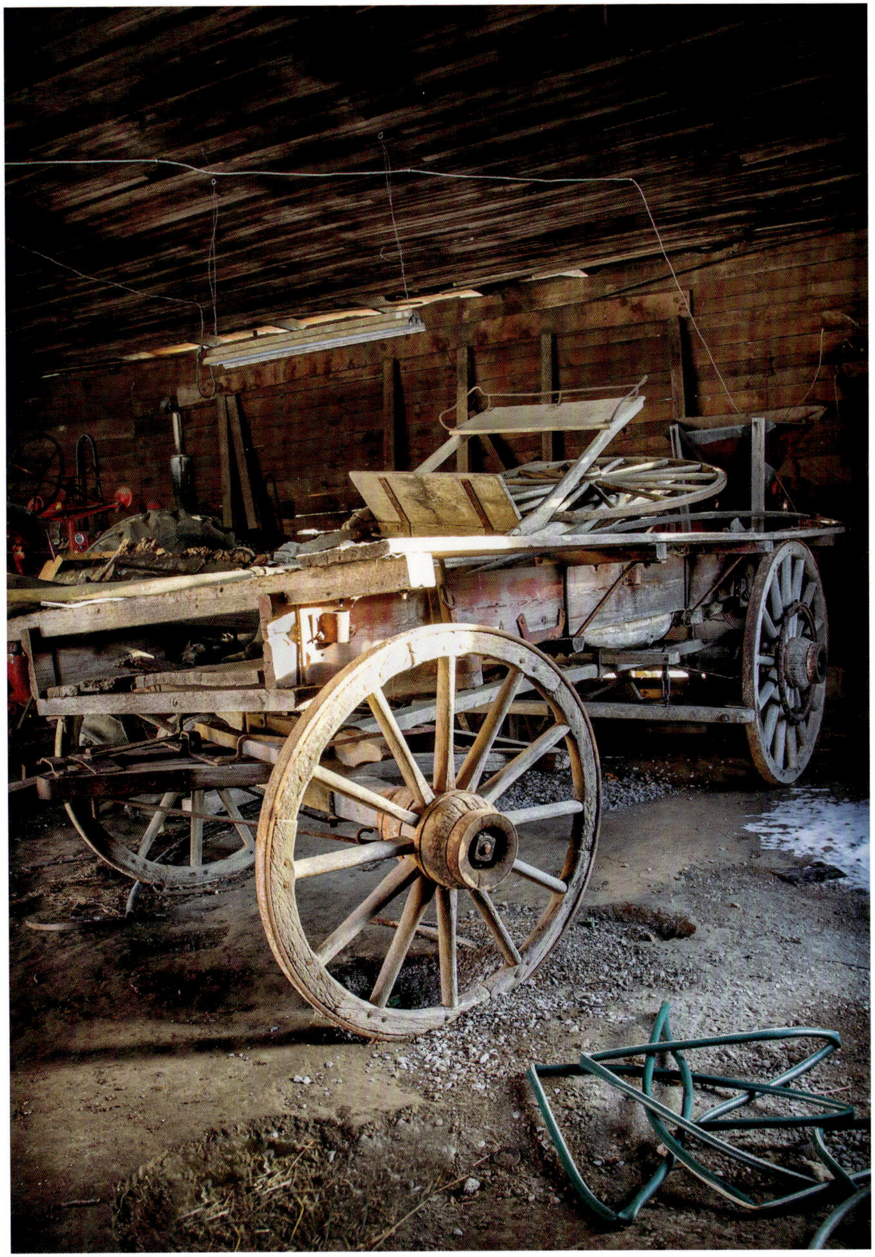

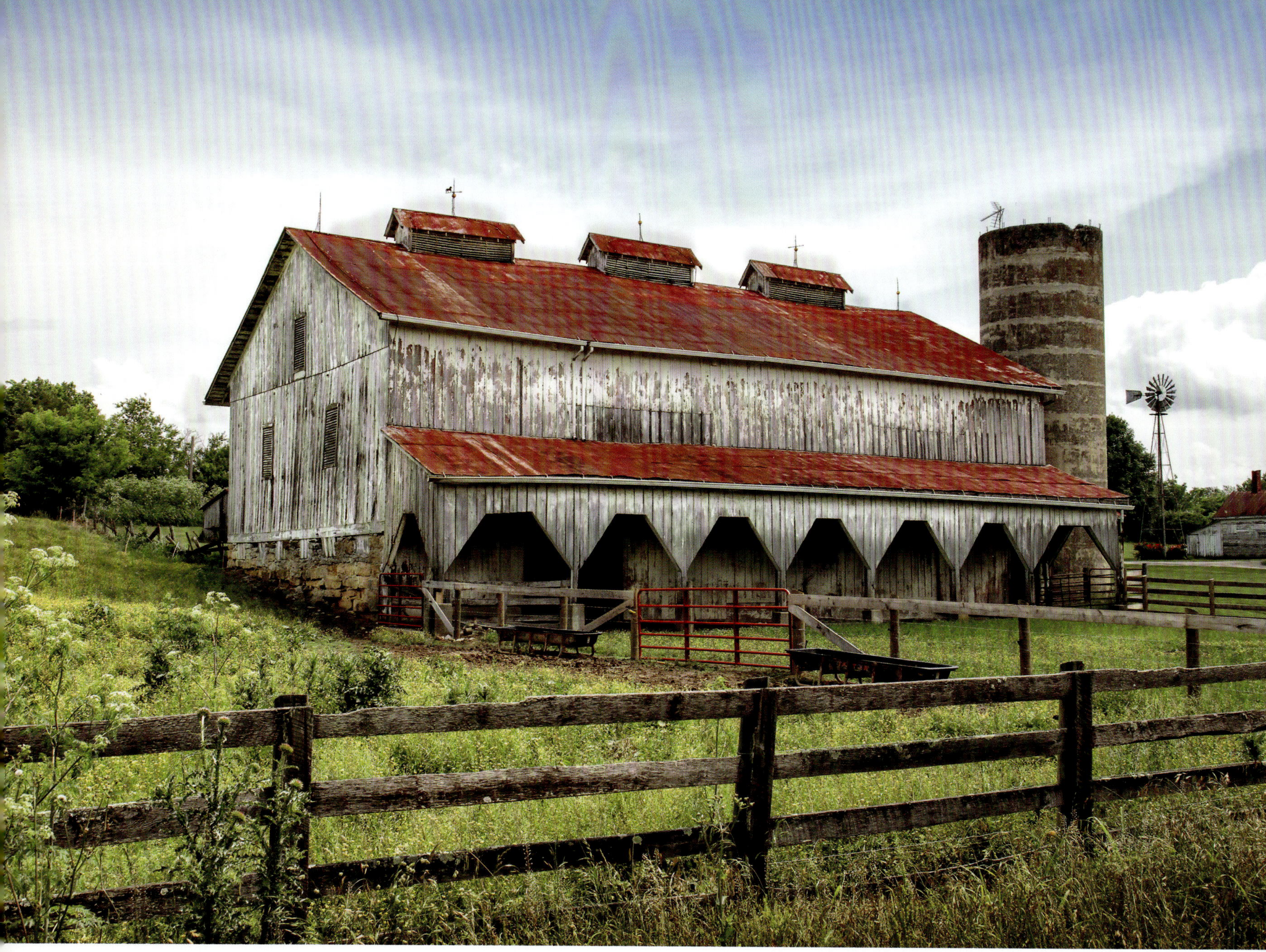

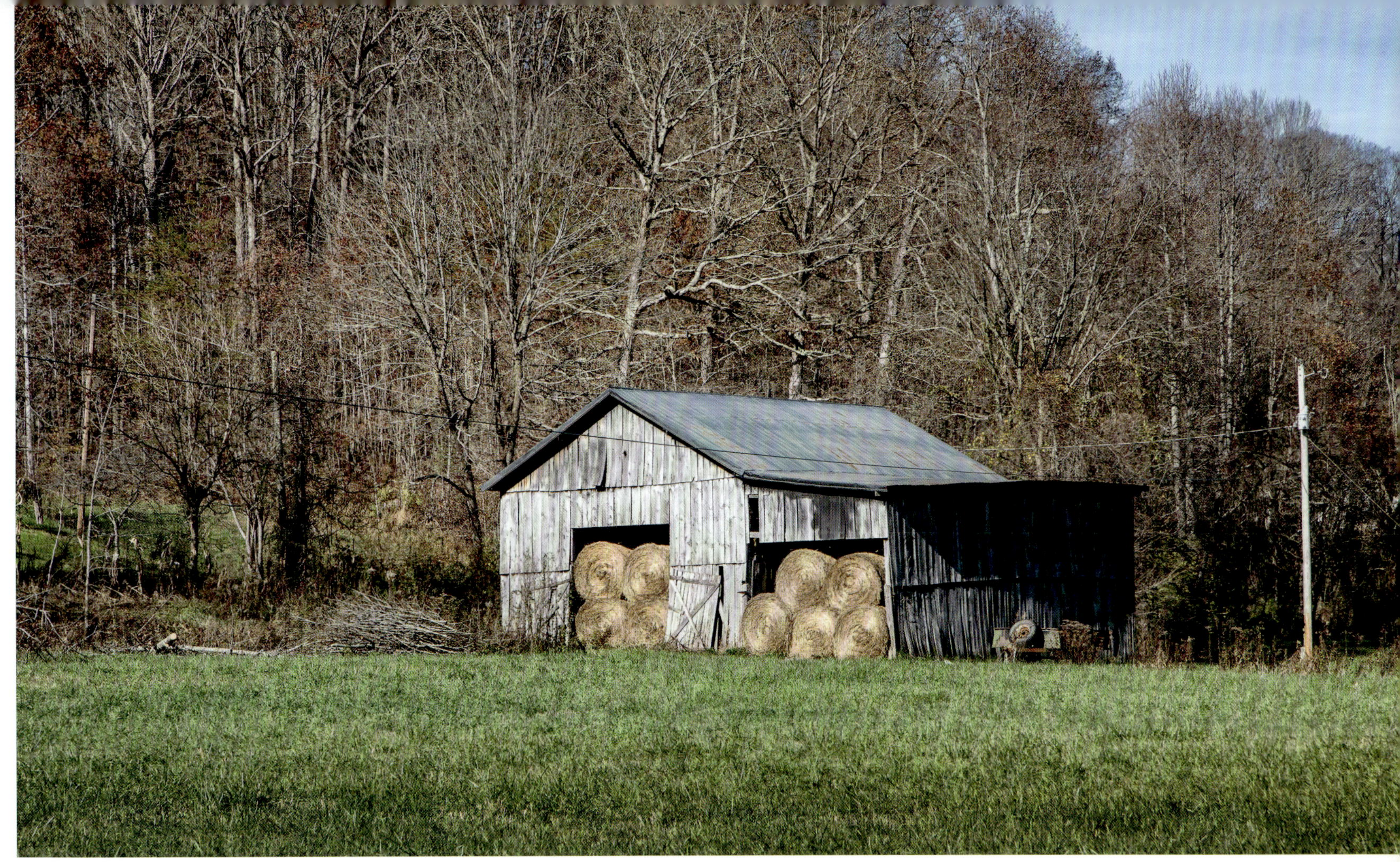

Lincoln County.

The Eaton Matheny barn was built by Elsie Chapel Eaton of cherry logs cut on the farm circa 1880. Elsie moved from Pennsylvania and built her new bank barn in the style she was familiar with from that region. The barn has stayed in the family continuously and is still in use.

Rockcastle County.

This barn, like others in the area, now houses rolls of hay rather than the stock animals it was built for.

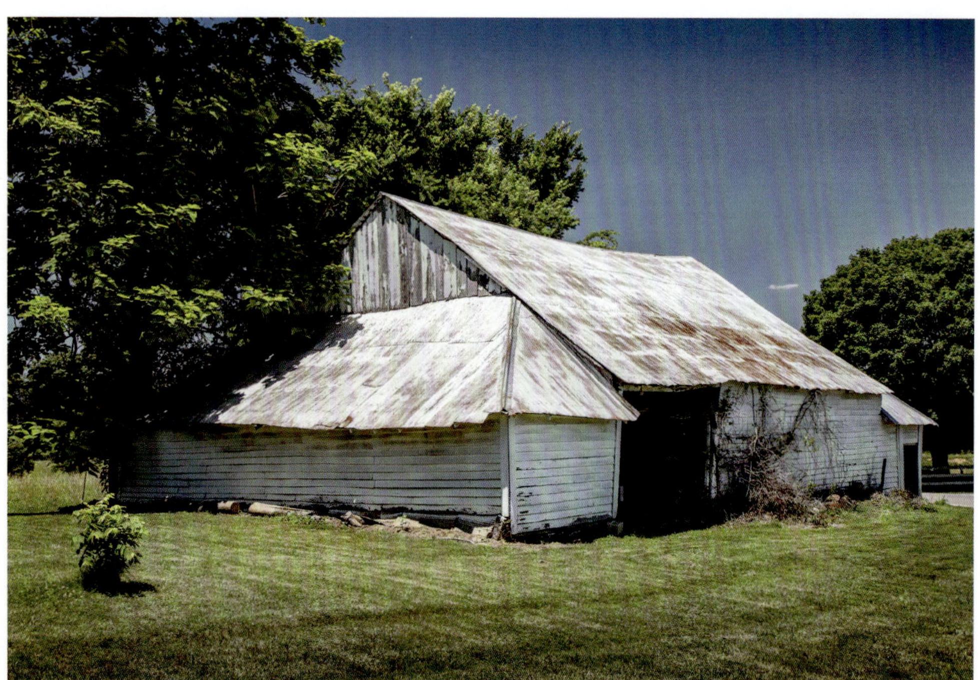

Union County.

The Shouse barn dates back to the late 1800s, remaining in the family. Inside is a log crib pen. More recent additions are visible on both ends.

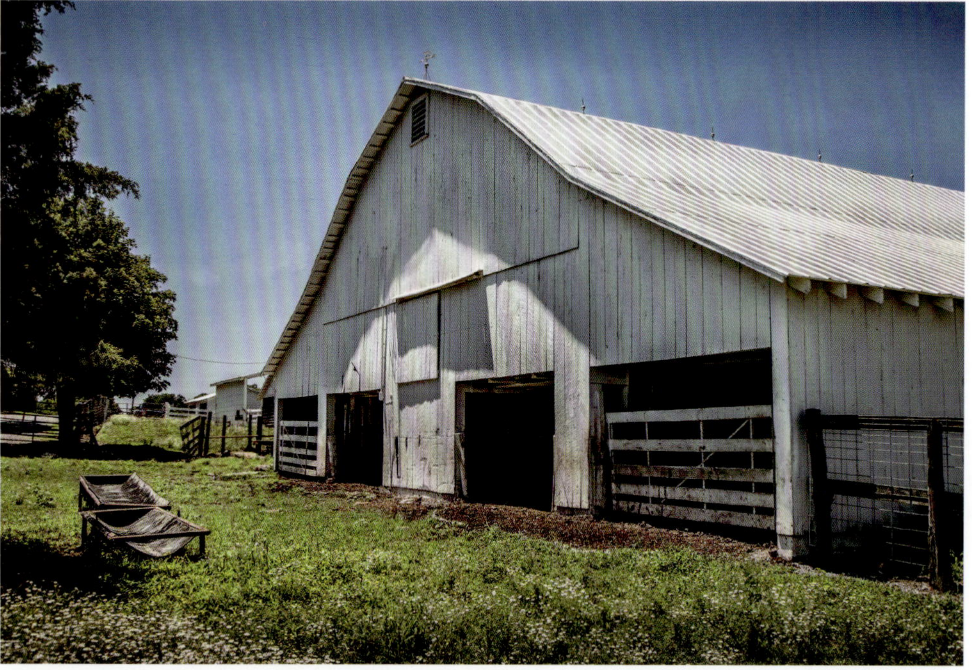

Union County.

Shelby Tucker built this barn seventy-five years ago, according to the current owner.

FACING PAGE

Casey County.

The Ellis family horse barn was built in the late nineteenth century. It was converted to hang tobacco in the loft area, retaining the stalls on the ground floor.

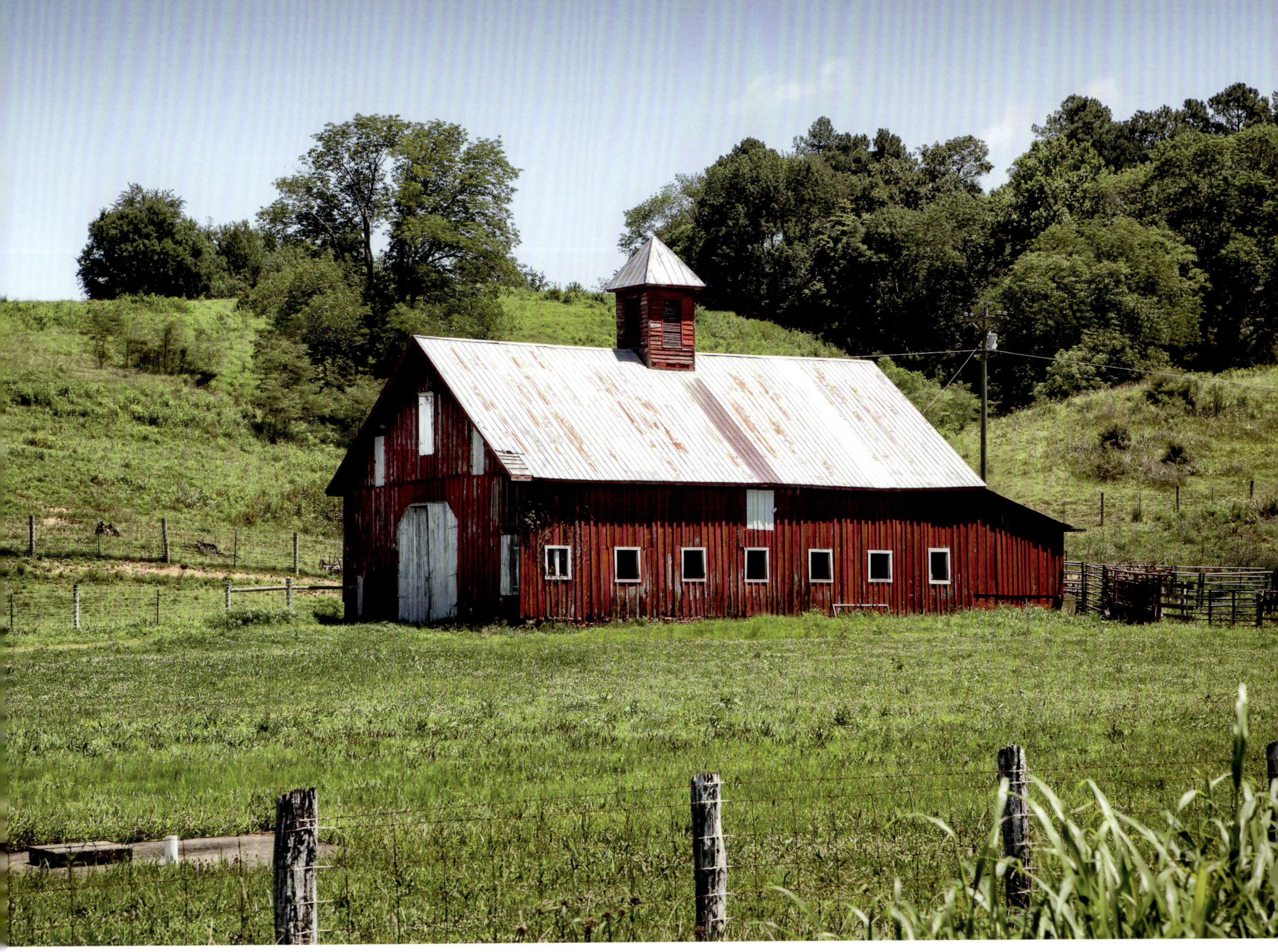

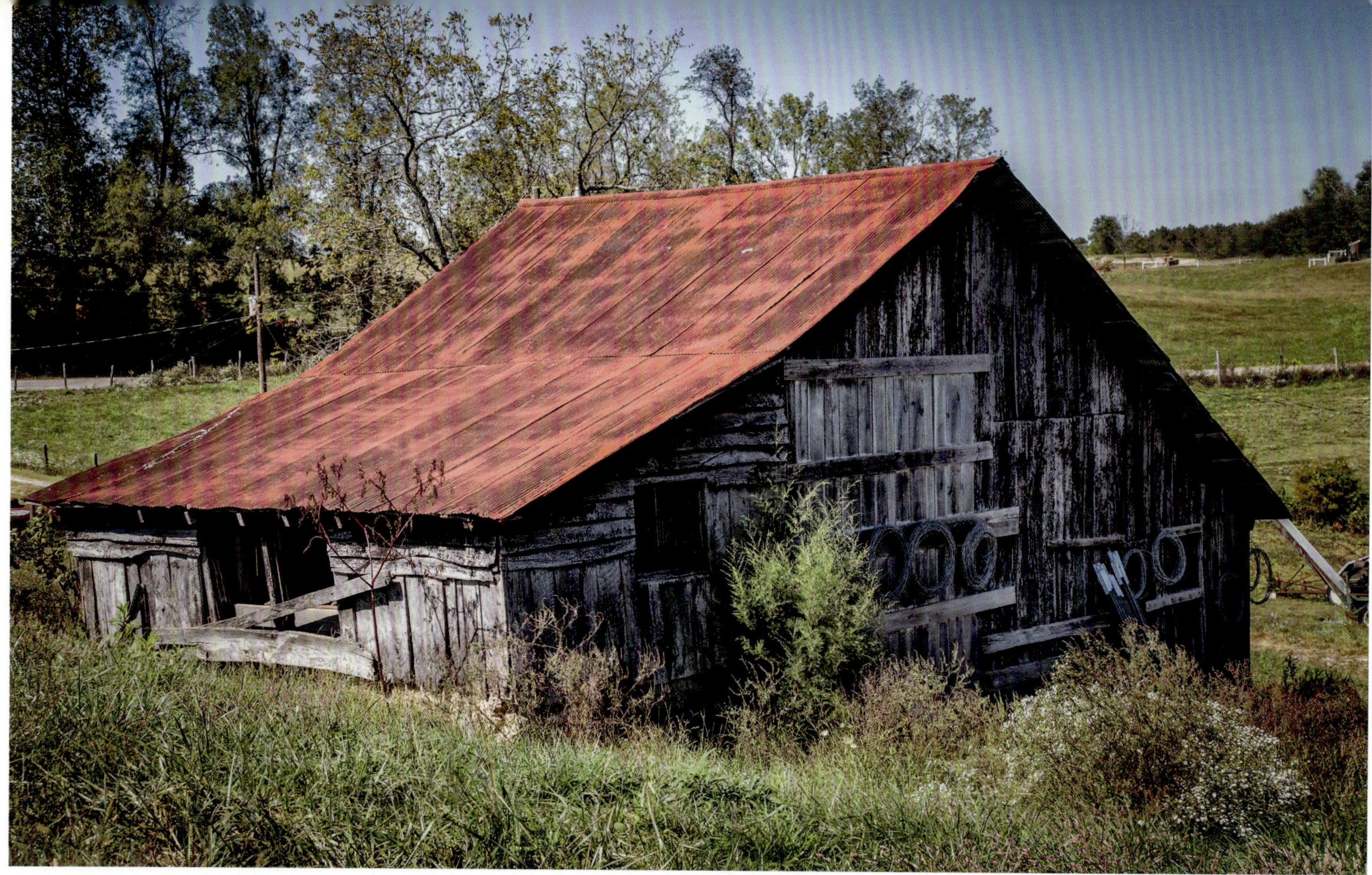

FACING PAGE

Lincoln County.

In the early 1880s, Lincoln County was the site of a settlement developed by Swiss immigrants. The town was named Ottenheim, and for years the German and Swiss heritage was maintained. Although many of later generations moved to neighboring counties, the area still retains buildings and the old footprint of the colony. This crib barn, built circa 1880, is on the Hartzog farm.

Caldwell County.

The University of Kentucky Agricultural Experiment Station at Princeton, Kentucky, was established in 1885. As part of its mission to serve the research needs of agriculture and rural citizens, the station has taken on the challenge of improving the quality of dark fired tobacco while keeping the nicotine content down. These images show an experimental dark tobacco firing barn exterior and interior, the product drying, and the packaging process.

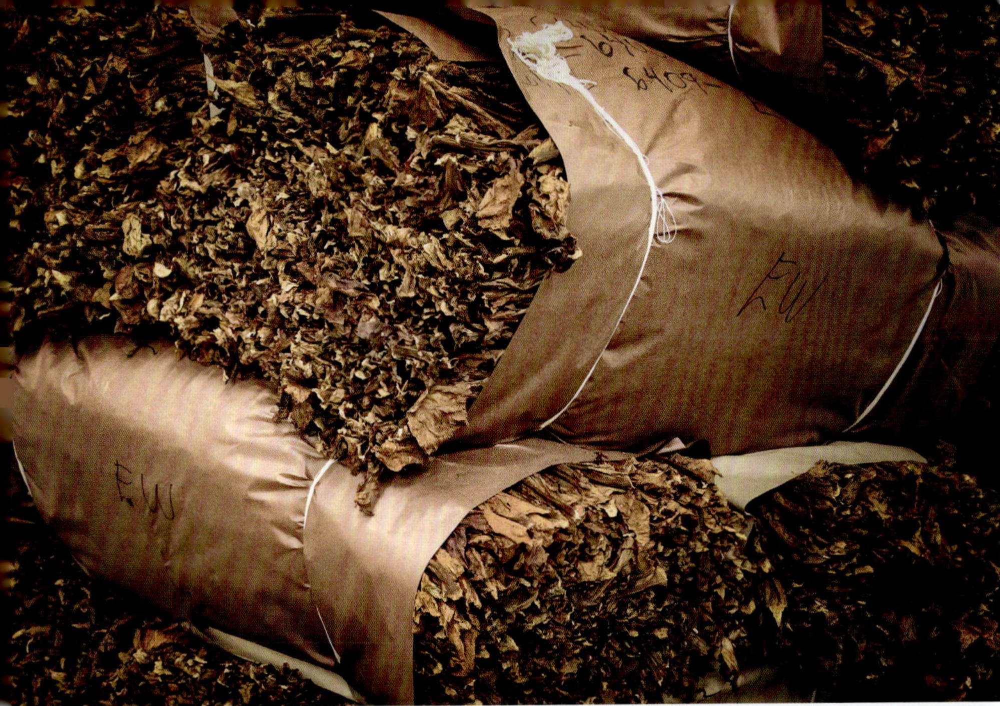

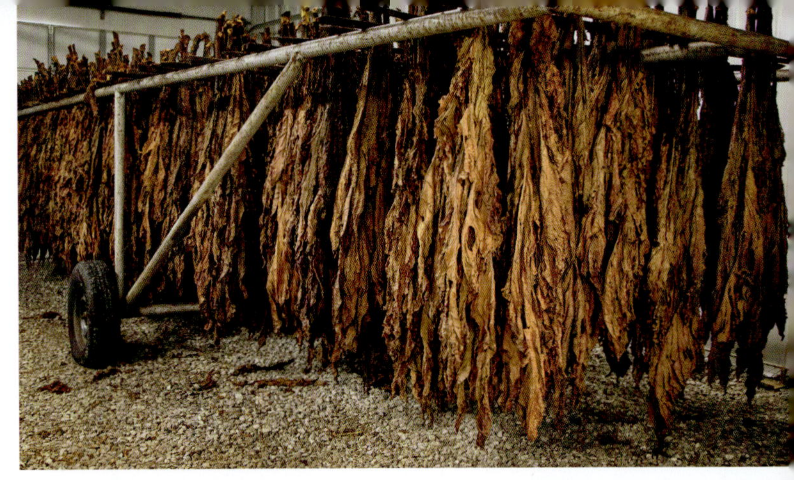

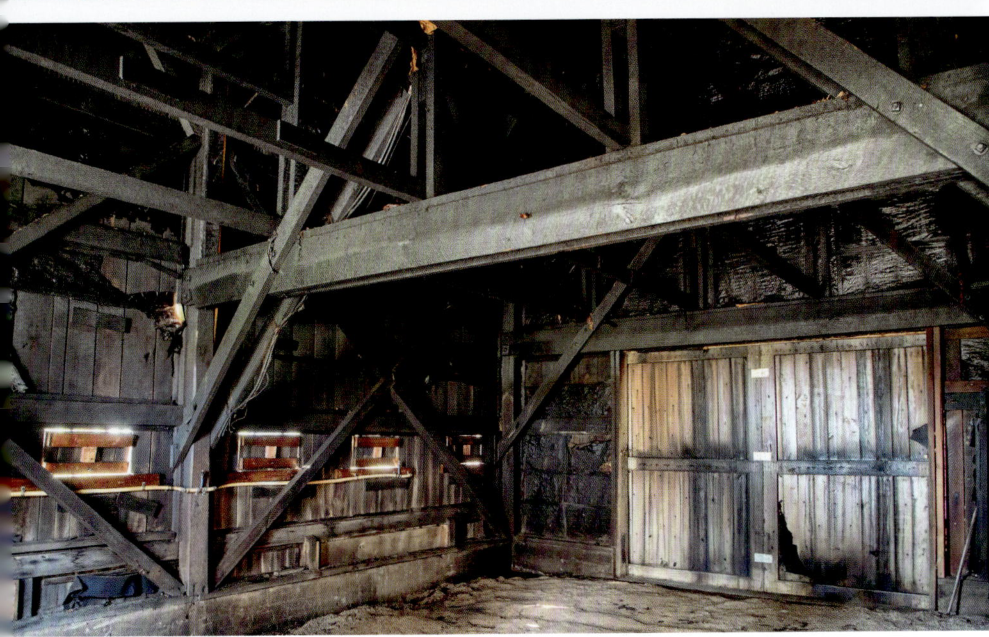

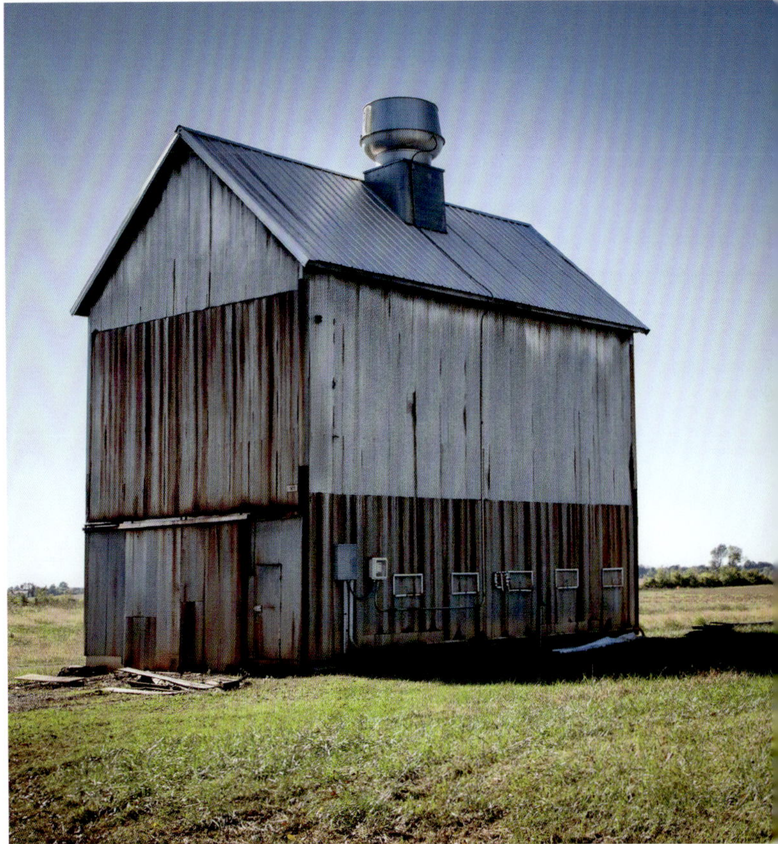

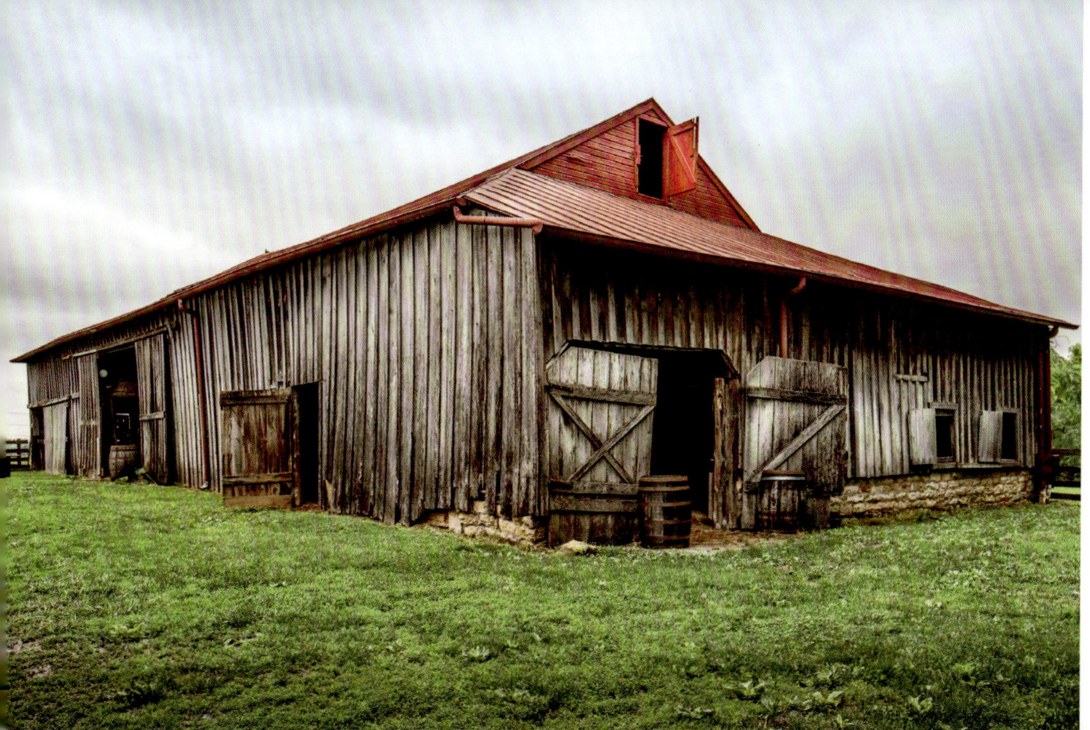

Jefferson County.

Blackacre Conservancy is on half of what once was the Moses Tyler settlement, established circa 1795. In addition to a stone house and springhouse, the Tylers built a double-pen log barn in the Appalachian style in that same year. Open to the public, the conservancy maintains the unique big barn for anyone to visit.

FACING PAGE

Bracken County.

Settlement of Rock Springs began in the early nineteenth century but had diminished by 1910. This barn shows German influence with a stone first floor, massive timber supports, and a timber second-story hayloft above.

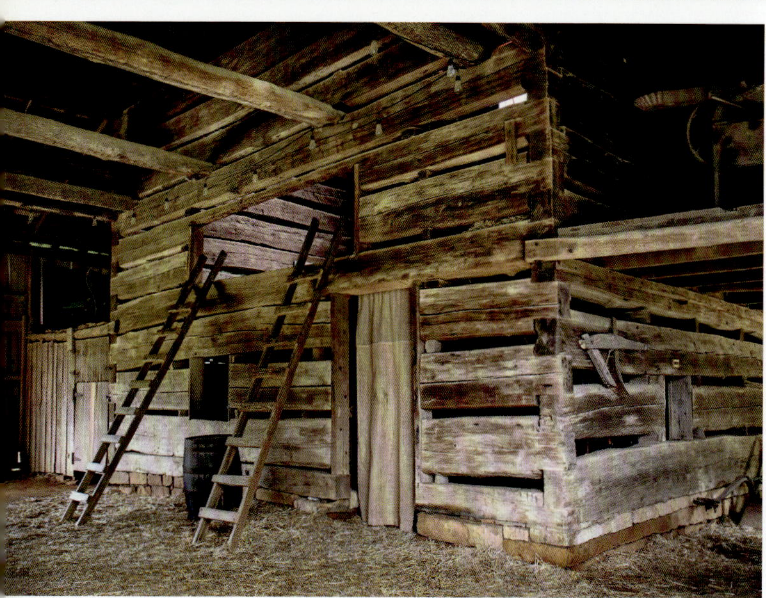

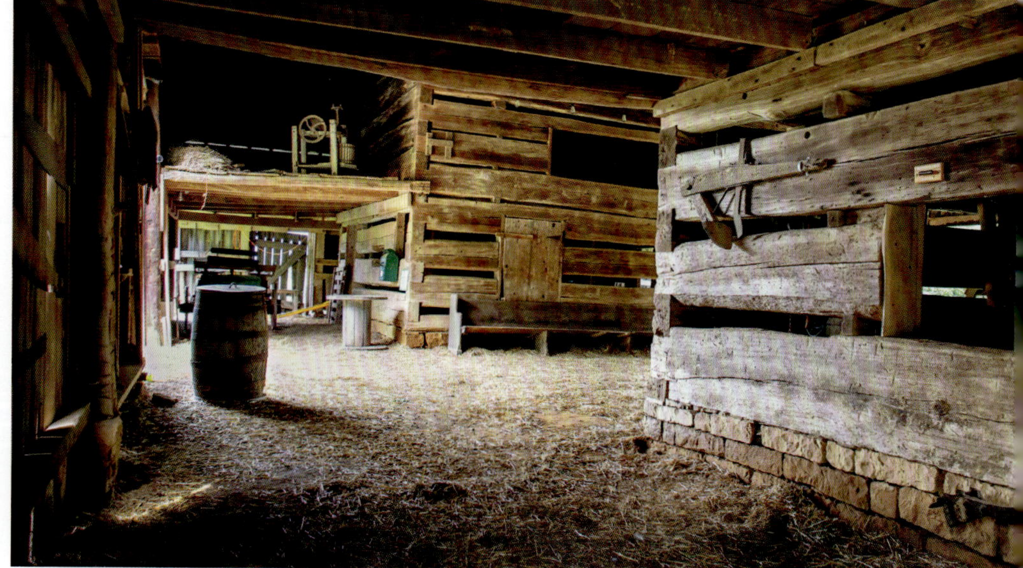

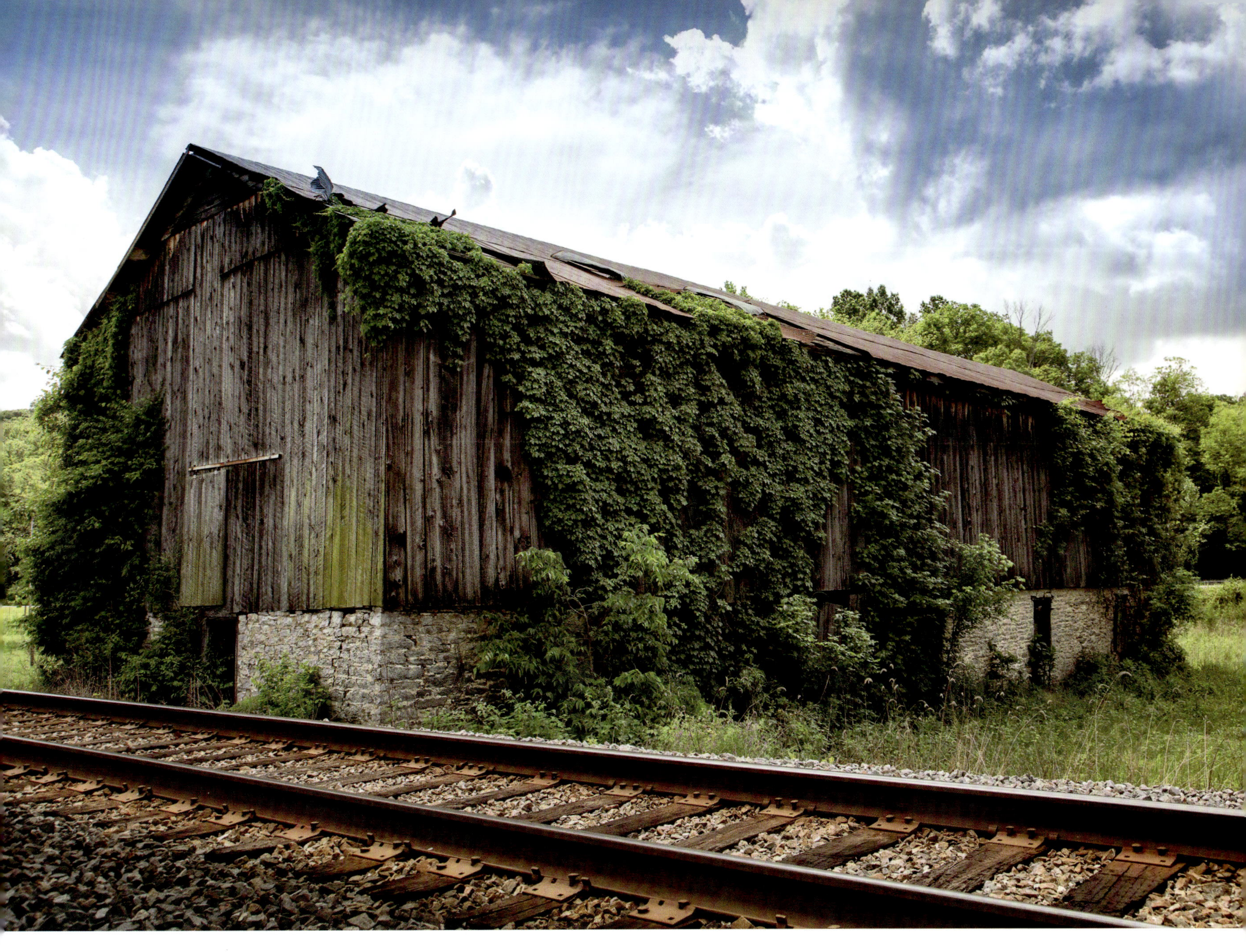

Lincoln County.

The Eby dairy barn, built circa the 1880s and located in the Swiss settlement of Ottenheim outside of Stanford, is no longer used for milking cows. Instead it houses hay for the new Mennonite owners.

Woodford County.

The first quilt barn trail was started in 2001, with forty-three of the US states and three Canadian provinces participating. The quilt squares draw attention to significant architecture or aesthetic landscapes.

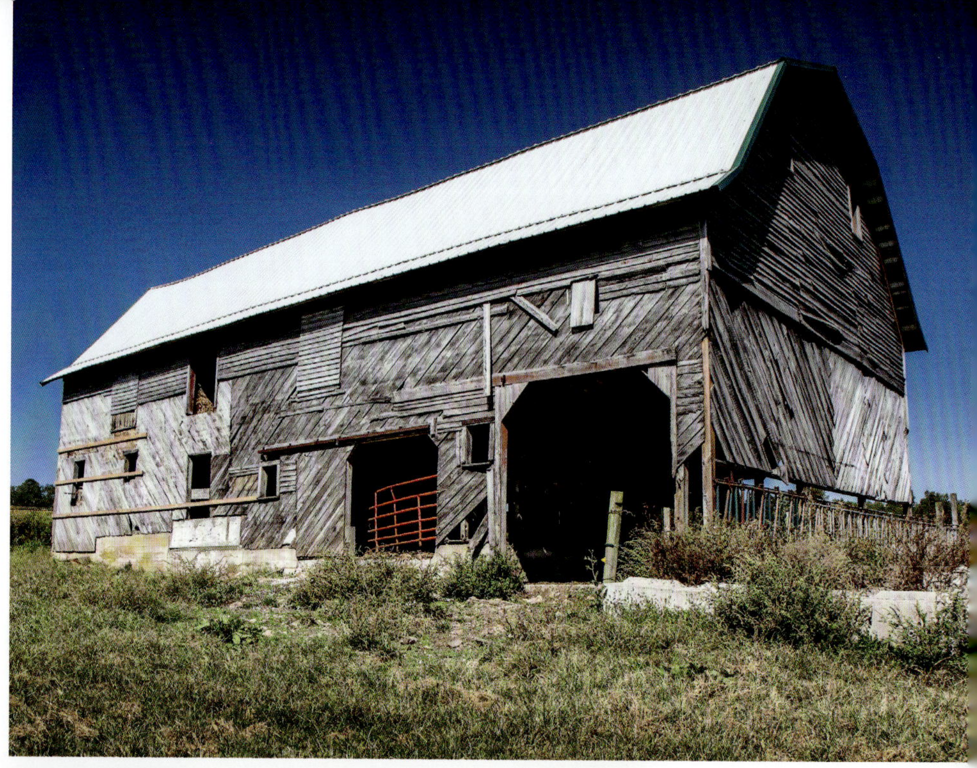

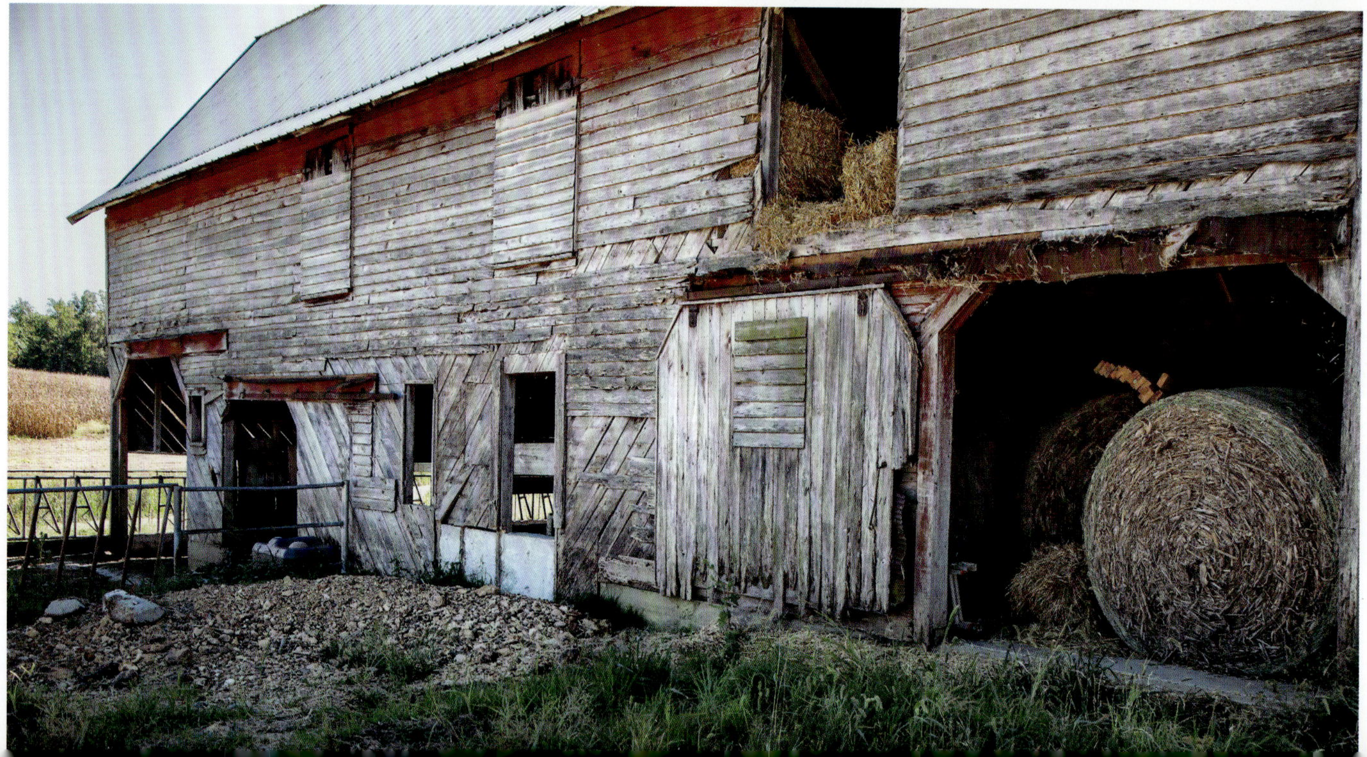

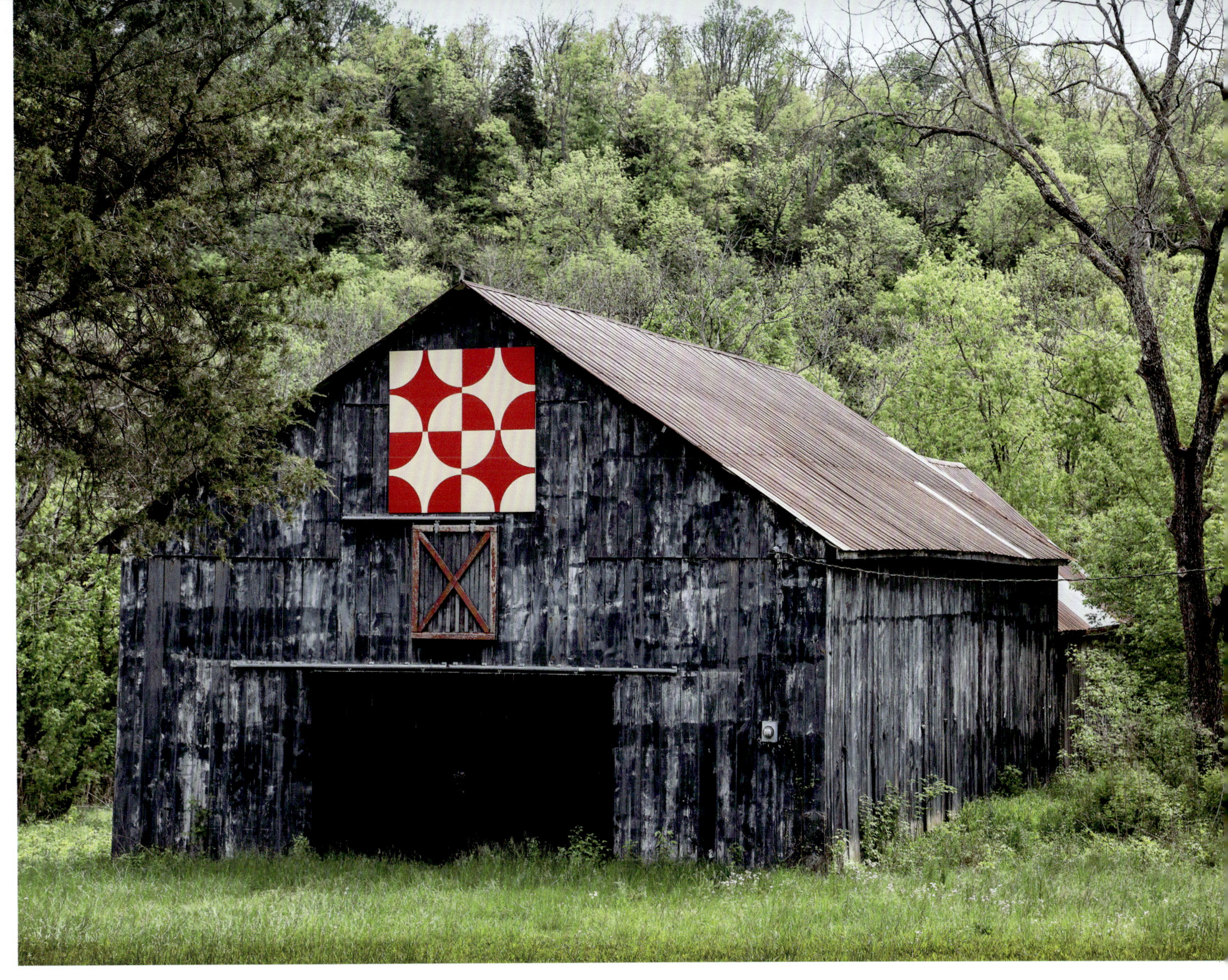

Bracken County.

Once a stone mill, this nineteenth-century building was turned into a livestock barn in later years.

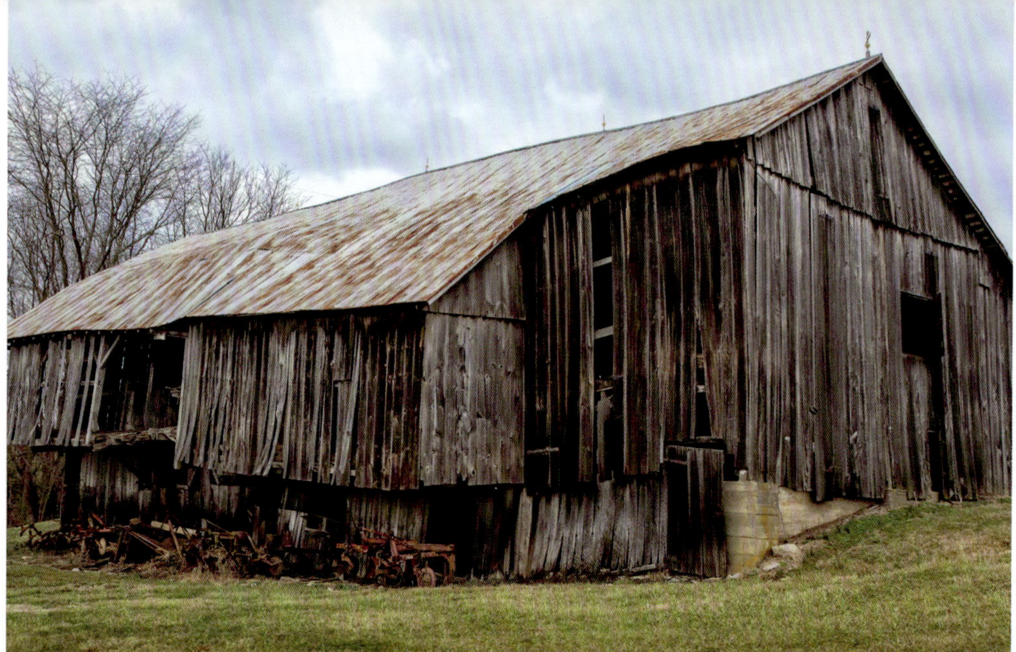

Mason County.

The Sweitzer barn has a cantilever fore bay and banked animal stalls. The style is unusual in Kentucky; it is German in origin and reached its peak of popularity in the late eighteenth century.

FACING PAGE

Bourbon County.

This tobacco barn typically has a connected stripping shed that was used for stripping the leaves off the stalks once they are dried. Early stripping sheds without electricity had windows for light instead.

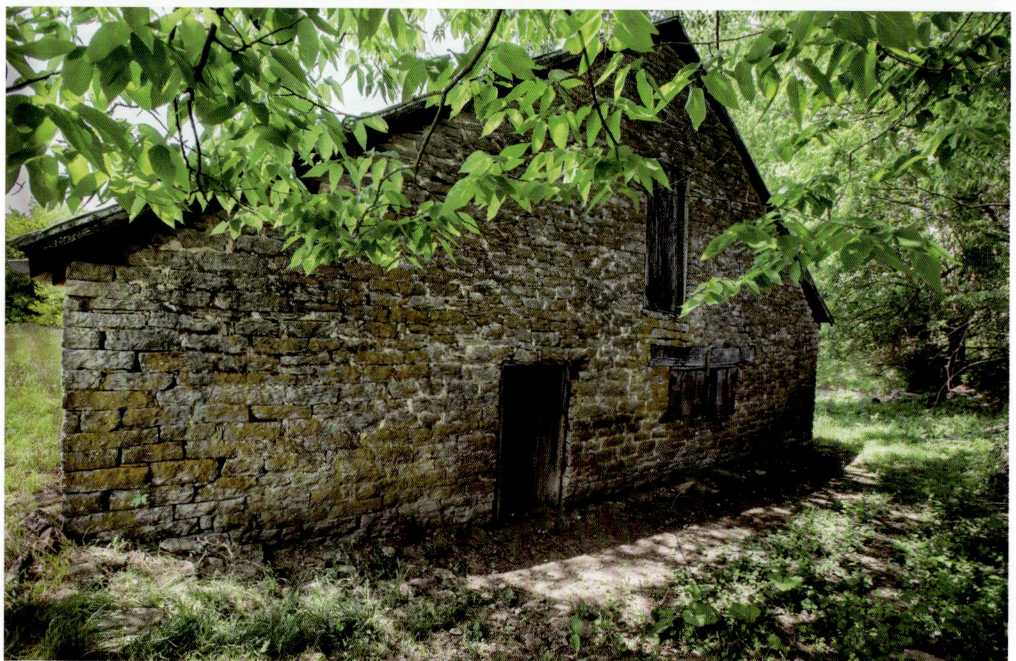

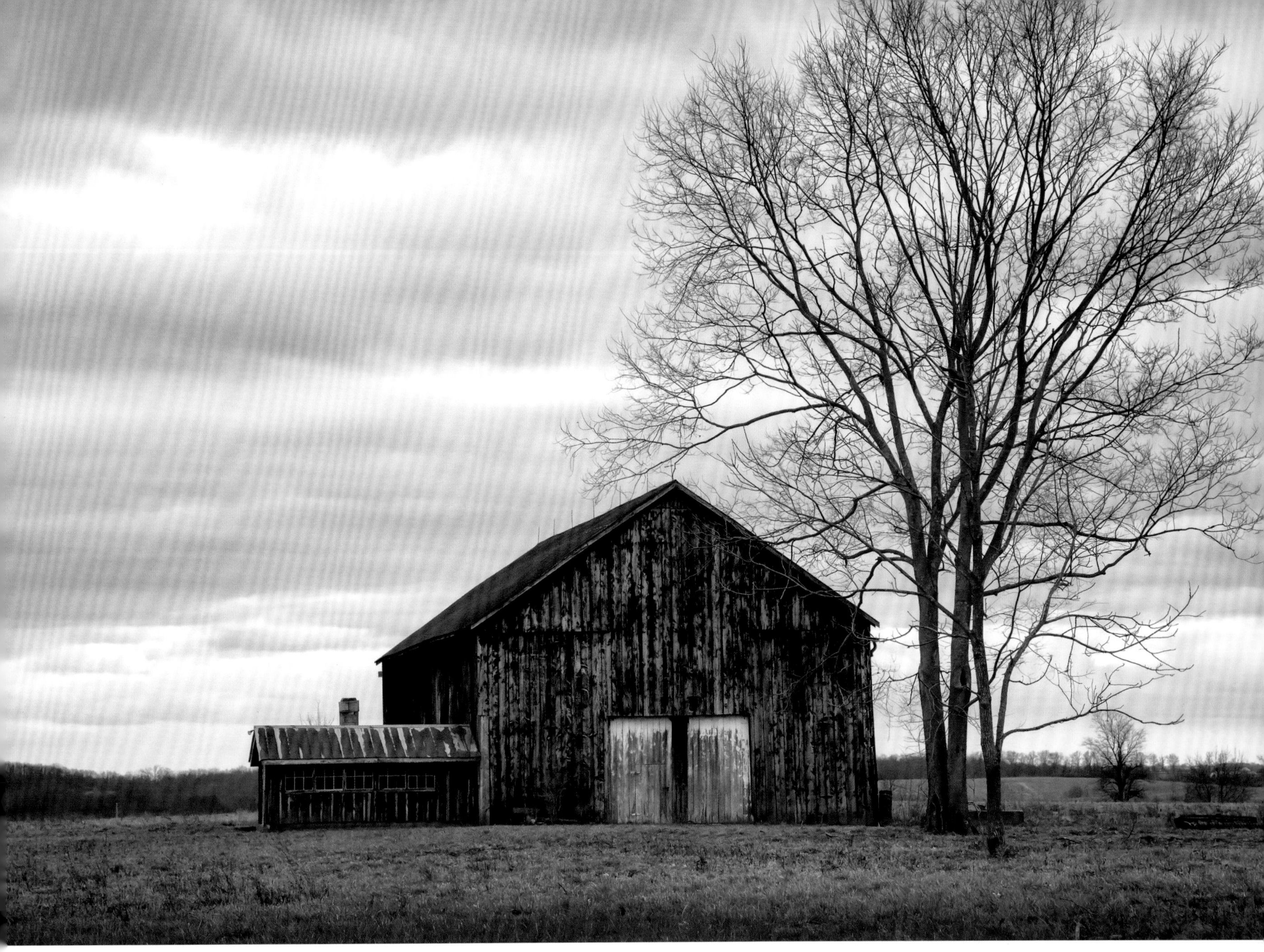

Madison County.

The Hanger family owned a large parcel of land
in the area near Eastern Kentucky University.
This fifteen-bent tobacco barn, with a tobacco
press inside, remains on the old farmstead
but sits in a housing development's path.
The stove is from a different barn but was a
typical means of staying warm in the barn's
stripping shed during chilly fall days.

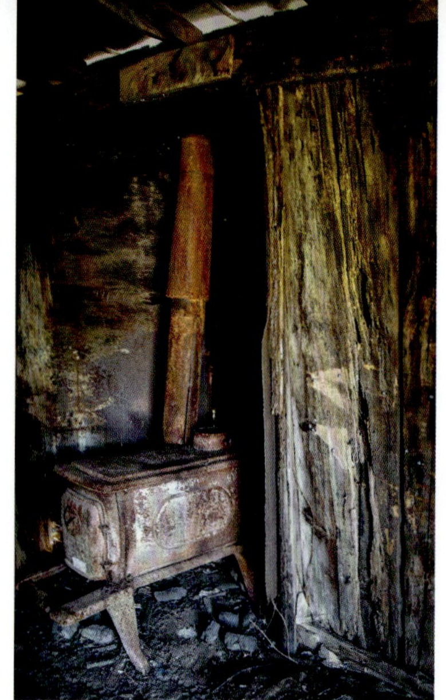

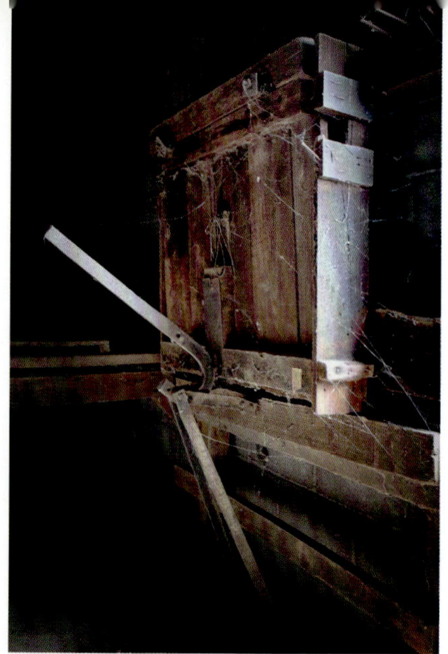

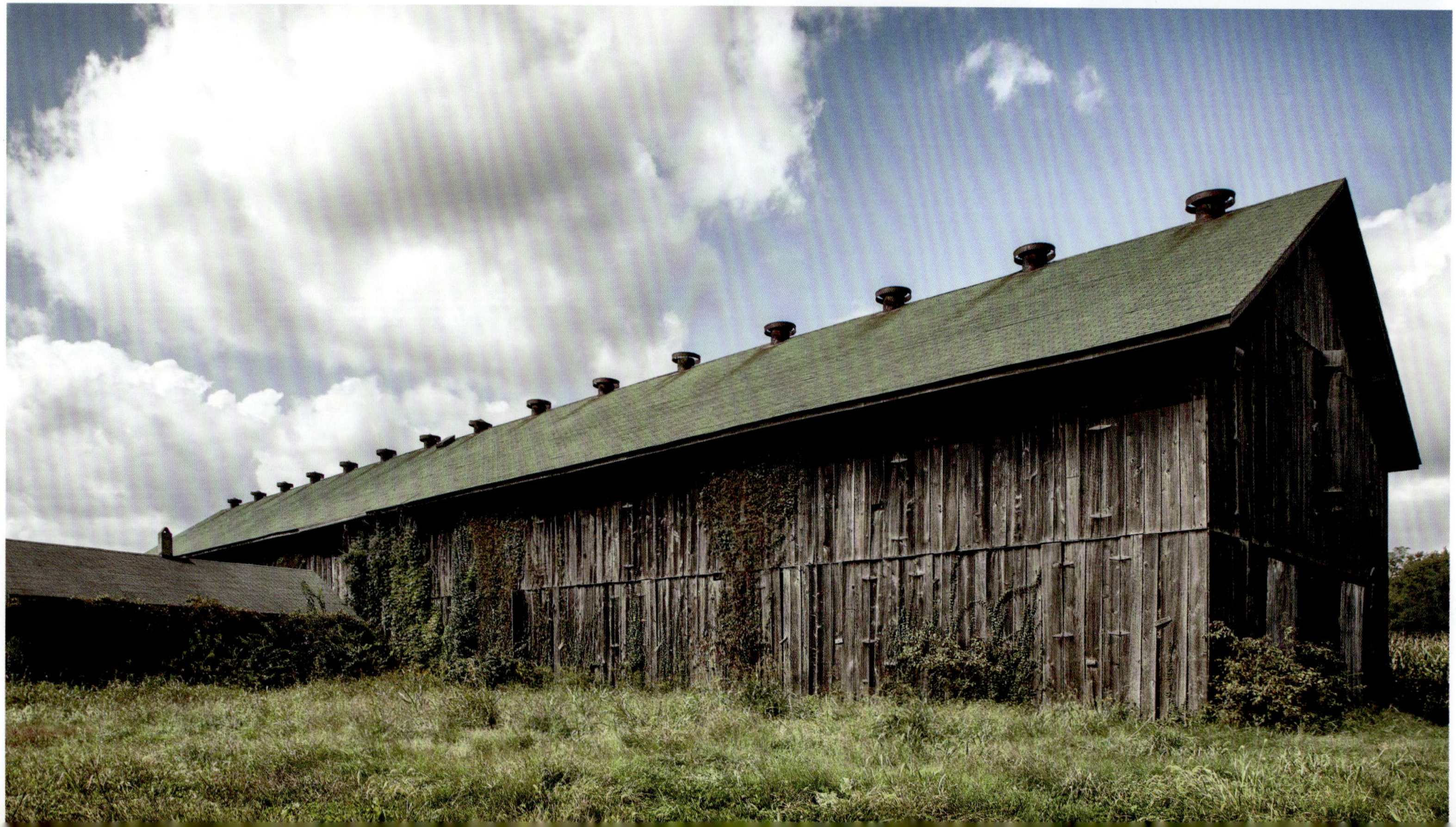

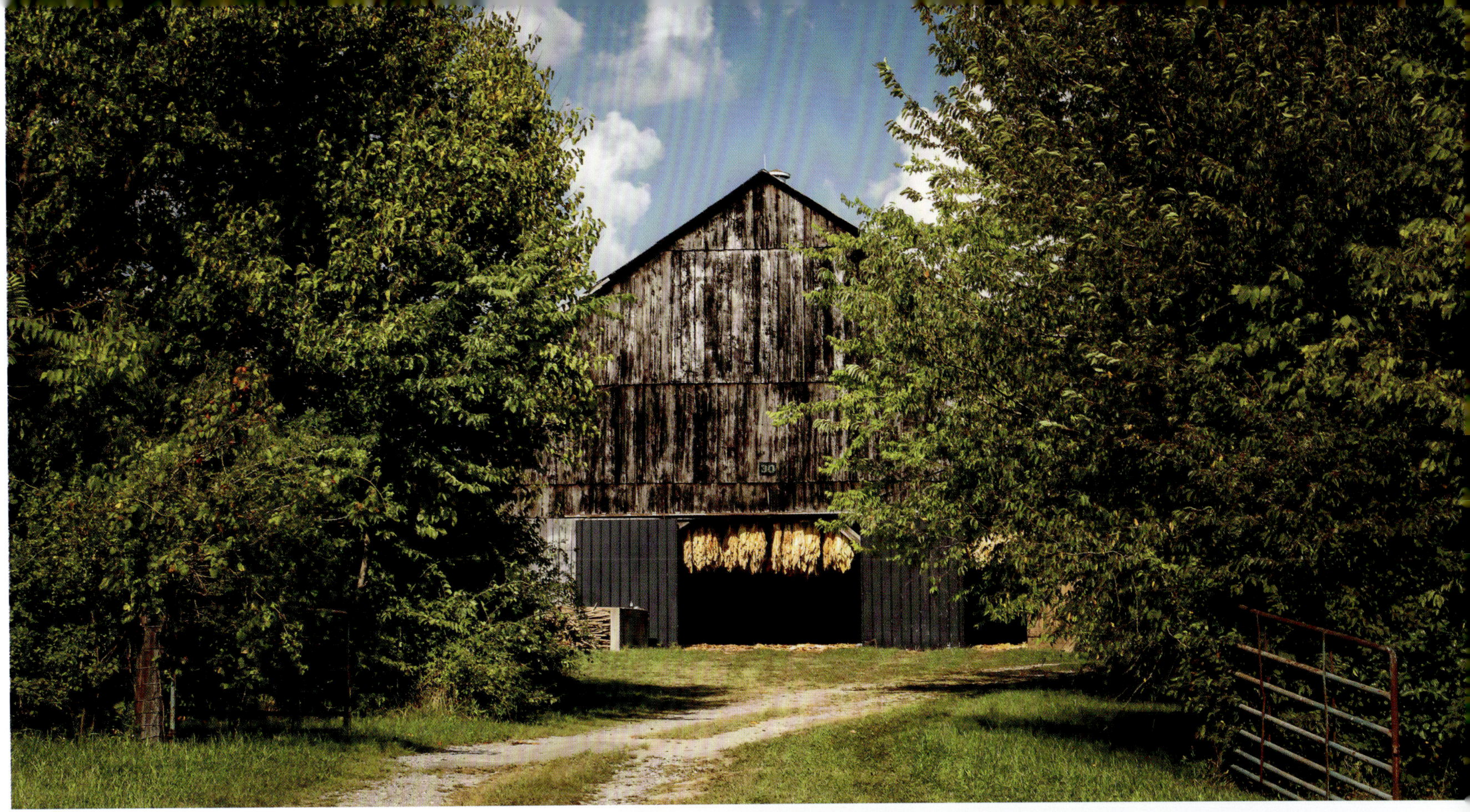

Boyle County.

Barns drying tobacco used to be a common sight in Kentucky. This barn is labeled barn 28, indicating that it was the twenty-eighth barn on the property. I visited 29 and 30, also leased to tobacco farmers.

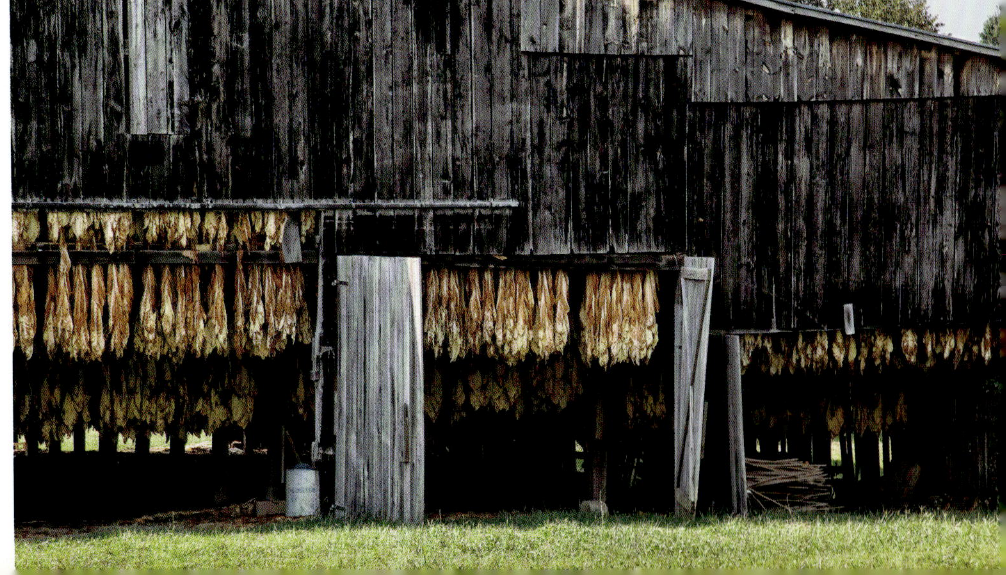

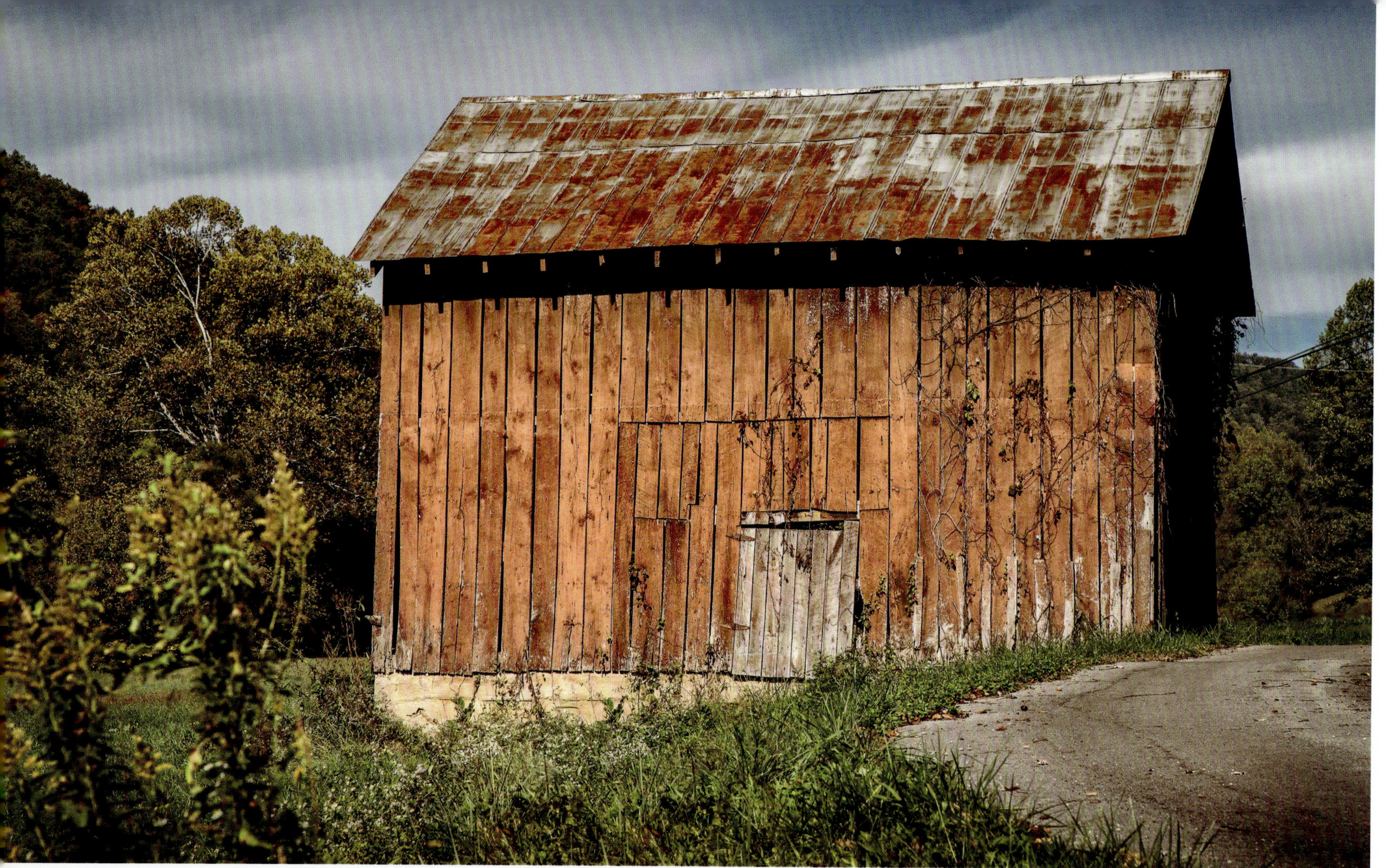

Bell County.

Called an English barn, a style notable because of the doors on the sides rather than ends, this barn is still in use as a stock barn.

FACING PAGE

Trigg County.

The Luther Thomas dark tobacco firing barn, east of Cadiz, was used for many purposes besides tobacco firing. At one time a family called it home after their house had been destroyed. Now it is empty and quieter.

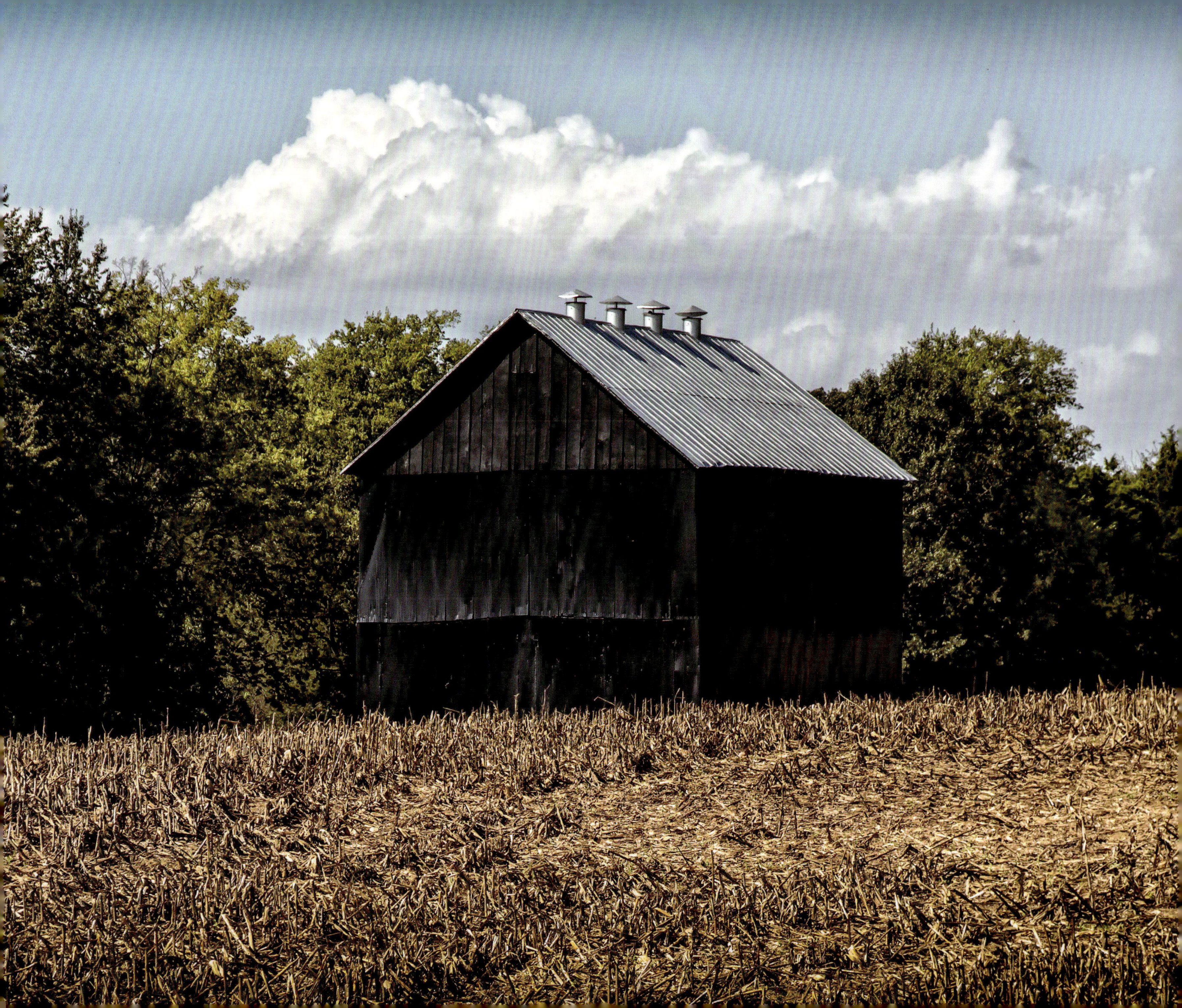

Todd County.

As technology and farming practices have changed, so have the needs of the farmers regarding their buildings. Soy is a common crop that has replaced tobacco in recent years, and it does not require a barn. This silo barn in Western Kentucky is probably used for equipment, if at all.

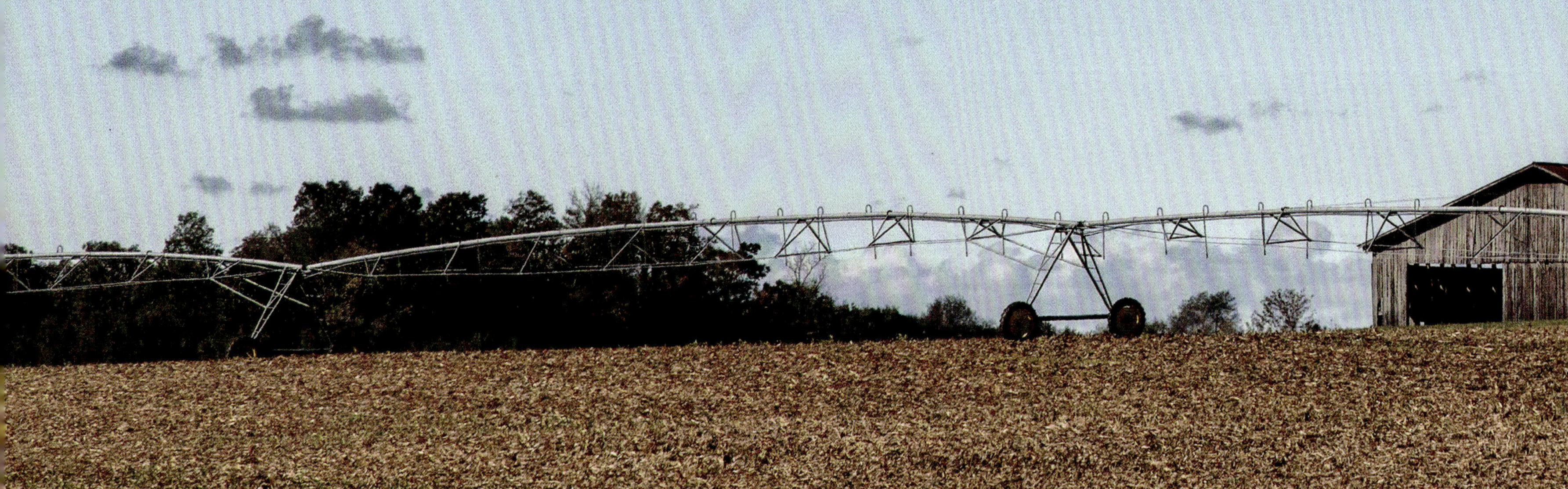

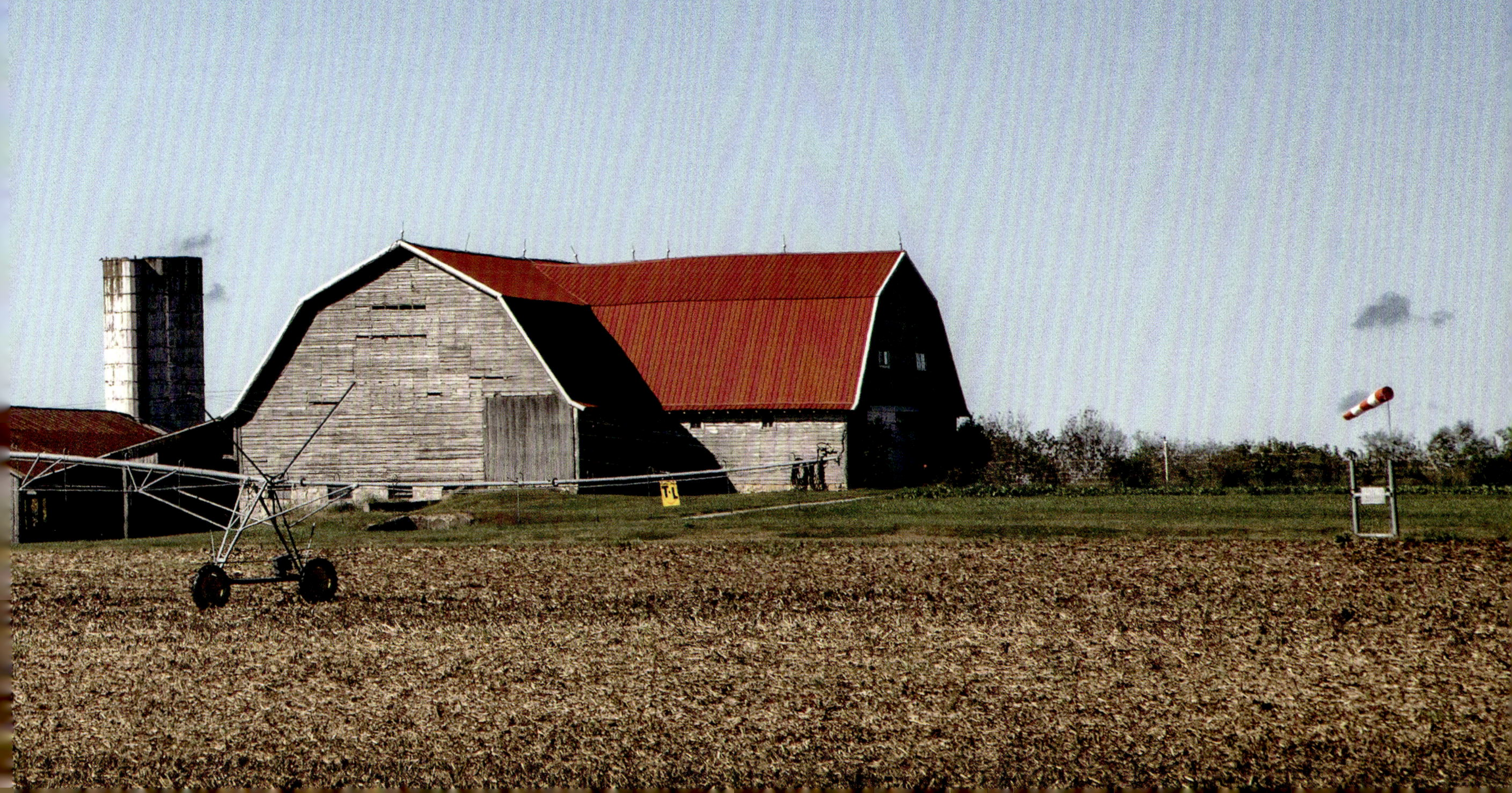

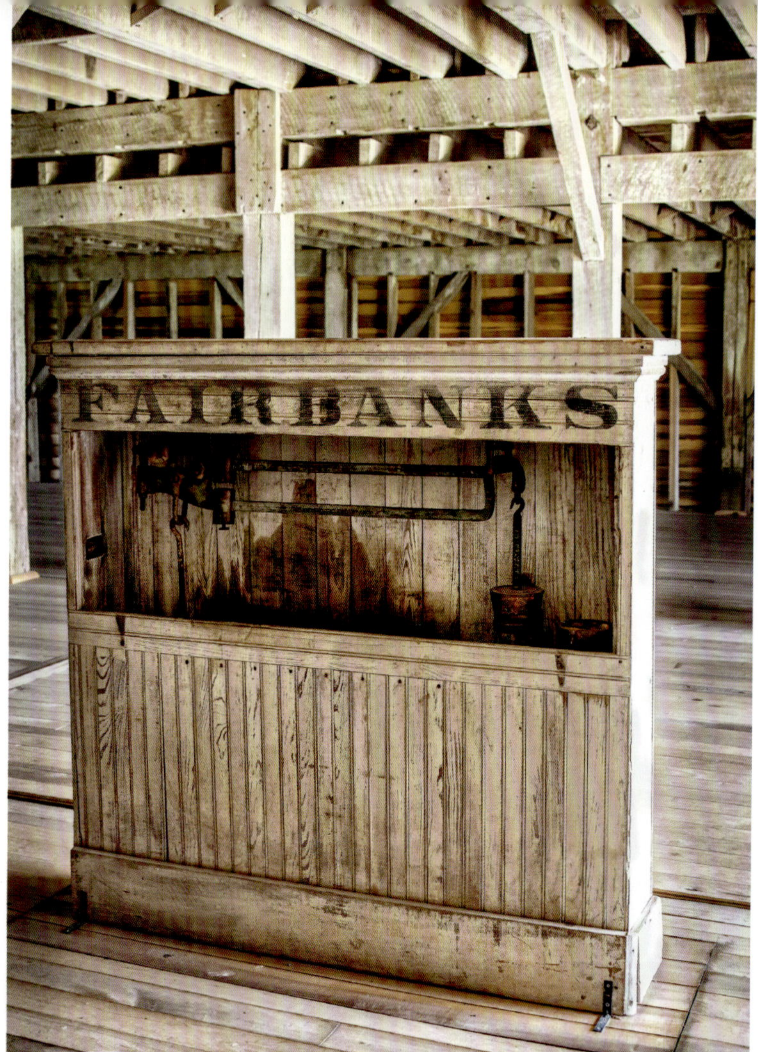

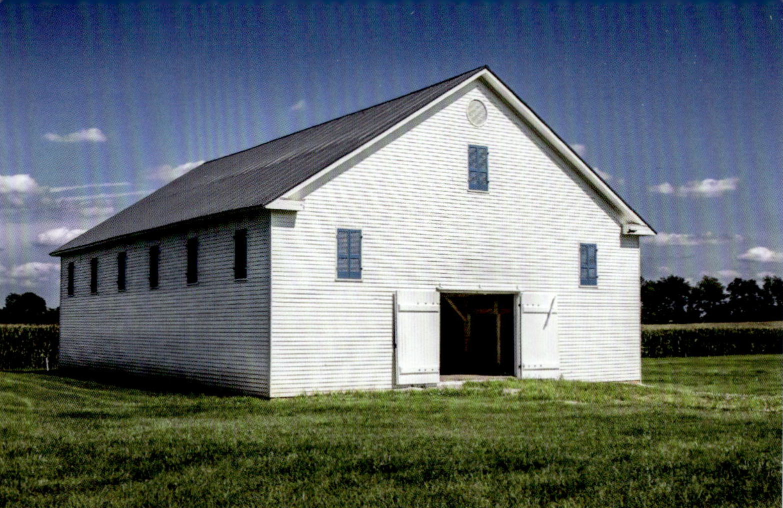

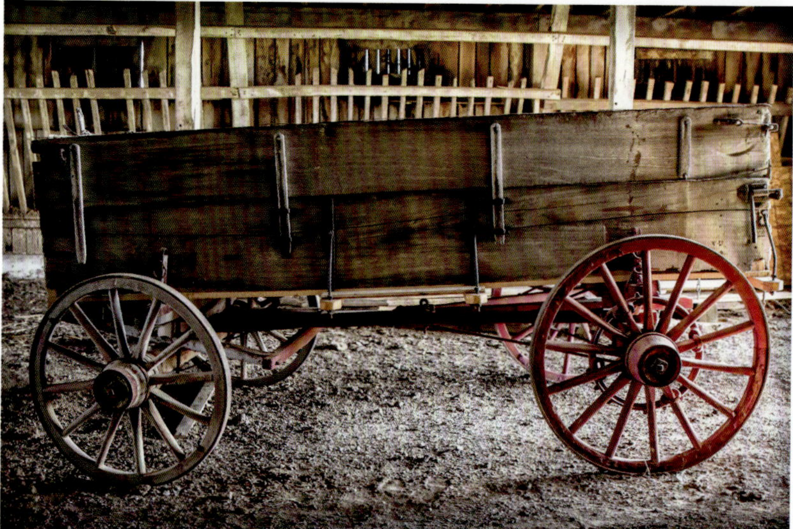

Logan County.

South Union was the home of a Shaker religious community from 1807 to 1922. Two barns exist on the site, including this one, built circa 1875 and recently renovated. Inside are a scale and some examples of carriages and carts.

Christian County.

Built circa 1930 by Robert Fritz, this unusual stone barn has been in continual use ever since, now mostly for storage.

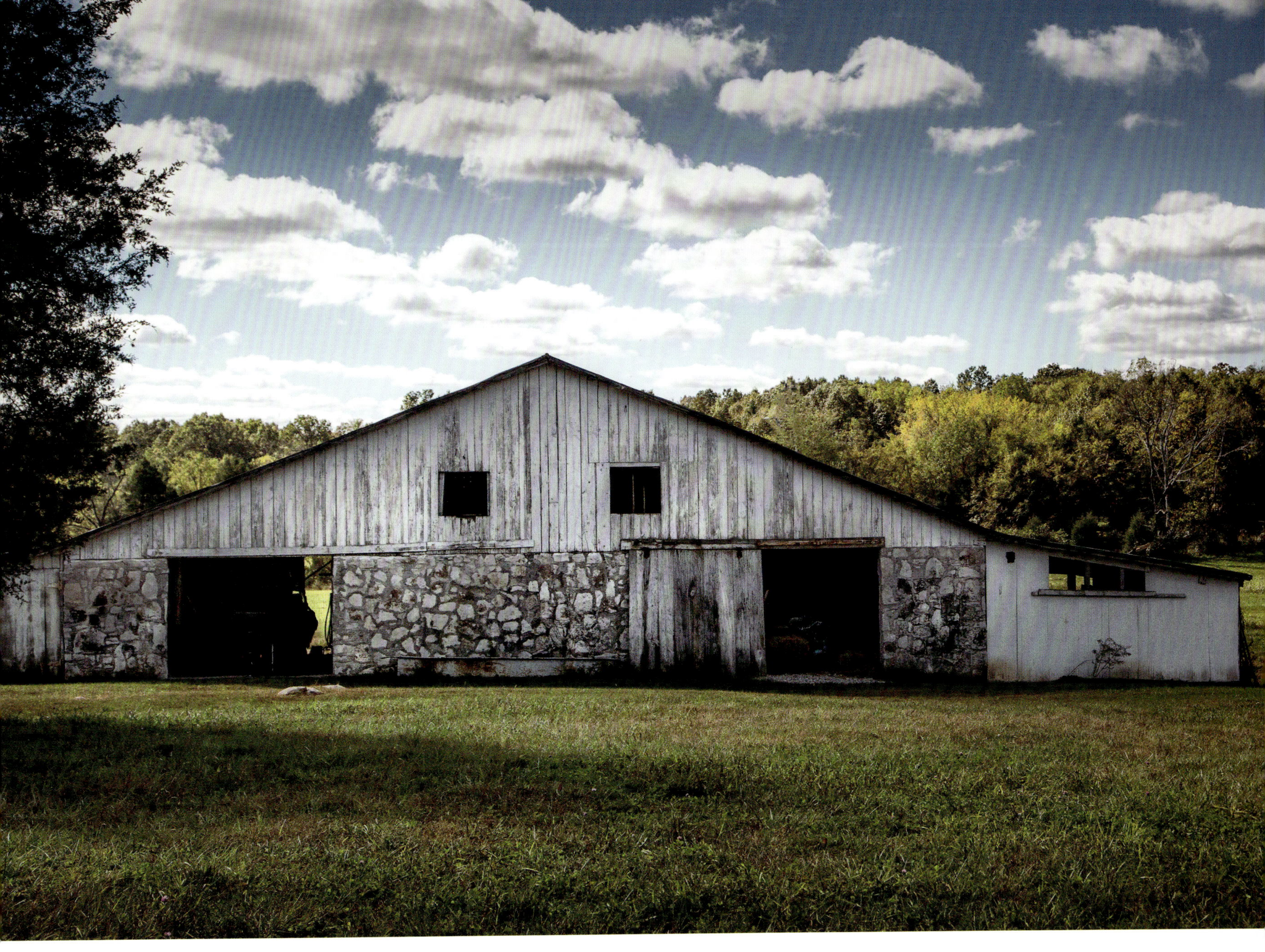

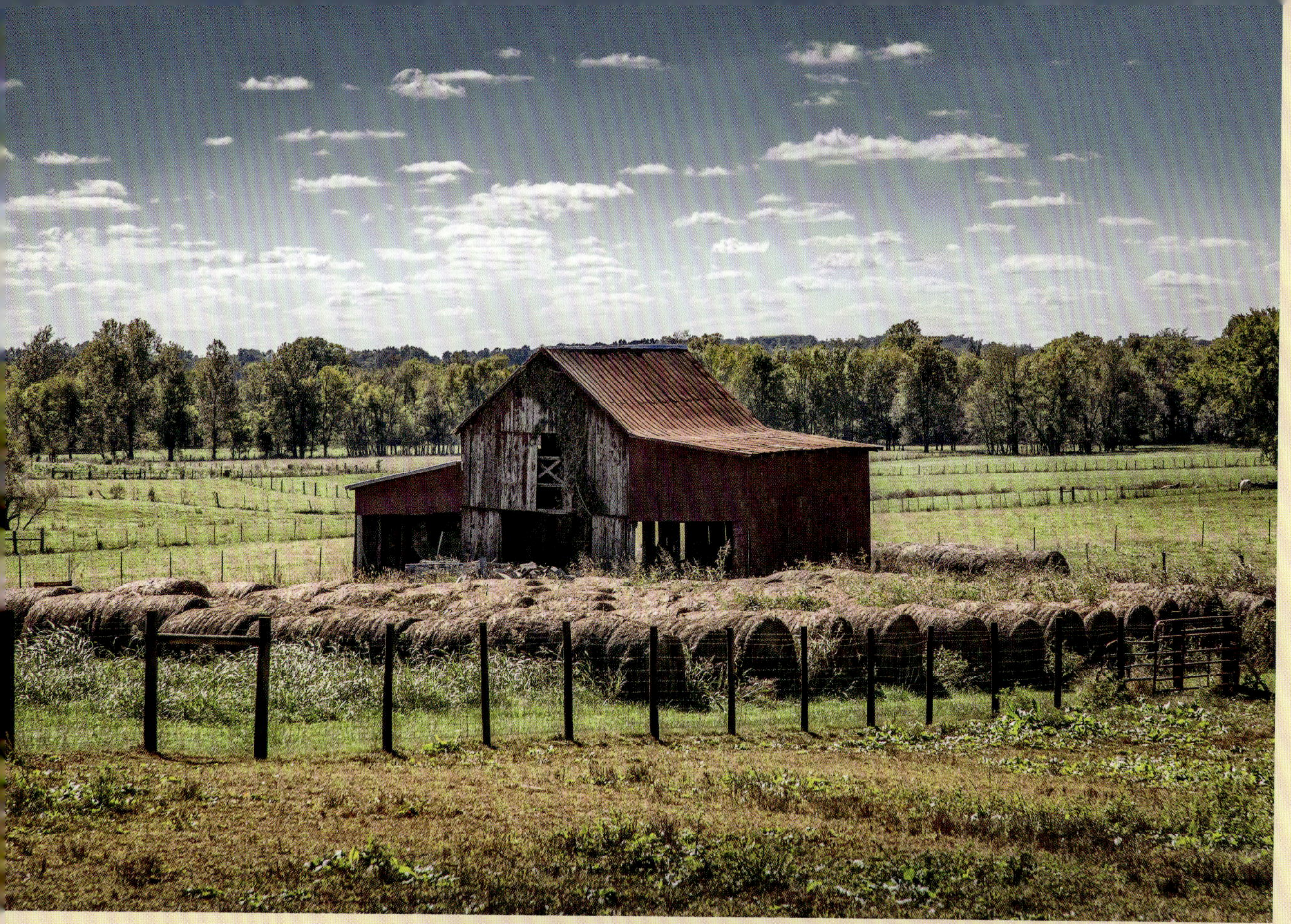

Christian County.

This two-story barn sits in a field and is used for hay roll storage. The yellow tine comes from a hay rake.

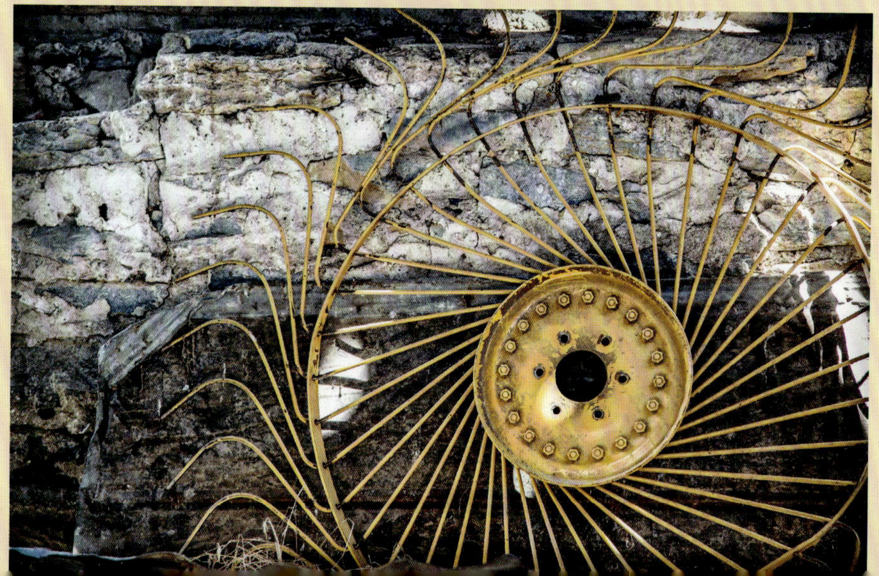

FACING PAGE

Madison County.

Once part of the McWilliams Revolutionary War land grant, this farm retains the historic eighteenth-century log crib and a nineteenth-century stock barn.

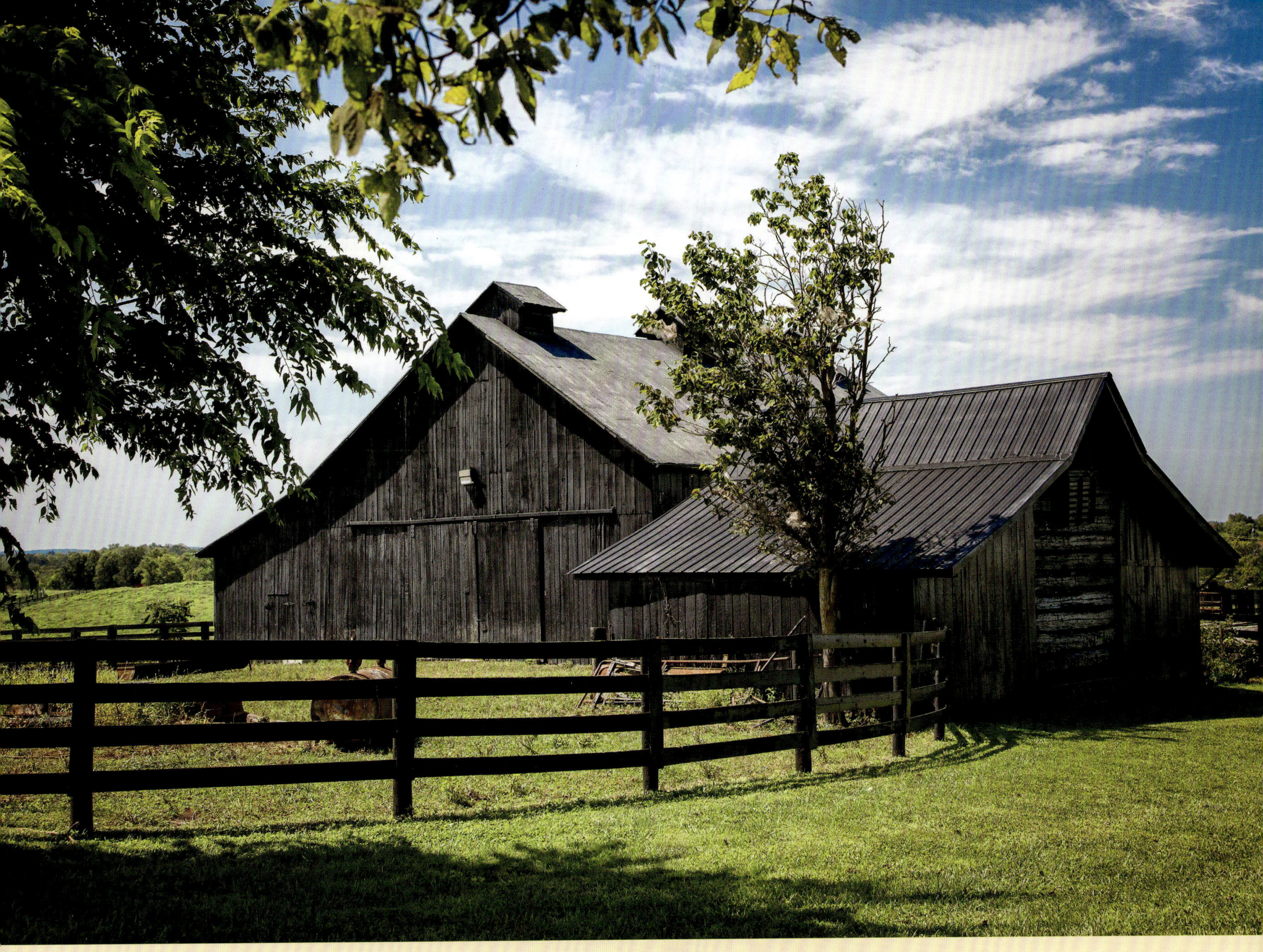

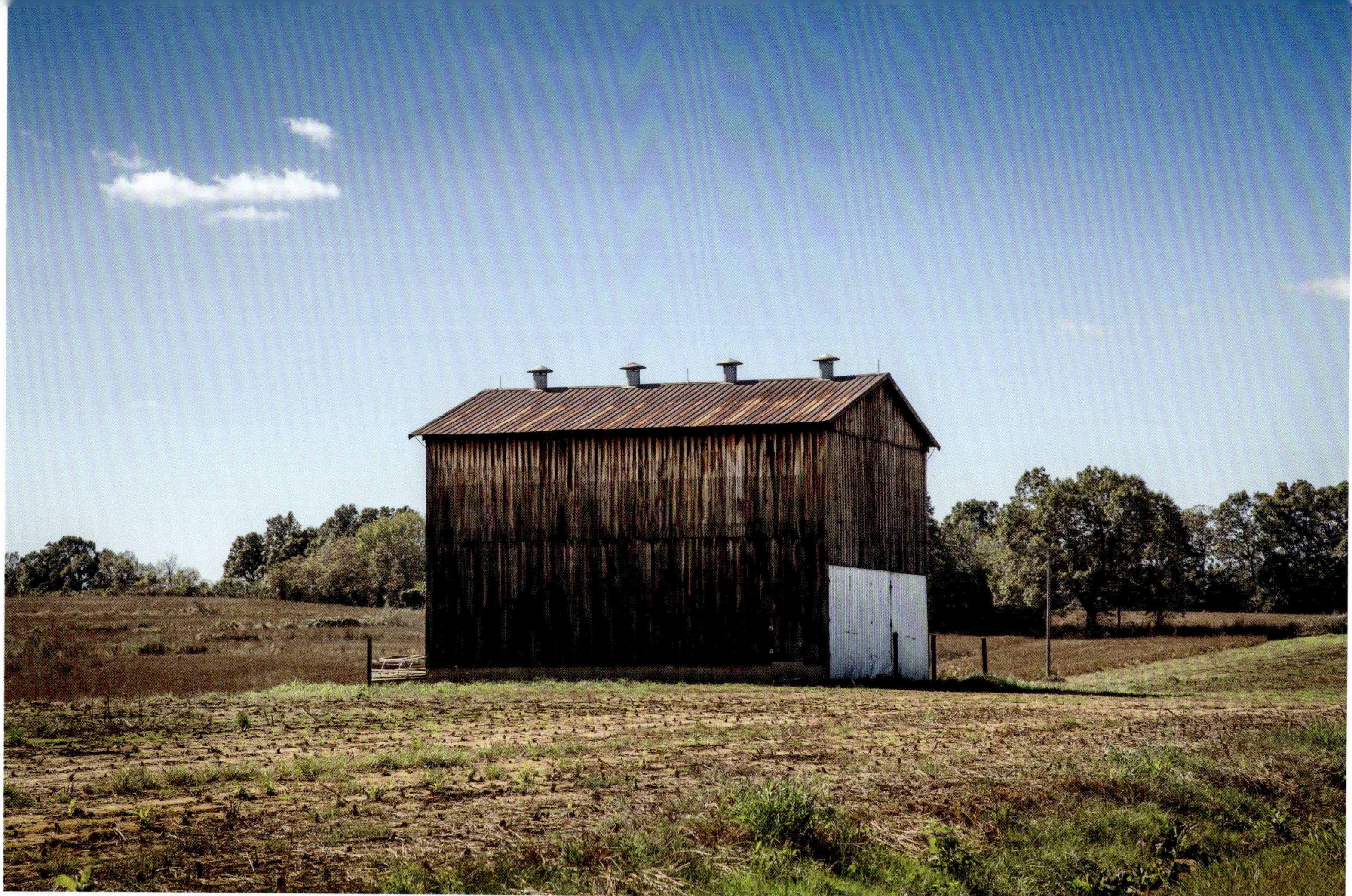

Christian County.

Along roads in Western Kentucky are many inactive wooden dark tobacco firing barns. Active farms are now larger and use metal buildings. Most old wooden barns are used for storage or abandoned altogether.

FACING PAGE

Rockcastle County.

The use of barns to advertise has a long history. Farmers added advertisements both by painting and by adhering signage to the sides of the barn.

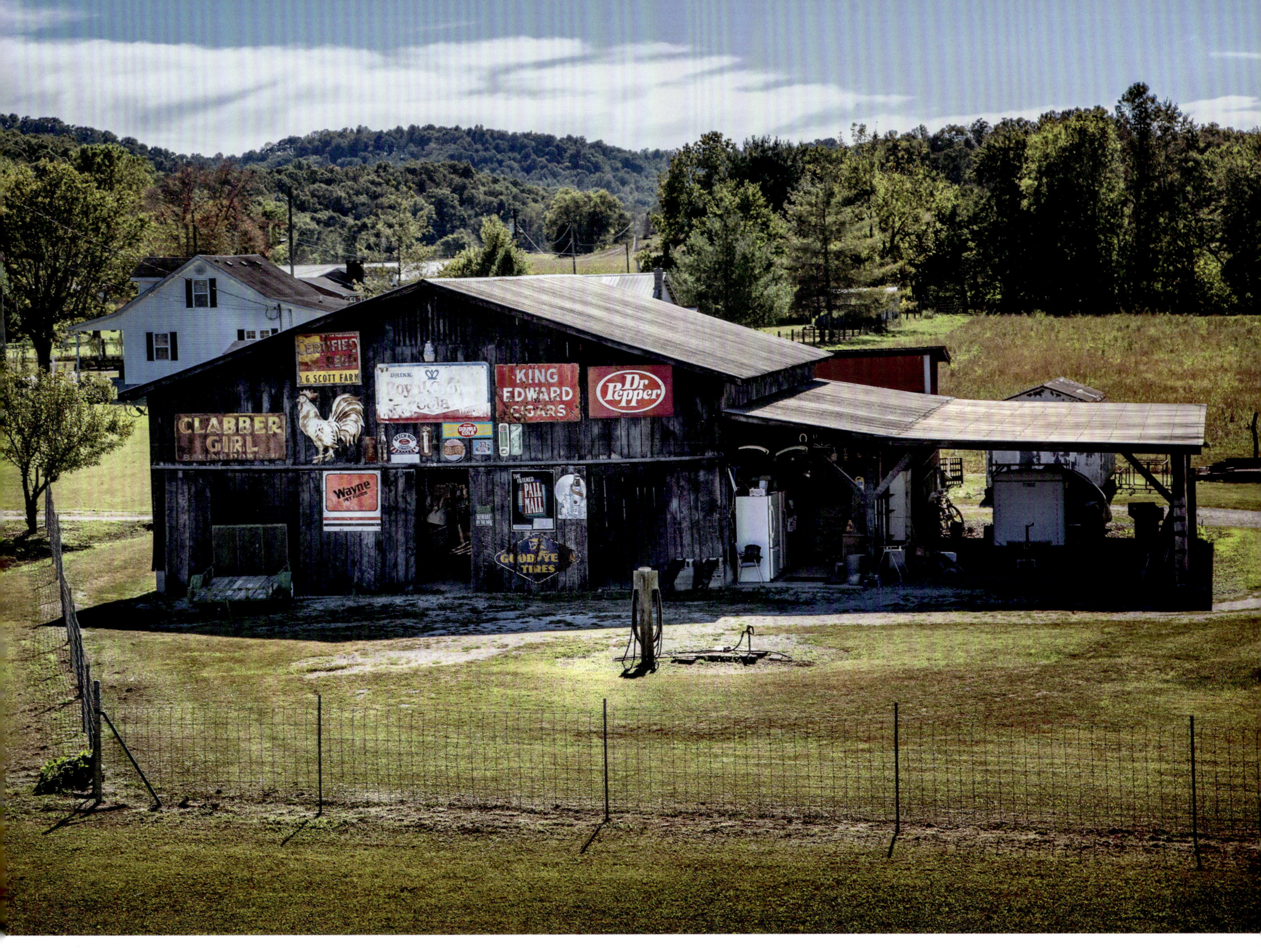
CERTIFIED FEED
D. G. SCOTT FARM
CLABBER GIRL
BAKING POWDER
DRINK
Royal Crown Cola
KING EDWARD CIGARS
Dr Pepper
DOUBLE COLA
Wayne
PET FOODS
THE FILTERED
PALL MALL
GOODYEAR TIRES

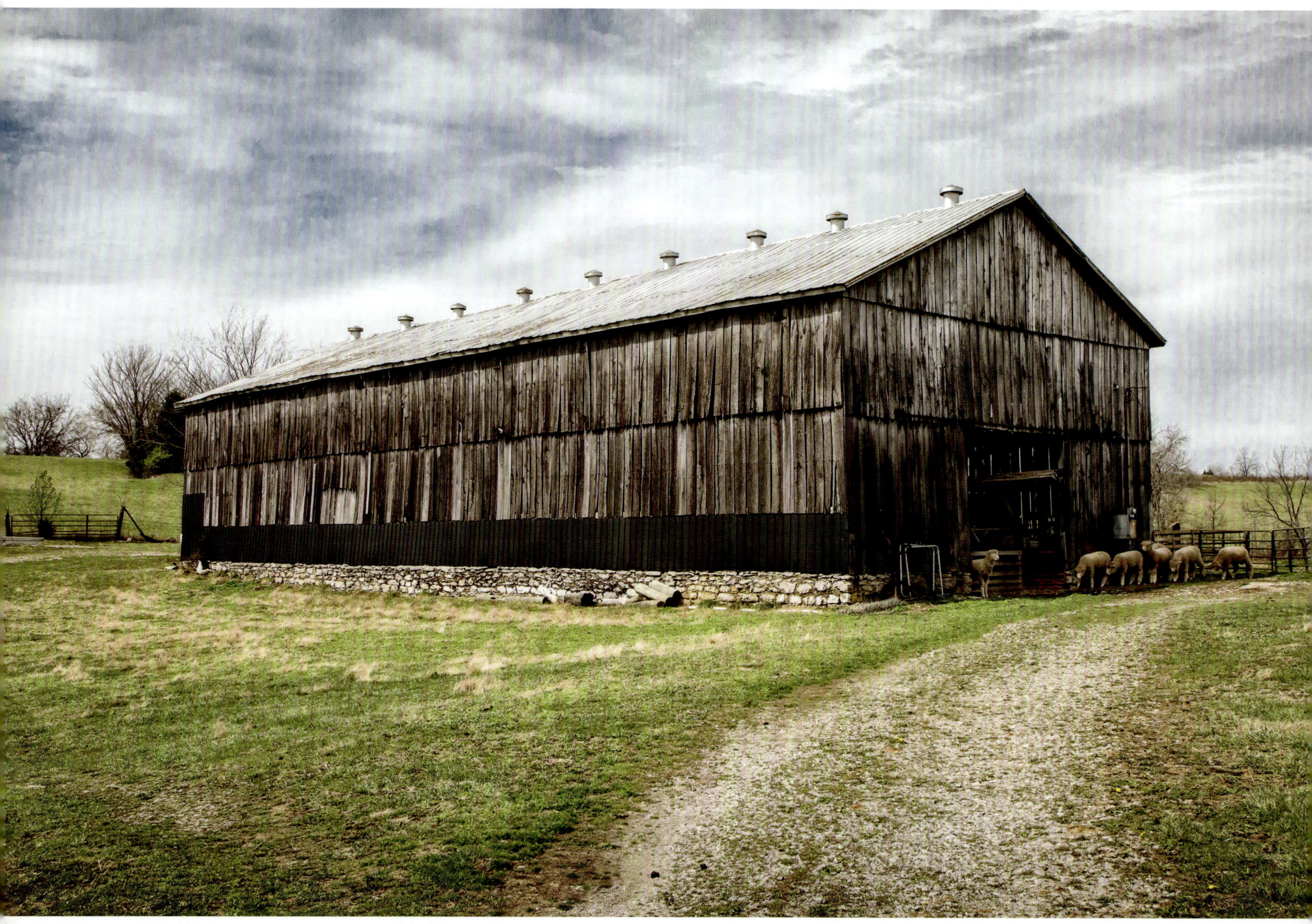

Mercer County.

This late nineteenth-century stock and tobacco barn has a stone foundation. The vents were added later for tobacco drying.

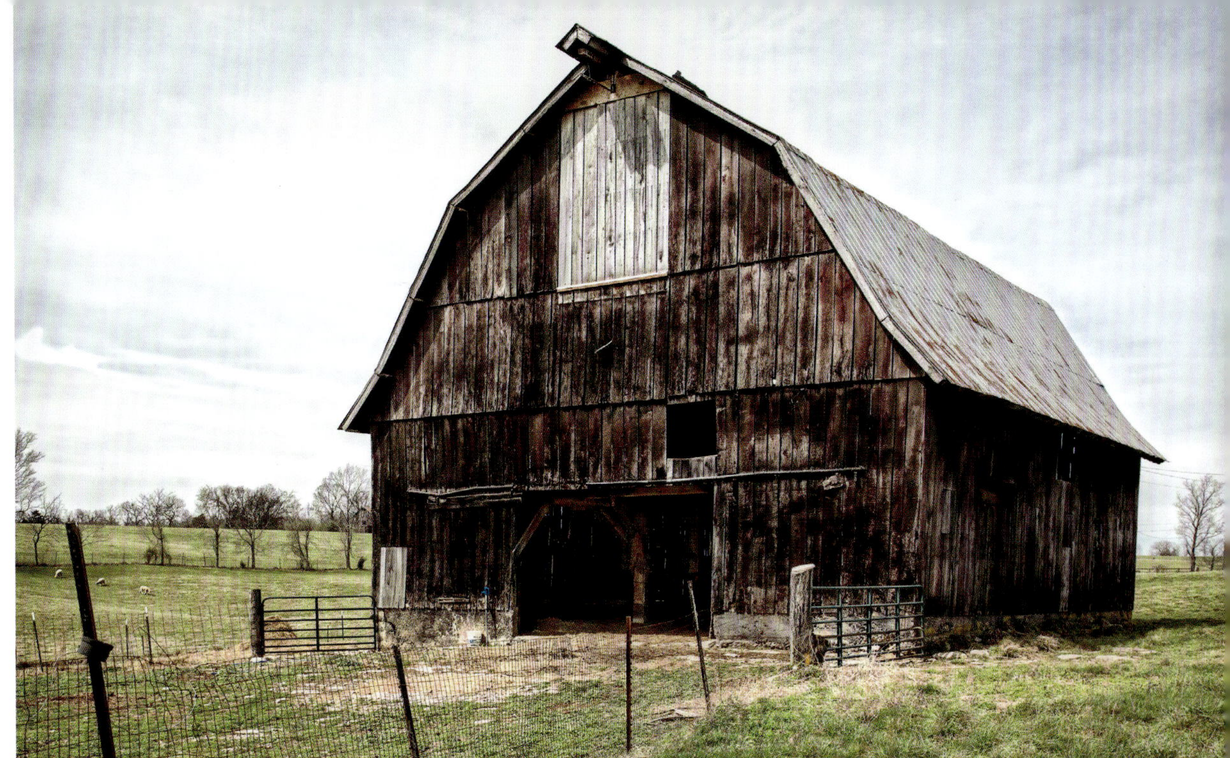

Mercer County.

A small sheep operation works out of this twentieth-century stock barn with a gambrel roof and small hay hood. Hay string drapes over the supports inside.

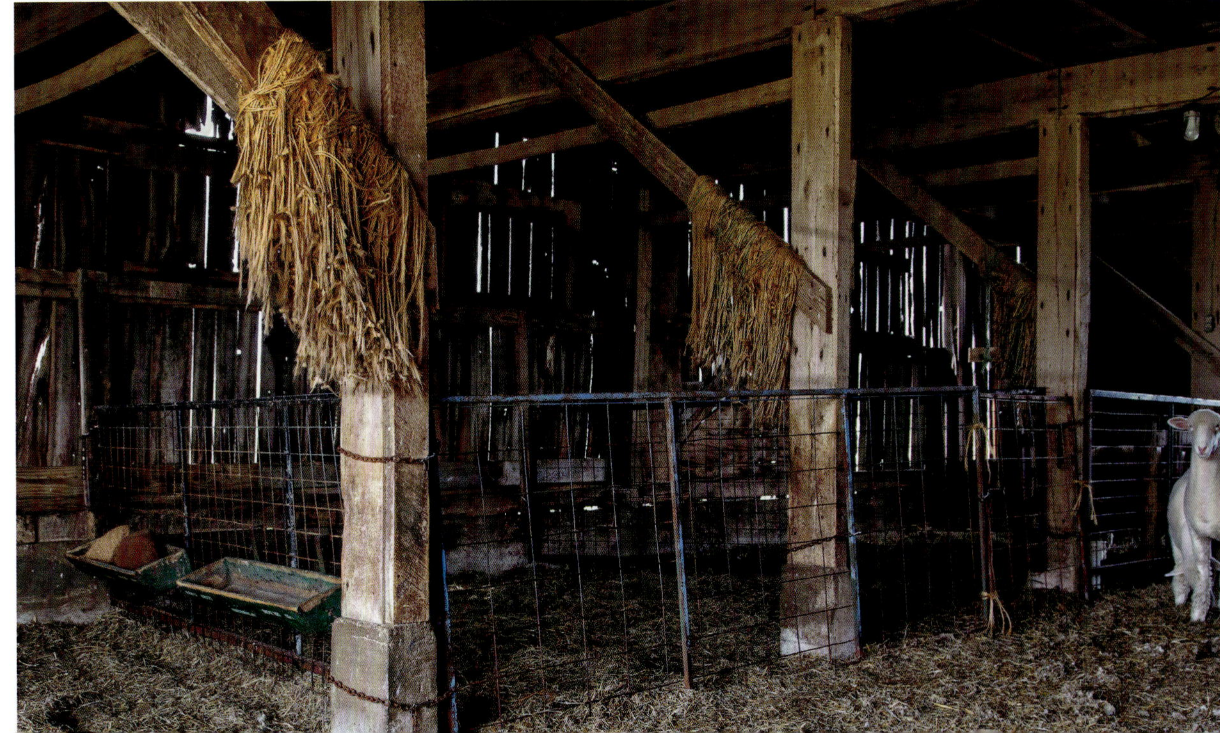

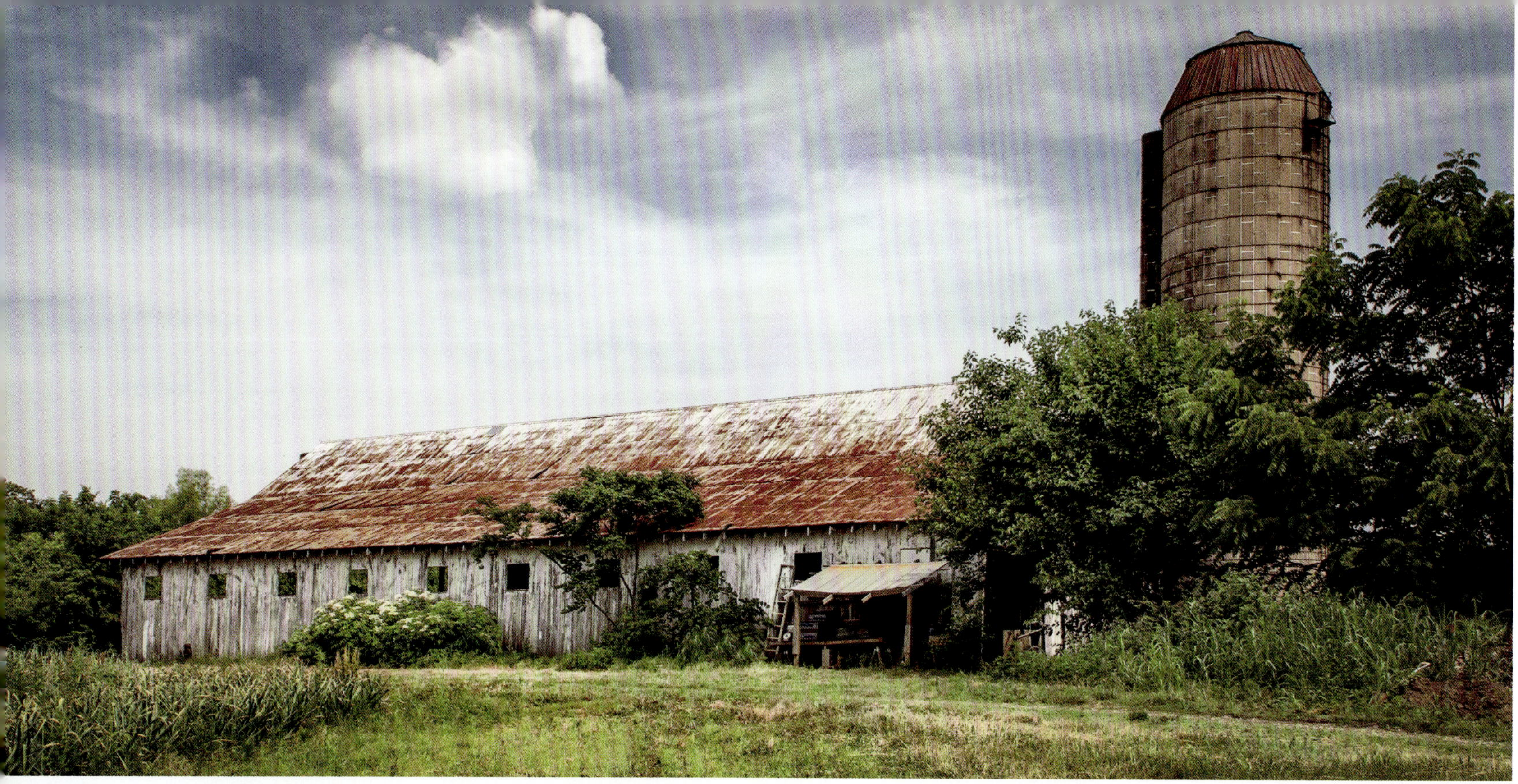

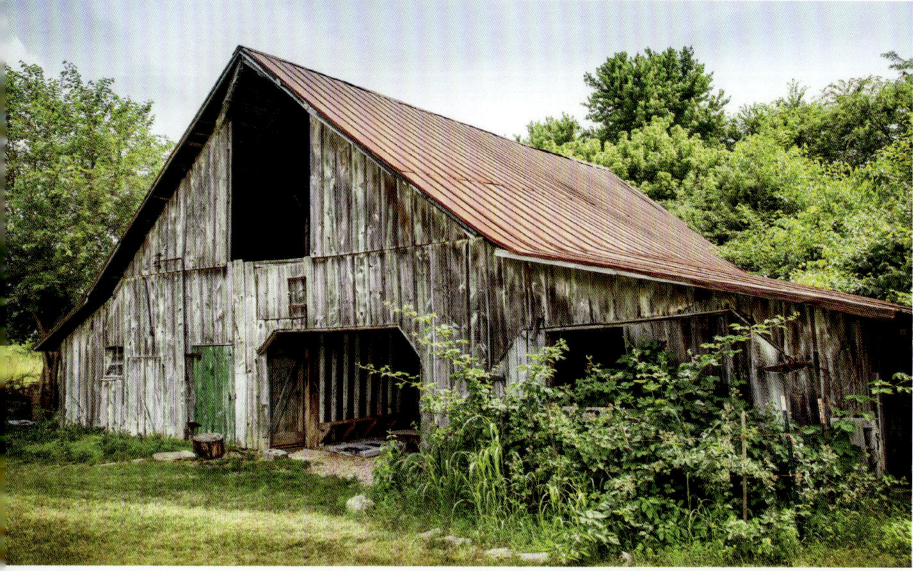

Jefferson County.

Tenant farming on Oxmoor, the Bullitt Family Estate, was an integral aspect of land use and farming activity during the twentieth century. Dairy, stock, and crops were the products of a farm located on this parcel of farmland, now used by the Food Literacy Project at Oxmoor Farm. According to their literature, the Food Literacy Project "invites youth and families to become active participants in a sustainable food system by planting, harvesting, tasting and cooking fresh vegetables." It was founded in 2006. The silo barn, built in the 1930s, houses equipment for the program. The carriage and crib barn contains the kitchen and dining area for the program, which emphasizes farm-based education, youth community agriculture programs, family engagement, and professional development.

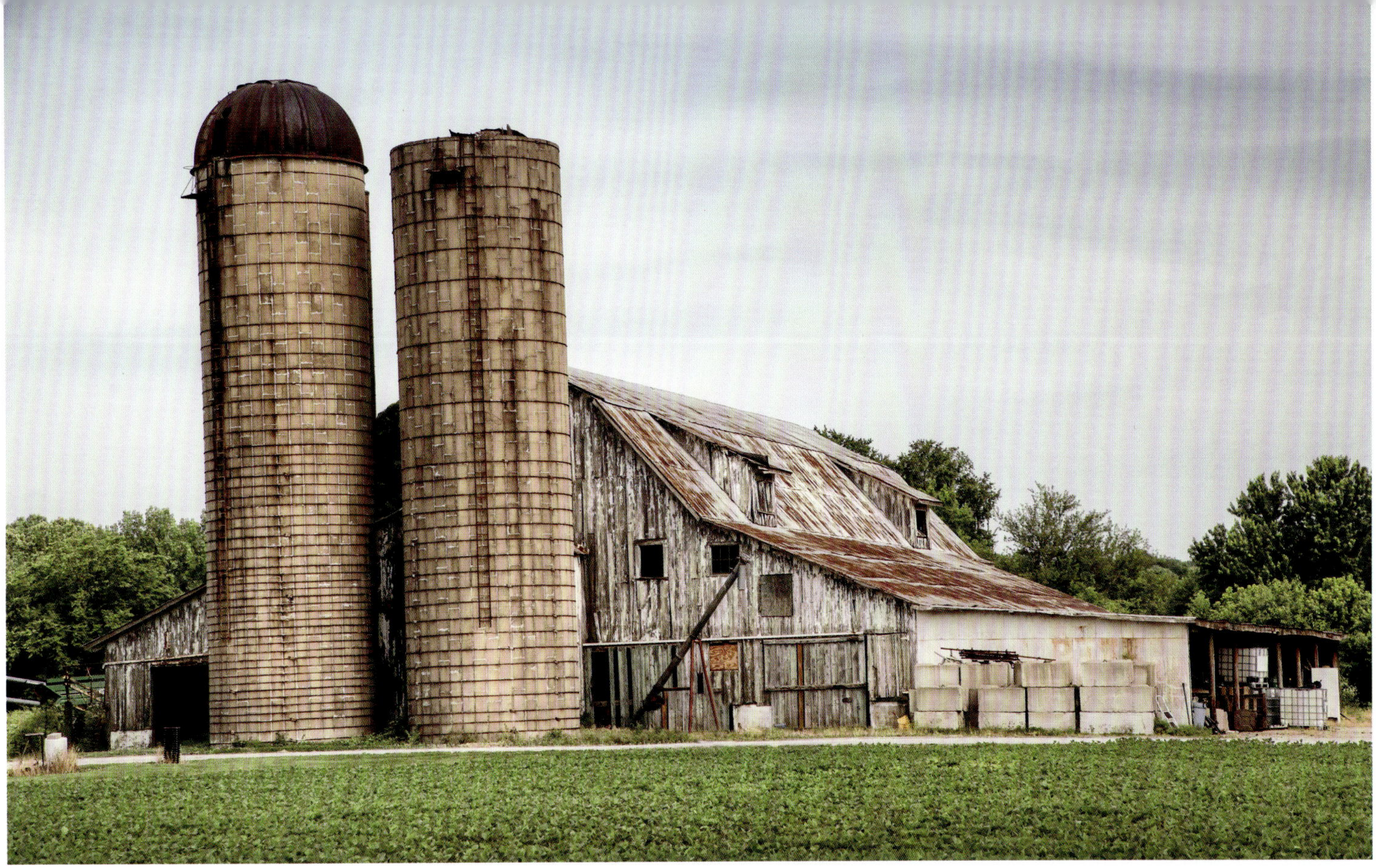

Jefferson County.

At one point in 1926, Mr. Bullitt turned over the acreage at Oxmoor to be farmed out by the Fidelity & Columbia Trust Company to run for him. The double-silo barn was part of that venture, used for stock, hay, and crops. The barn has a wonderful extended bay on one end and a side extension that was added later.

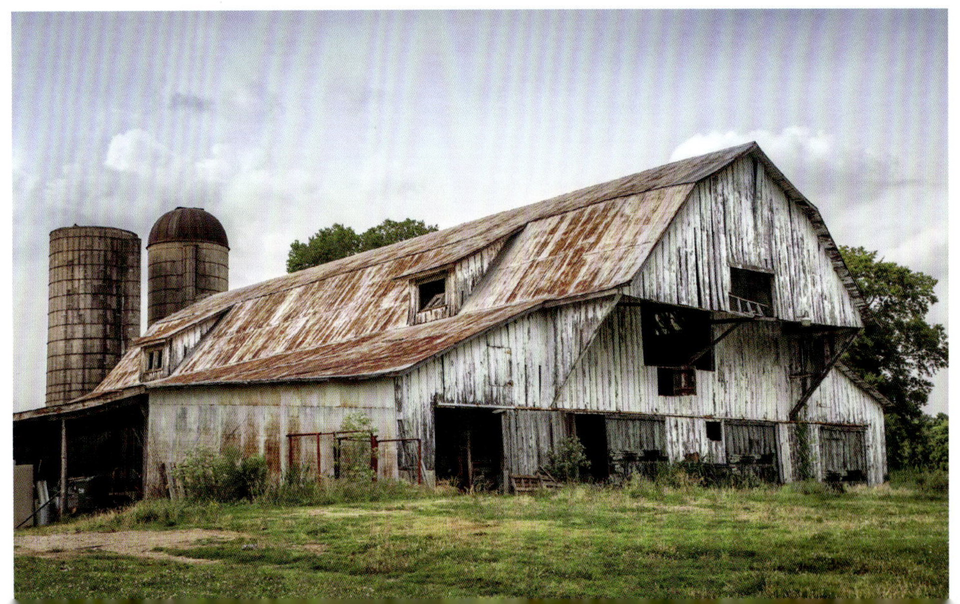

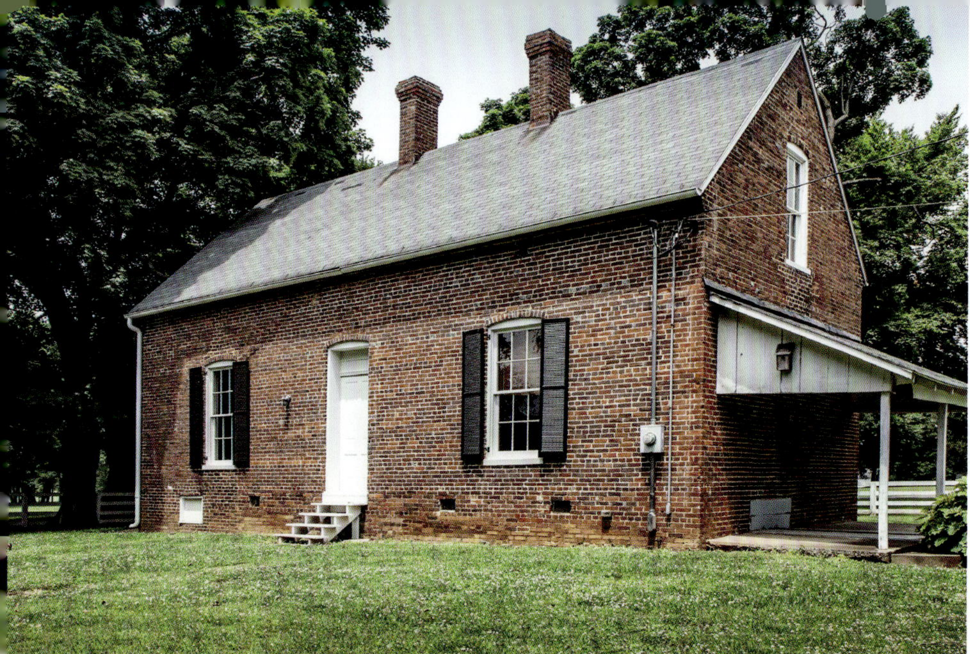

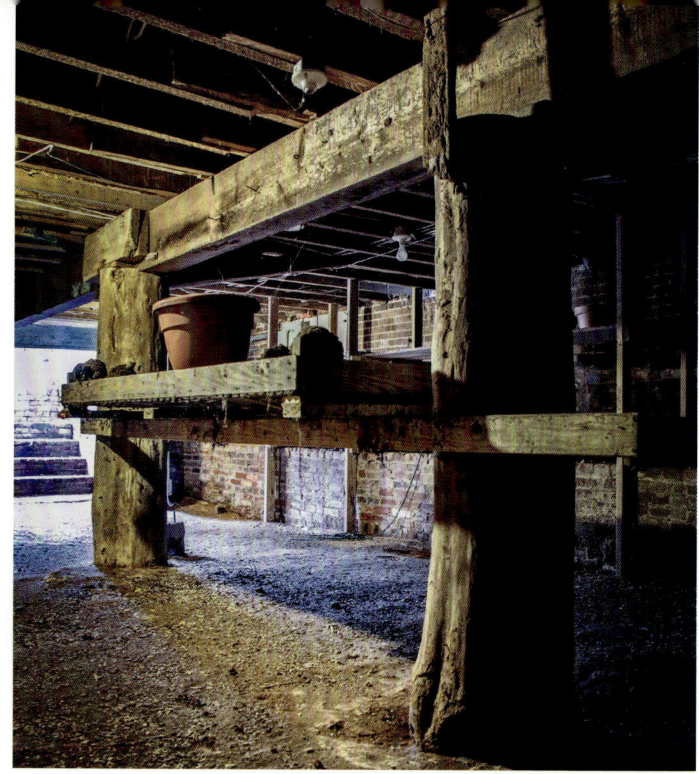

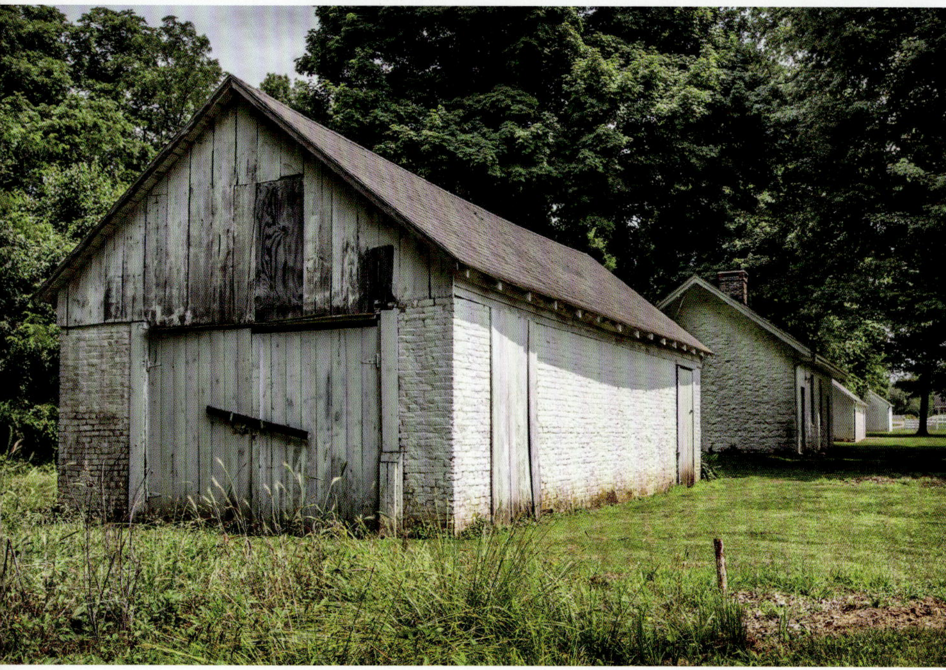

Jefferson County.

The hemp barn at Oxmoor Estate was constructed around 1840. It was renovated into a tenant farmer's residence in 1890. Huge tree trunks in the cellar support the long center beams that support the building. On the other side of the estate, a row of slave quarters remains, built circa 1843–1858. One of the cabins was converted into a barn later and still houses nineteenth-century farm equipment.

FACING PAGE

Jefferson County.

Horses became an integral part of Oxmoor Estate in the twentieth century. The stables were built in 1931, and an annual steeplechase commenced in 1940. The steeplechases continued until 2002.

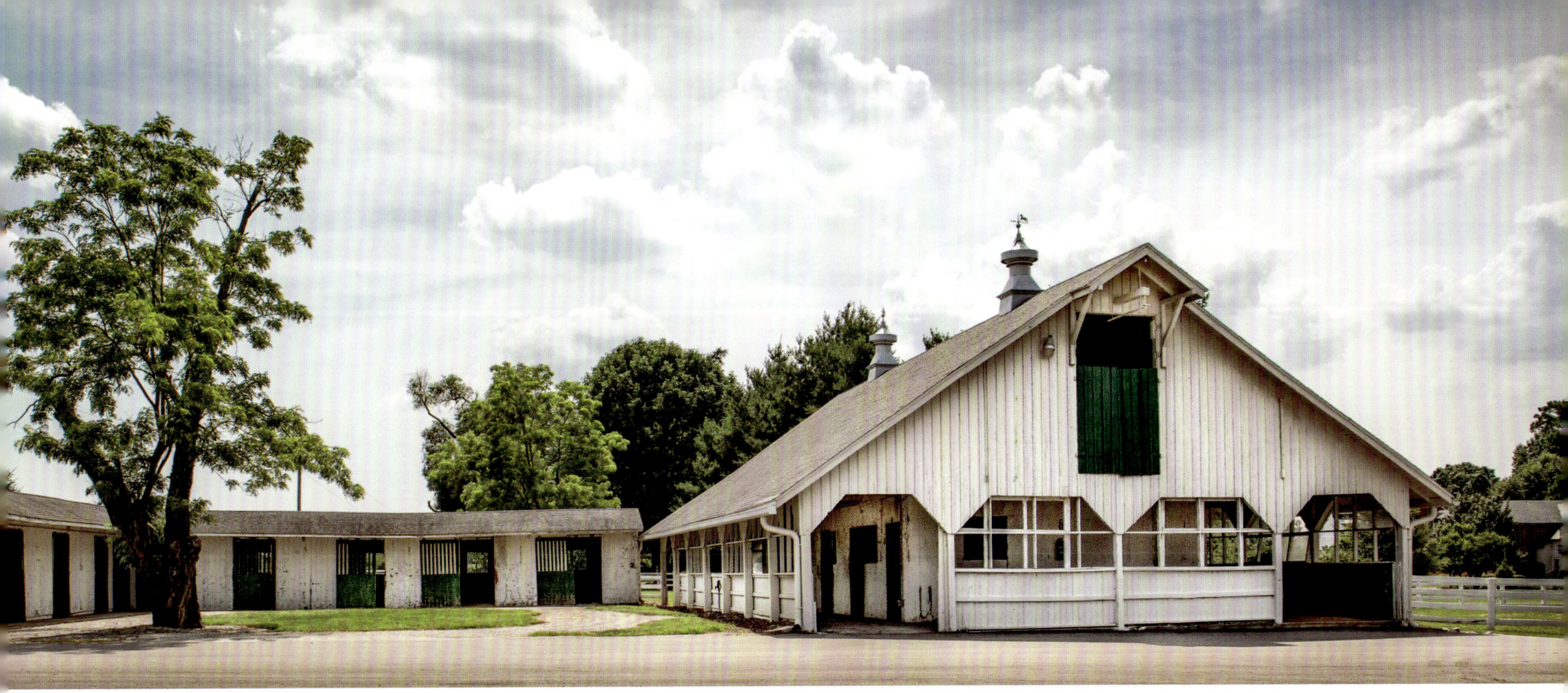

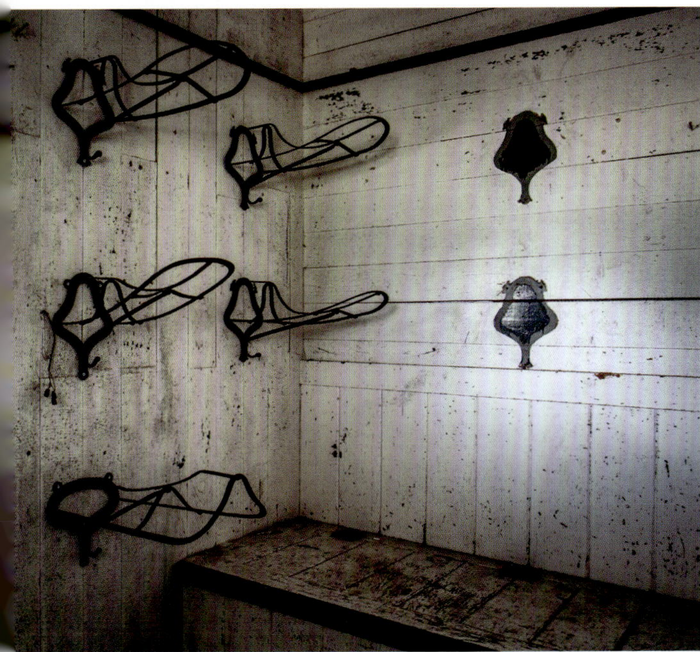

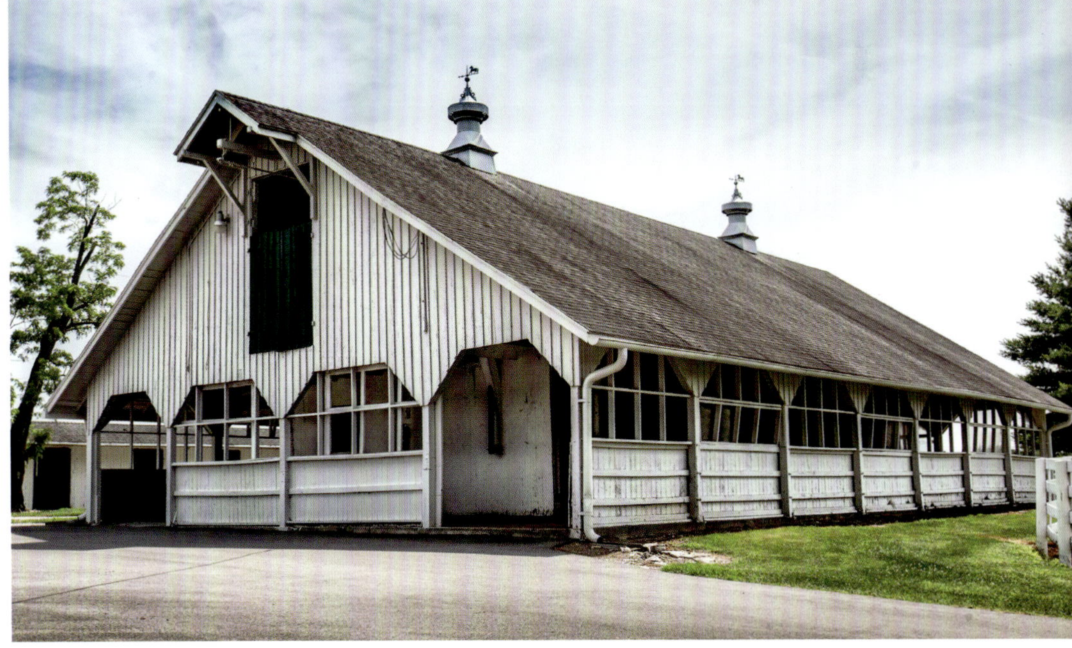

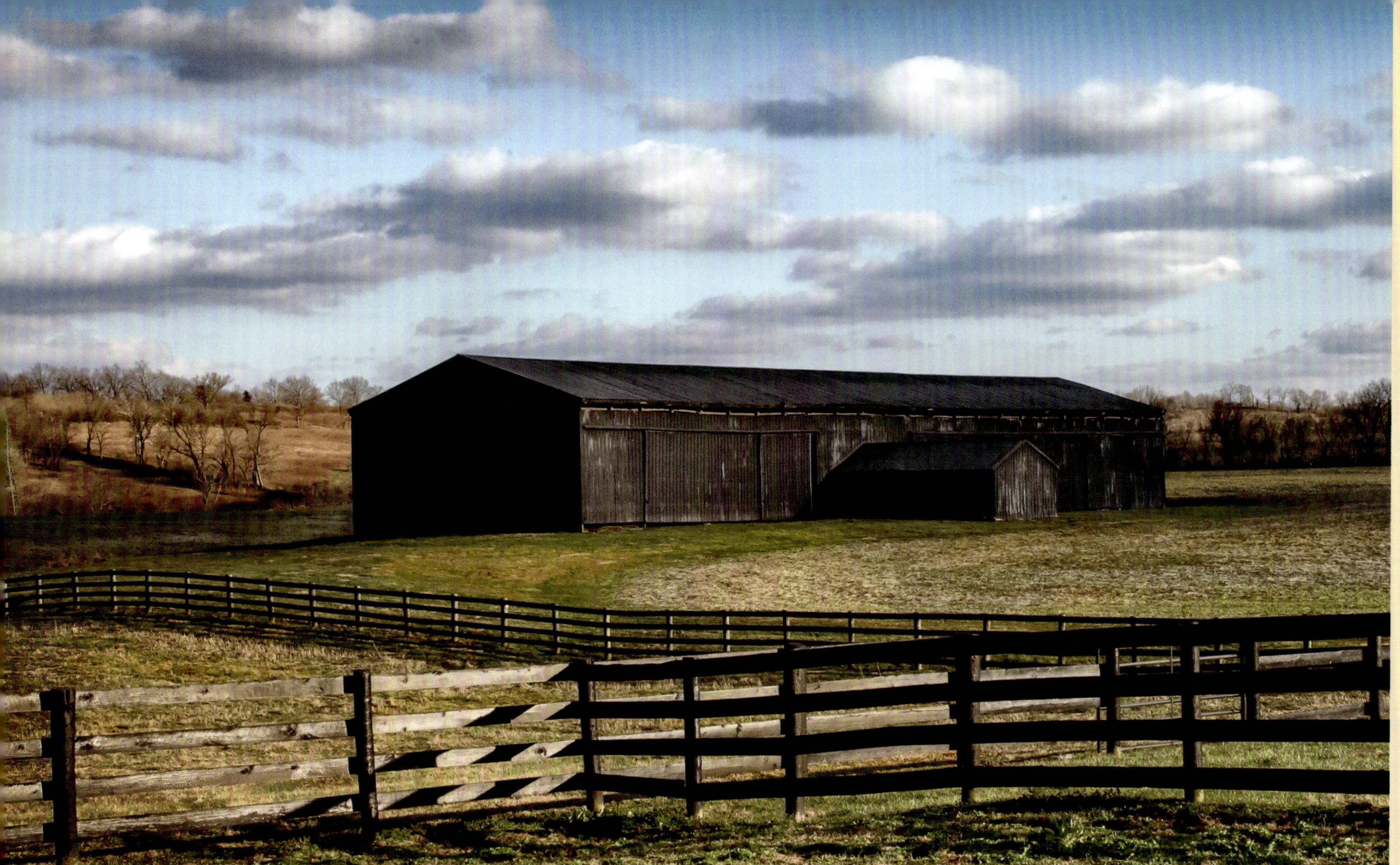

Mercer County.

There are still landscapes that suggest how Kentucky looked when farming was the predominant livelihood.

Mercer County.

The Pleasant Hill Shaker Community began in 1806 in Kentucky, and by the mid-1800s, it was known for its farming products. Barns original to the community are painted red, such as this nineteenth-century barn on the West Lot.

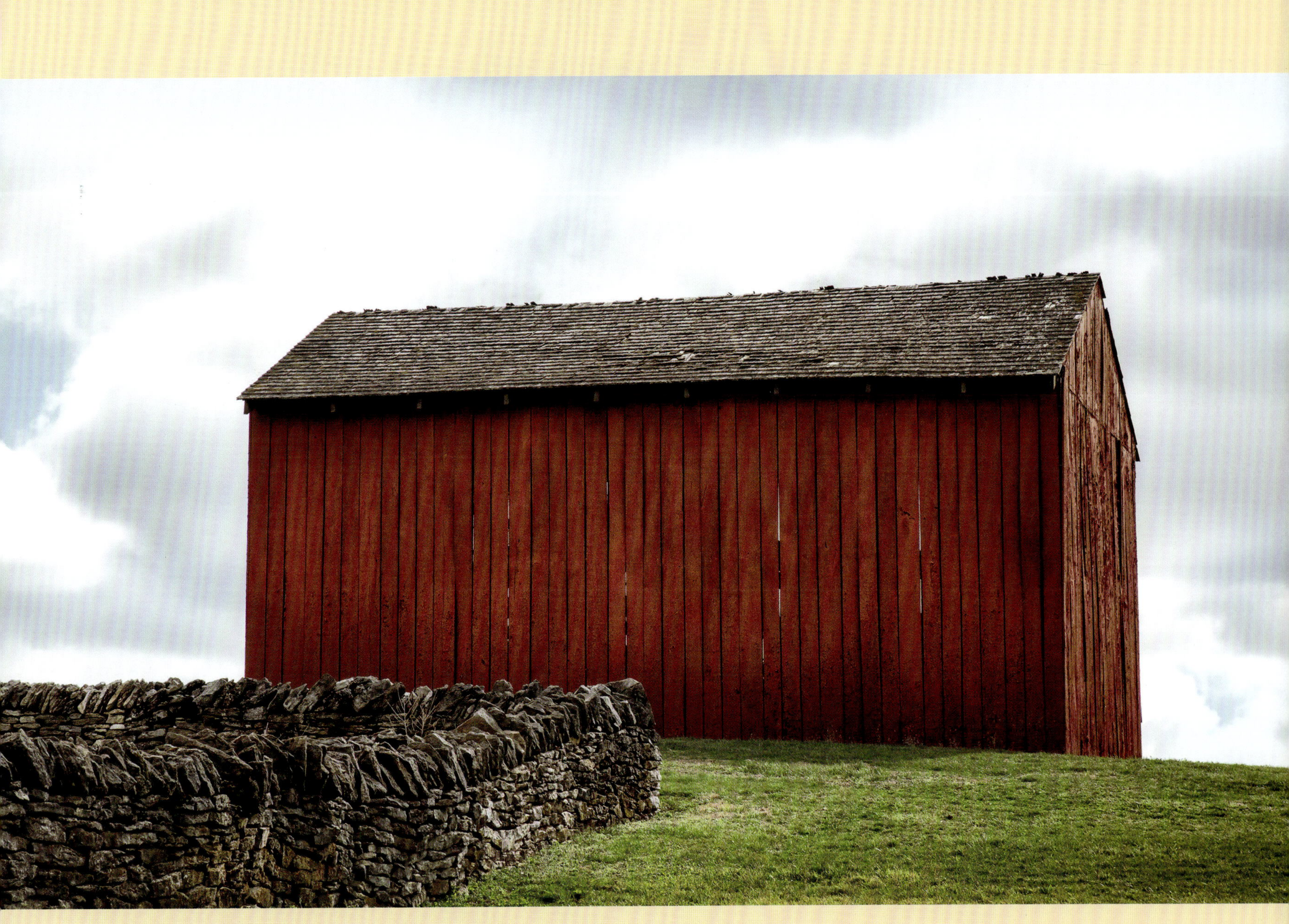

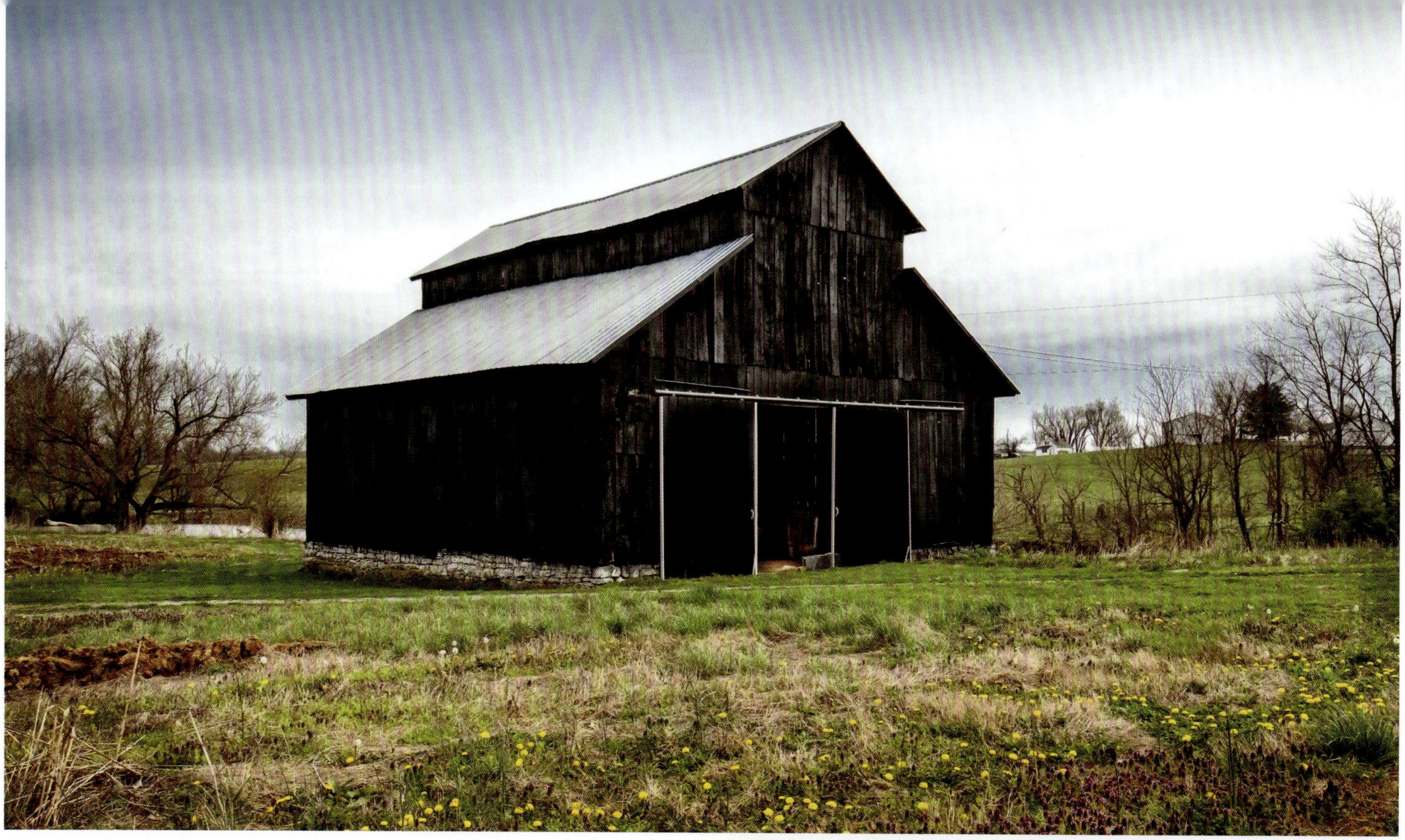

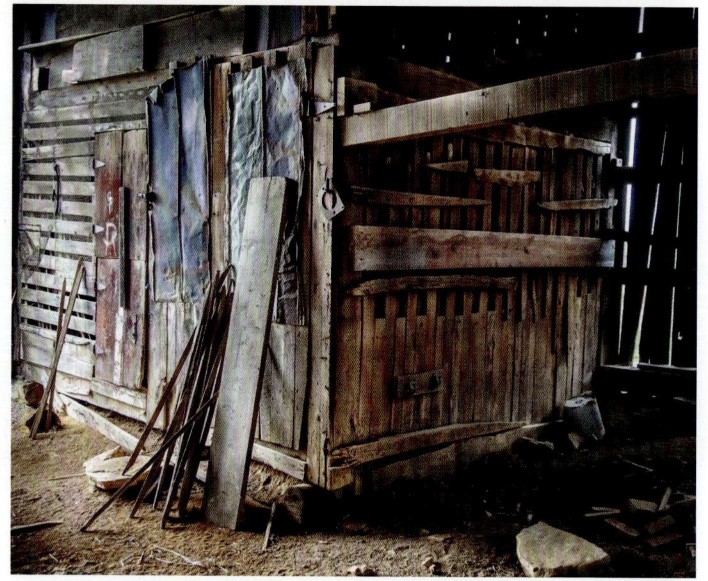

Mercer County.

This nineteenth-century barn has remained in the Shewmaker family for generations. Like many barns, it has had several reincarnations, from stock barn to tobacco barn and milking barn. The corncrib inside is original, now used for storage.

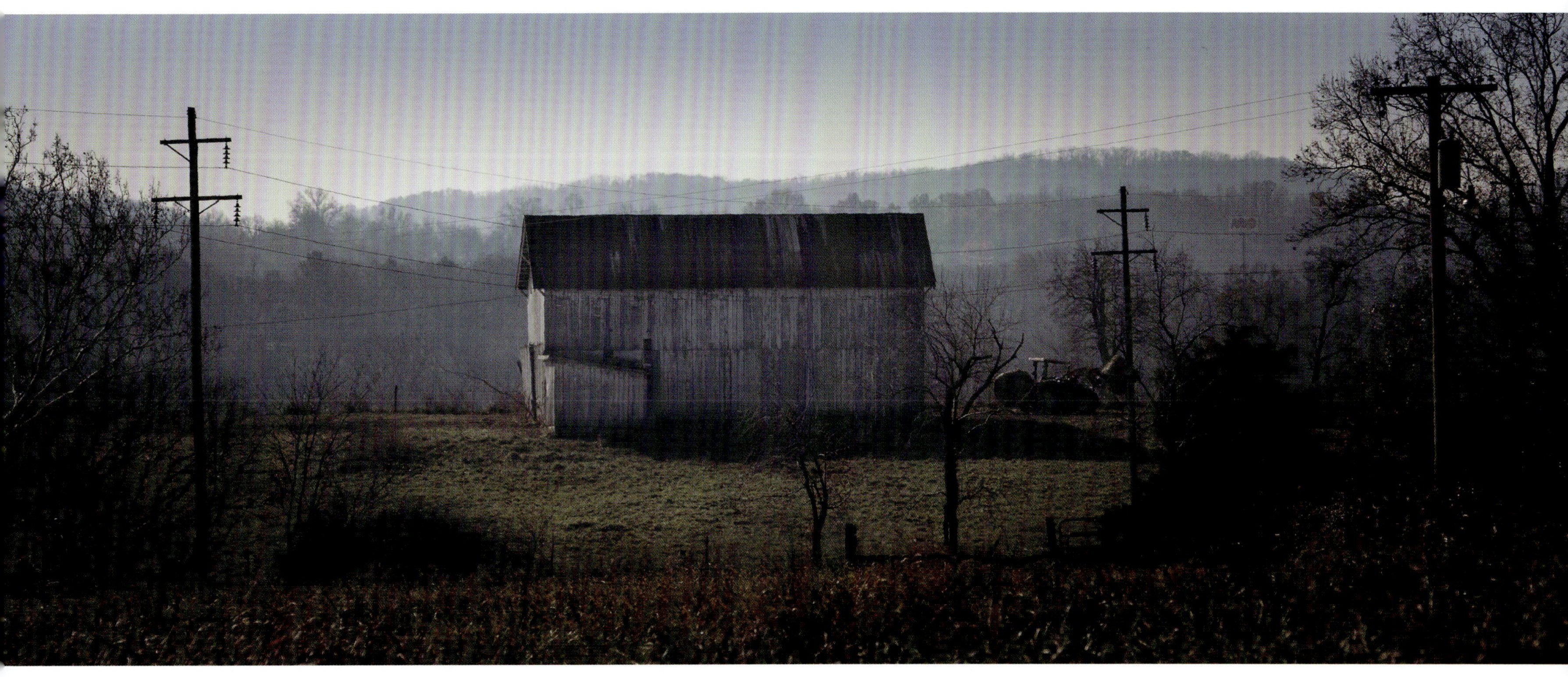

Madison County.

Twilight sets in, metaphorically and literally, on the Kentucky barn.

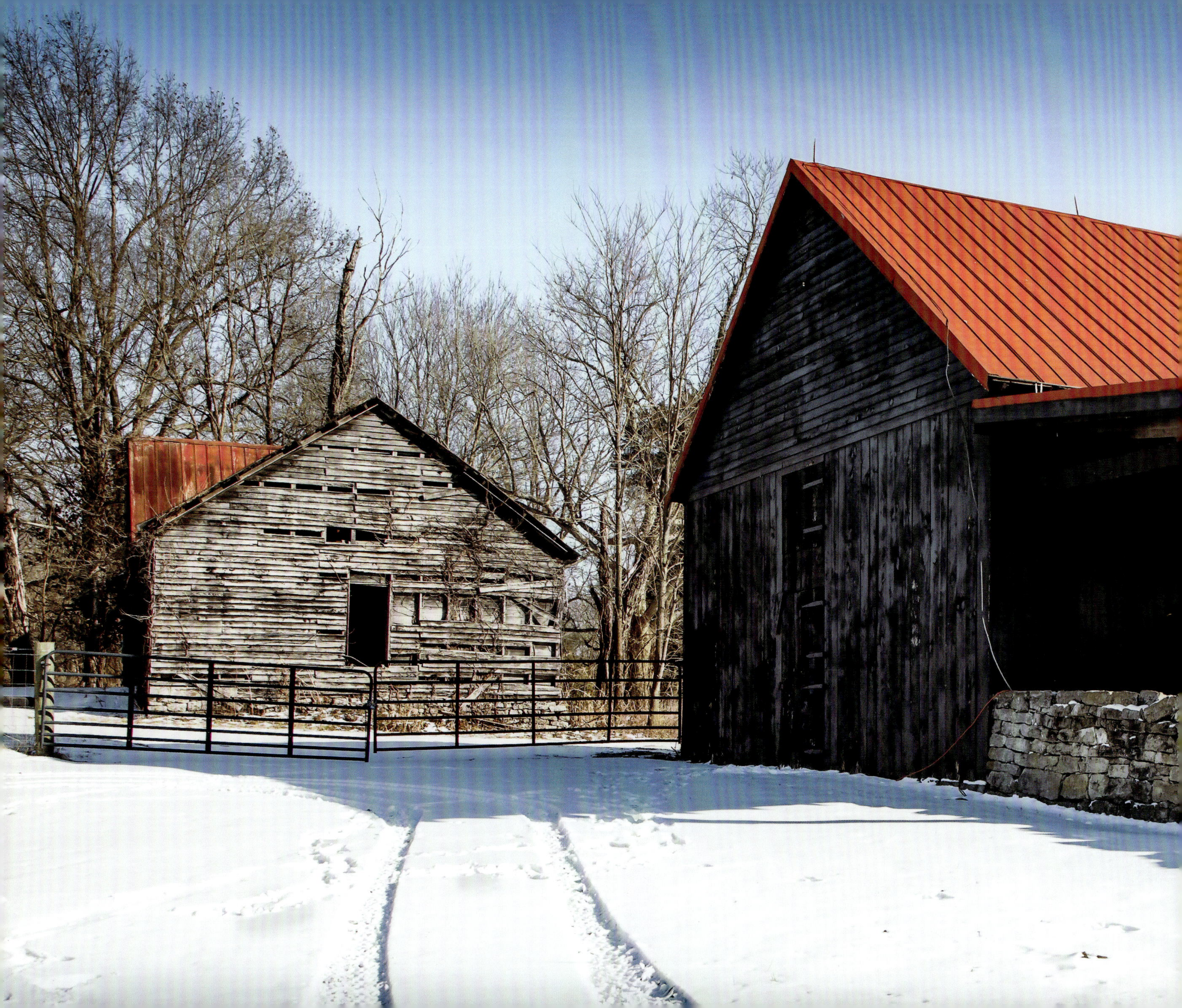

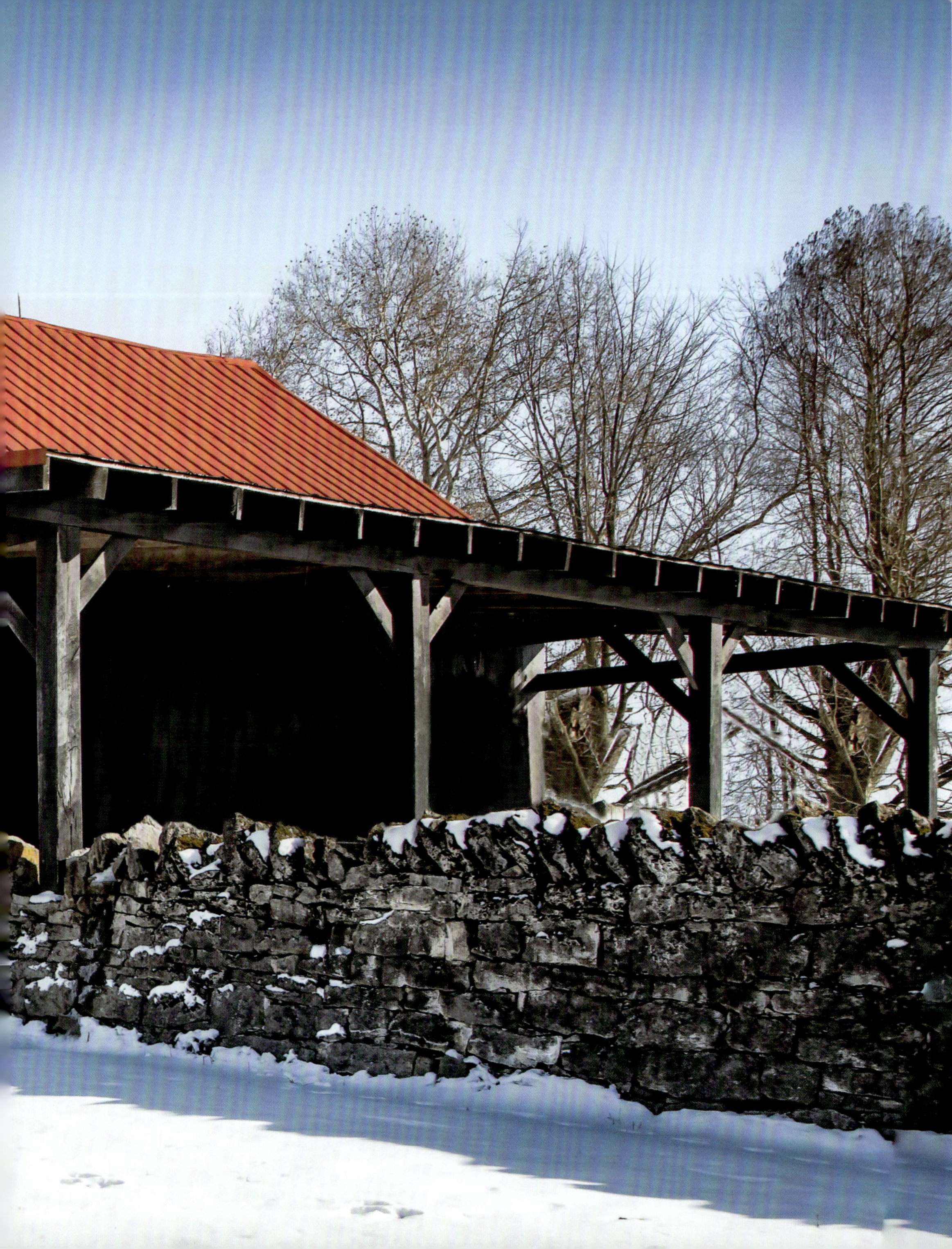

Bourbon County.

The mule barn at Auvergne, the homestead of historic statesman Brutus Clay and his descendants, was built circa 1841. The carriage house of the same age is in the background.

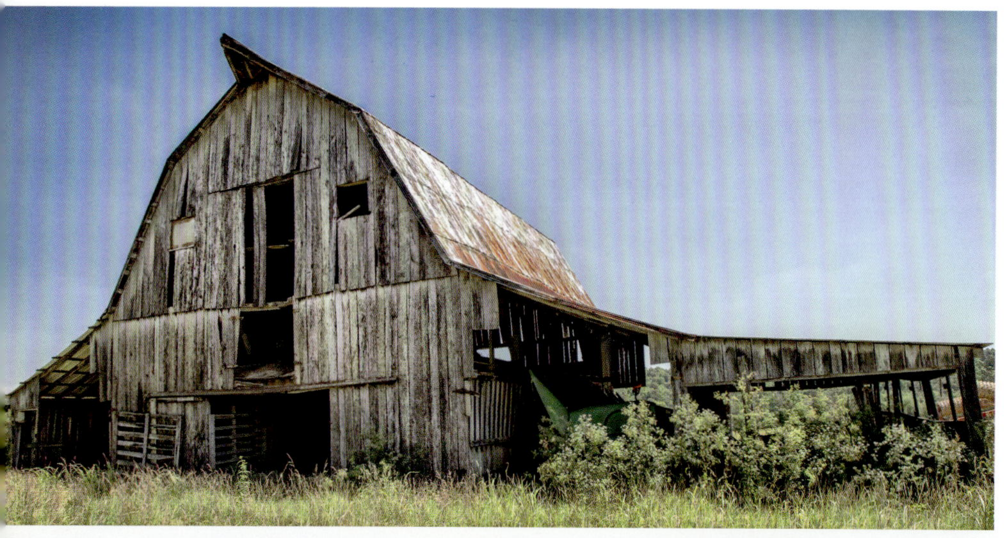

Crittenden County.

Two separate barns with similar architecture are located on opposites side of the road; these were probably on the same farm at one time.

FACING PAGE

Crittenden County.

Barns are often adapted over time to meet changing crops and farmers' needs, sometimes resulting in large barns like this one.

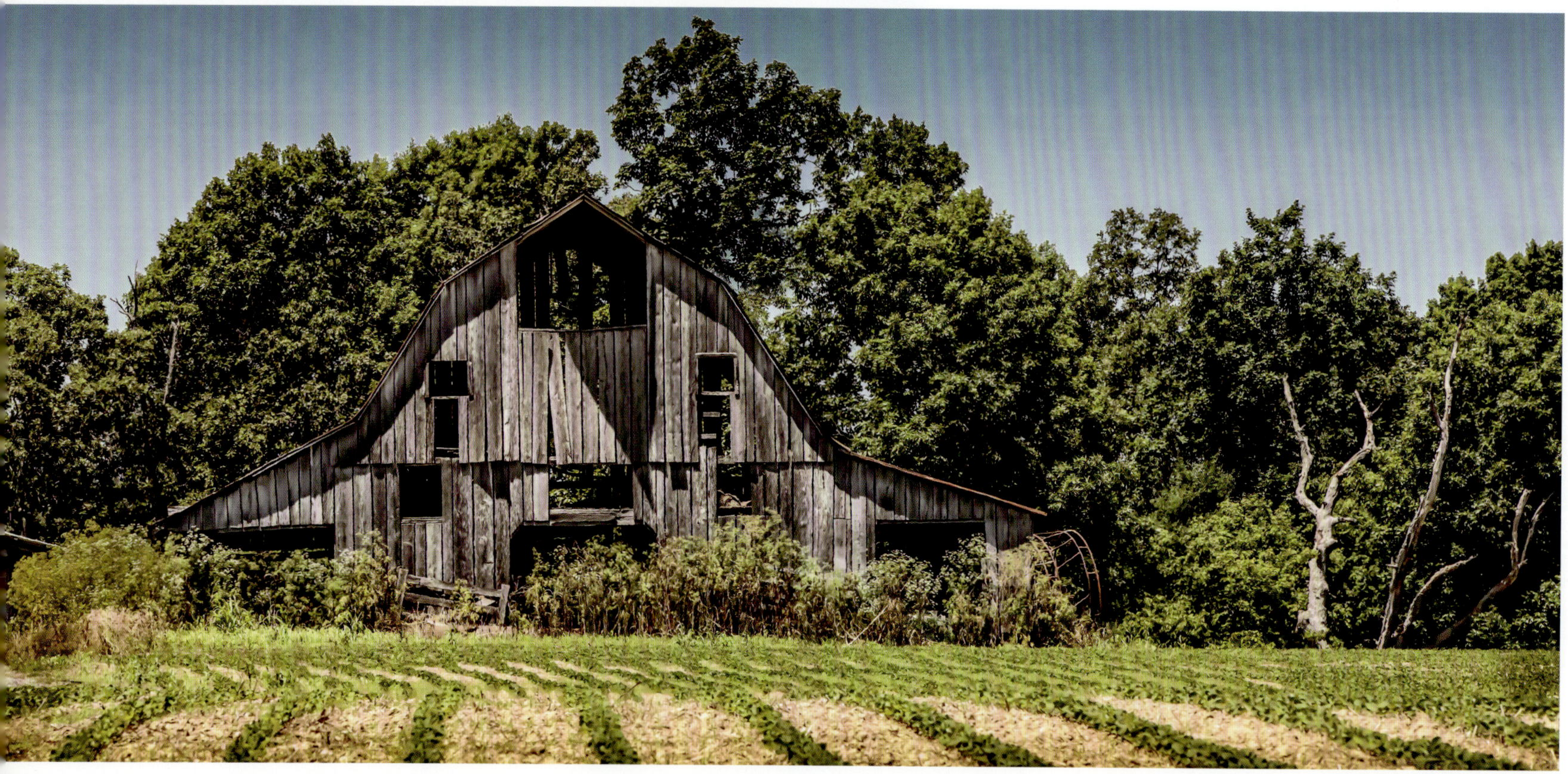

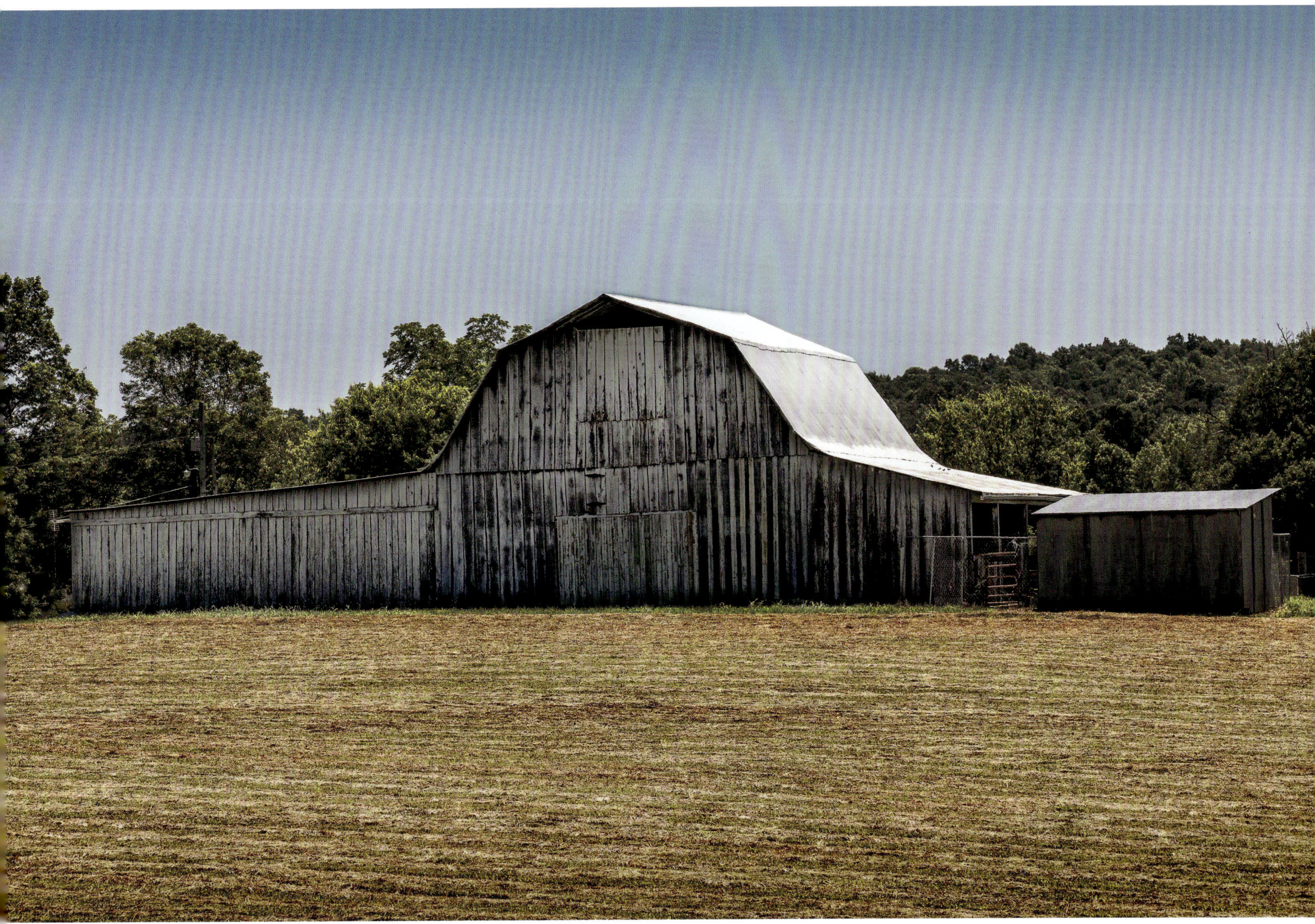

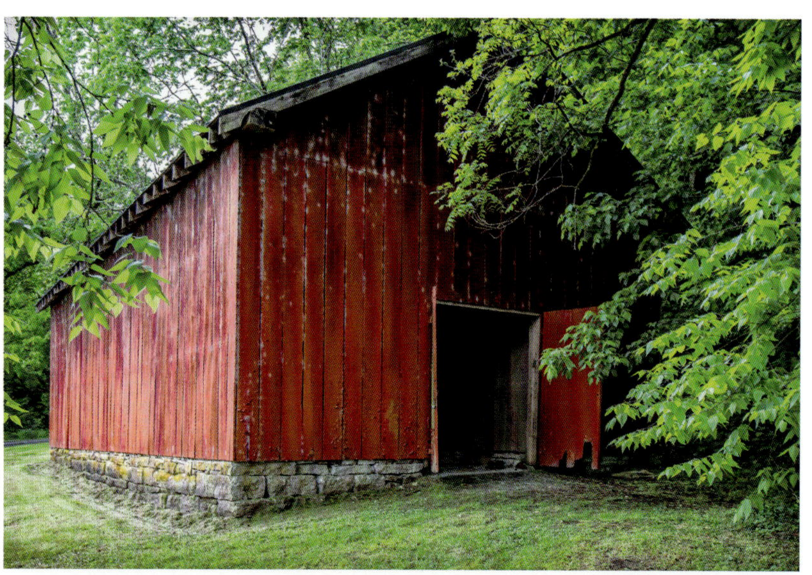

At the Pleasant Hill Shaker Community, barns were built down by the Kentucky River as well as around the main site. This restored nineteenth-century barn was used to warehouse goods coming and going on the river.

Henry County.

The bow-truss roof, introduced to North America by Dutch settlers, has a large storage space. The roof on this barn was probably the influence of Dutch colonists who migrated into this area from Mercer County in the nineteenth century.

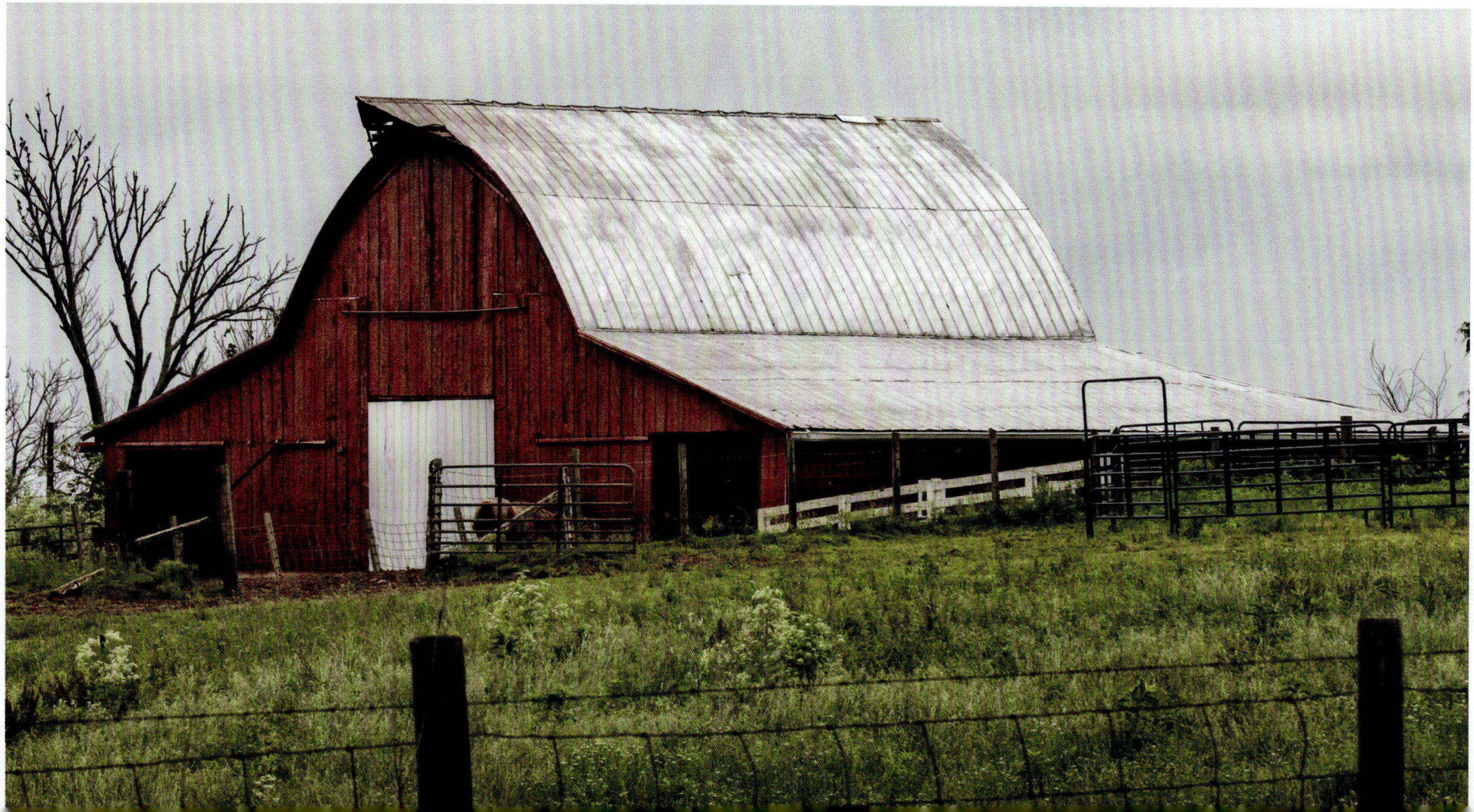

Henry County.

The Six Mile Meetinghouse was constructed circa 1824 as a religious center for the Low Dutch colony that settled here. Later the framework of the church was converted to a barn on the List family farm. Sam and Joann Adams obtained the timbers and reconstructed the meeting house in 2003–2005, preserving the wood and architectural construction.

Henry County.

Henry County was the site of a settlement of Low Dutch immigrants who had first settled in Mercer County from Pennsylvania in the early 1800s. Many families from those immigrants remained in the area, influencing architecture in the surrounding area. The extended bay is called a threshing bay, an architectural feature from New England.

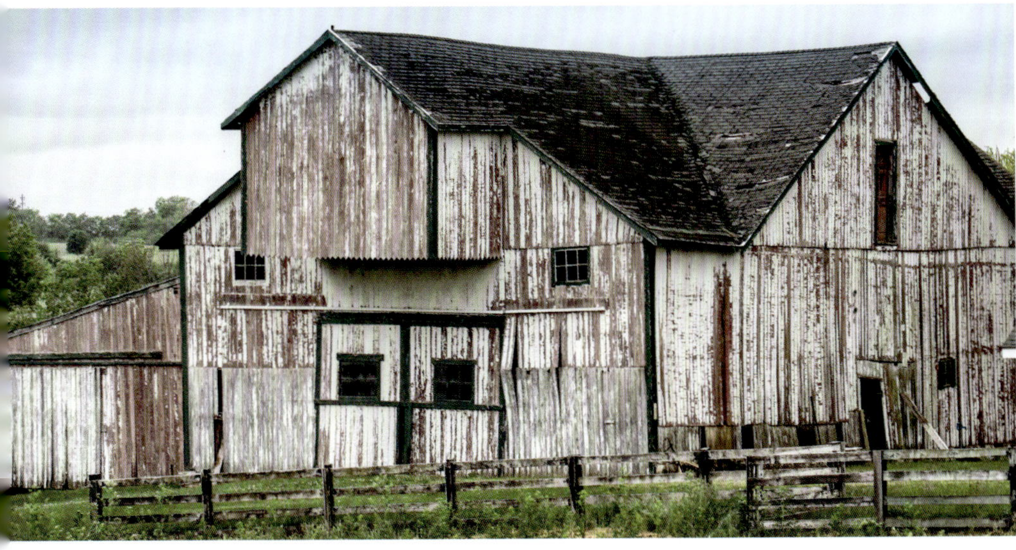

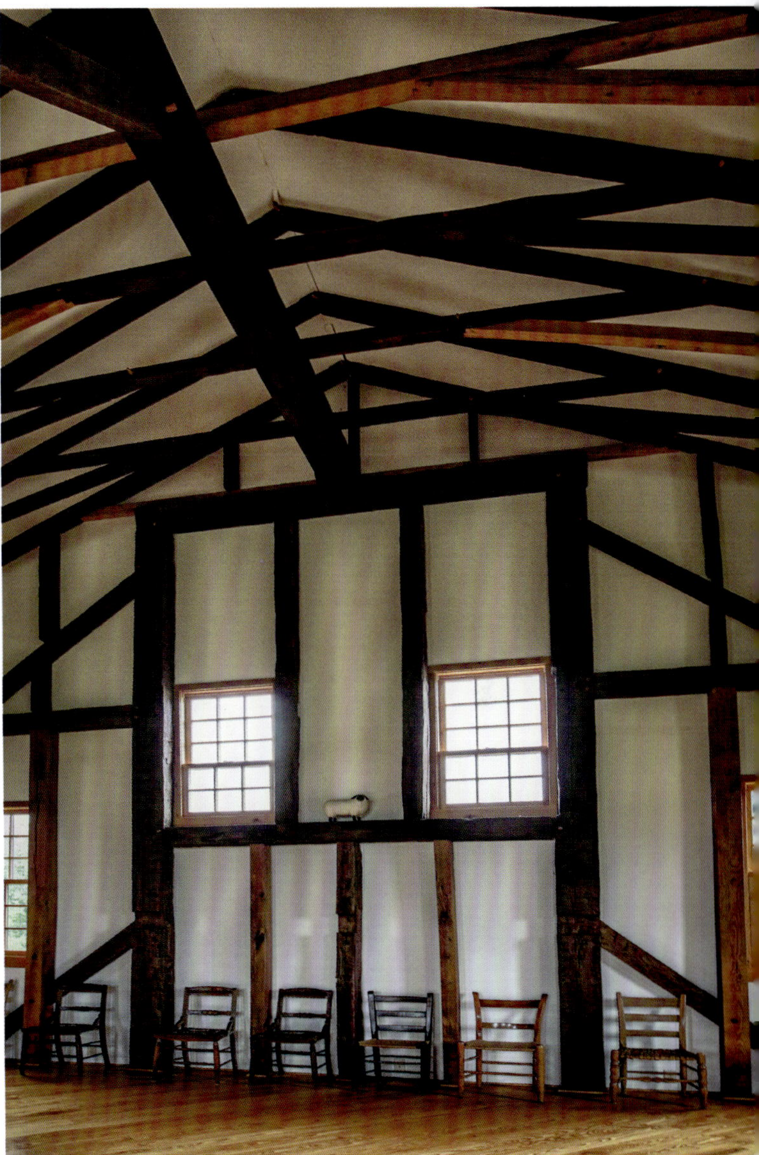

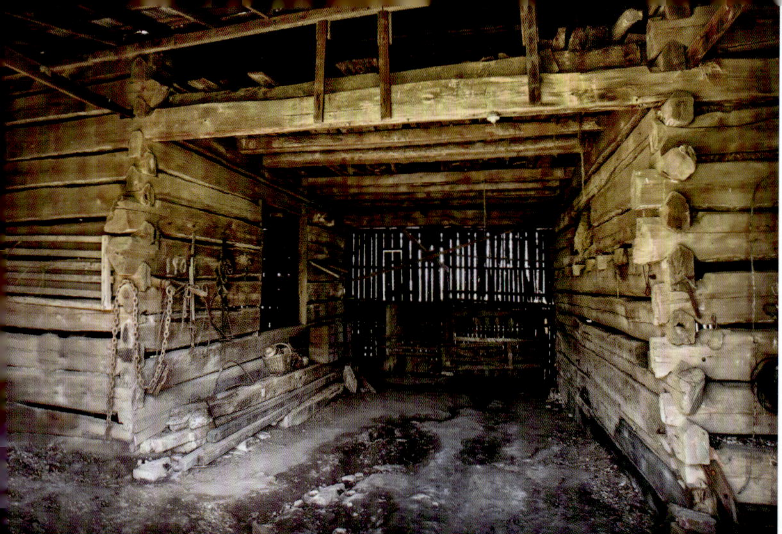

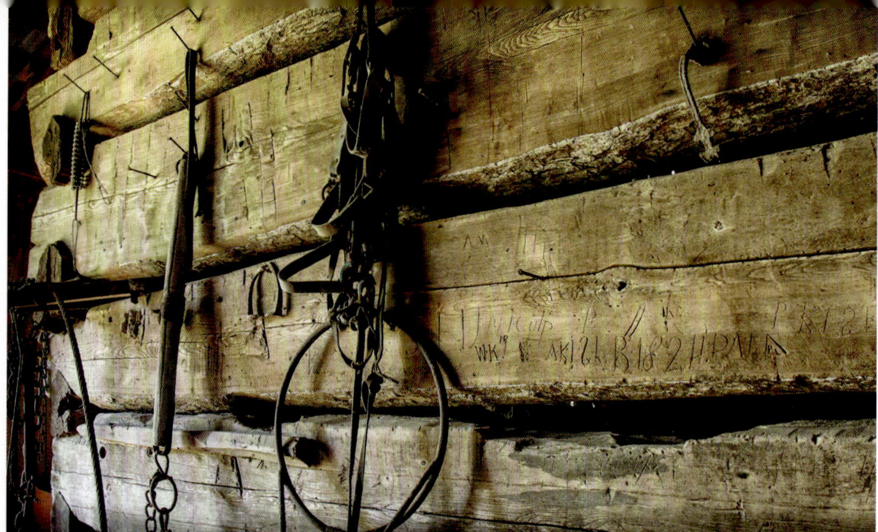

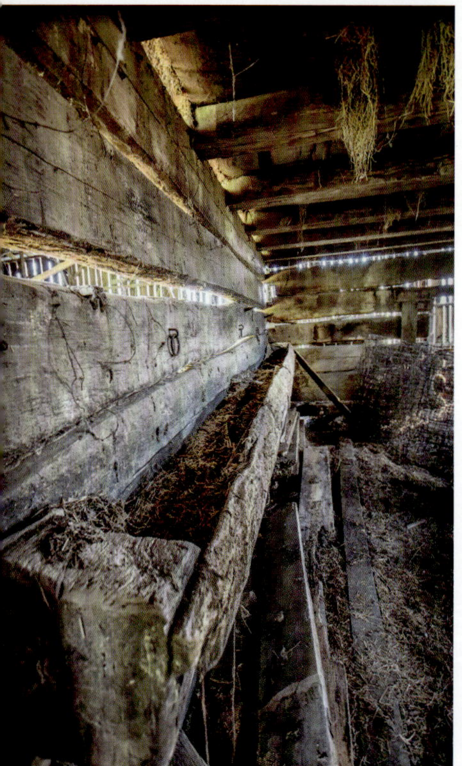

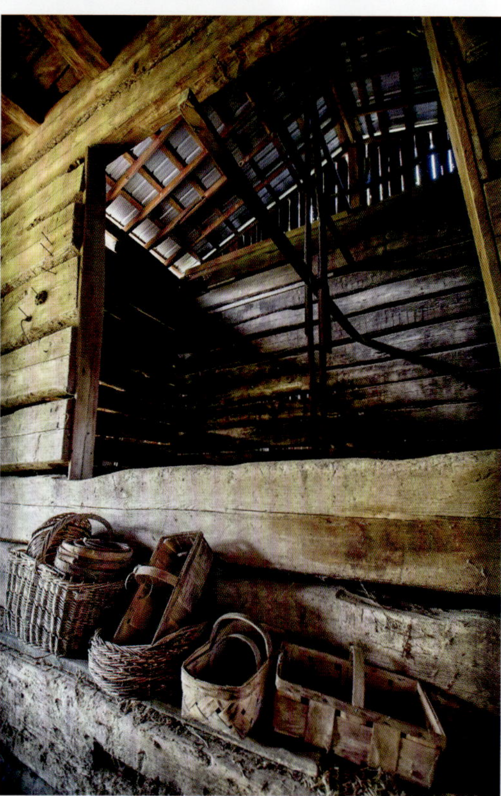

Bourbon County.

John Kiser, a prominent early settler, established Kiser Station circa 1781, including a home and this barn. The double-pen log barn shows V notching and a solid beam connecting the two pens. Inside one pen is a solid log trough with iron rings for livestock. The pen on the other side was converted for tobacco at one time.

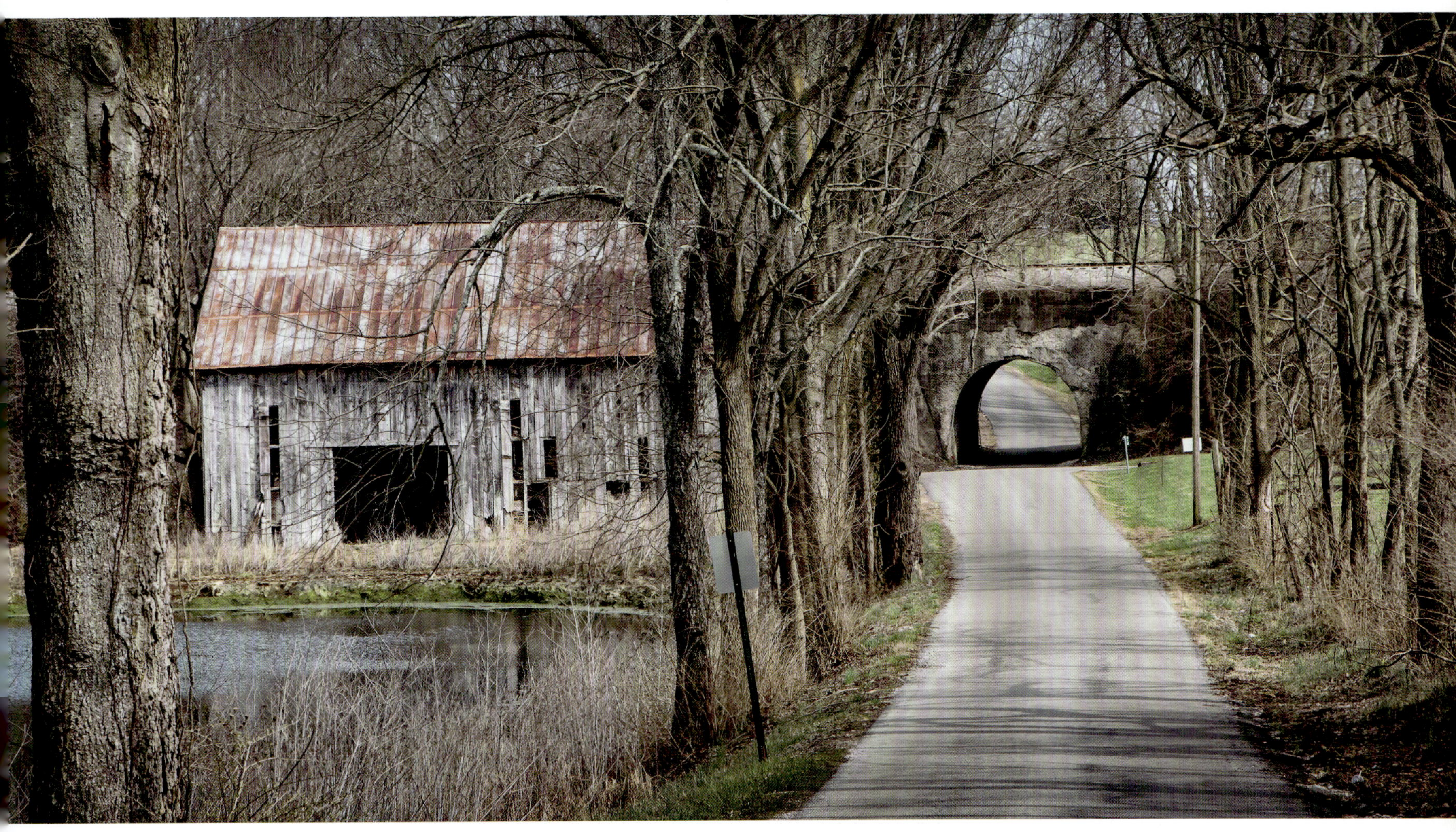

Bourbon County.

A drive along Bourbon County roads reveals this classic Kentucky scene.

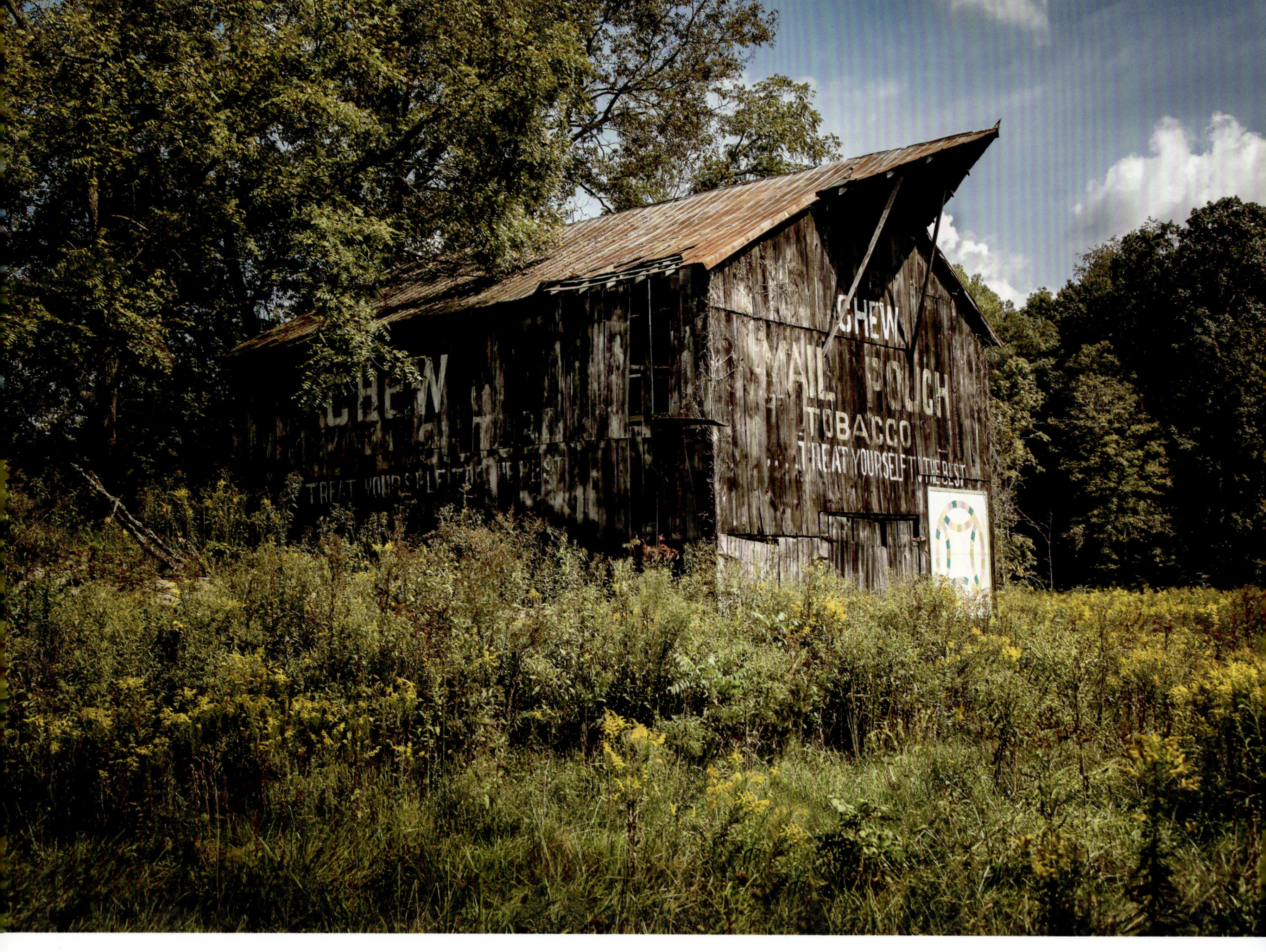

CHEW
CHEW
MAIL POUCH
TOBACCO
...TREAT YOURSELF TO THE BEST
TREAT YOURSELF TO THE BEST

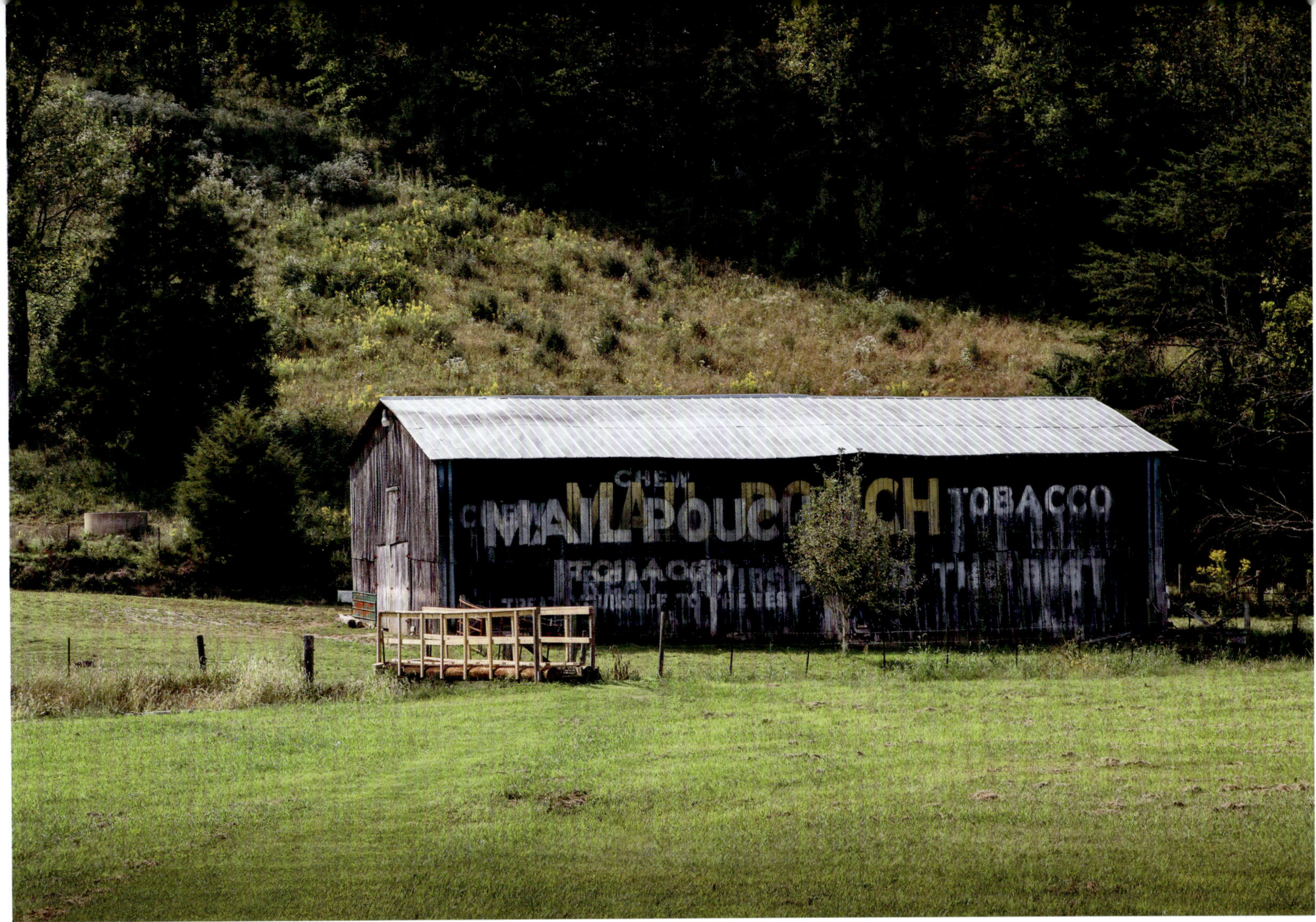

Carter County.

This abandoned Mail Pouch barn sports a hay hood and a quilt plaque. Old lettering can be seen coming through the newer application.

Carter County.

Advertisers might revisit Mail Pouch barns several times to refresh the paint. Several applications are visible at once on this barn.

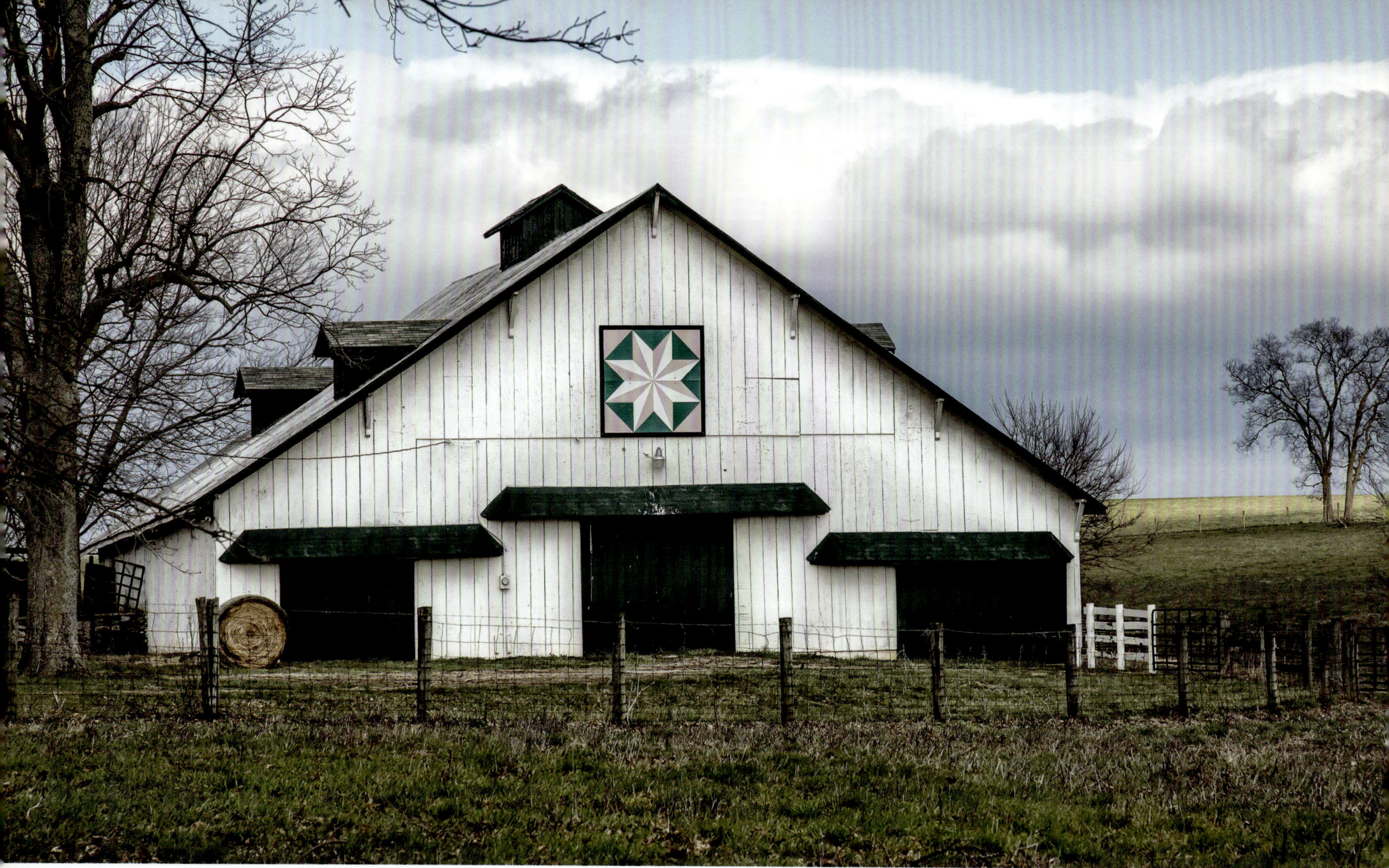

The quilt pattern on a barn may be traditional or newly created.

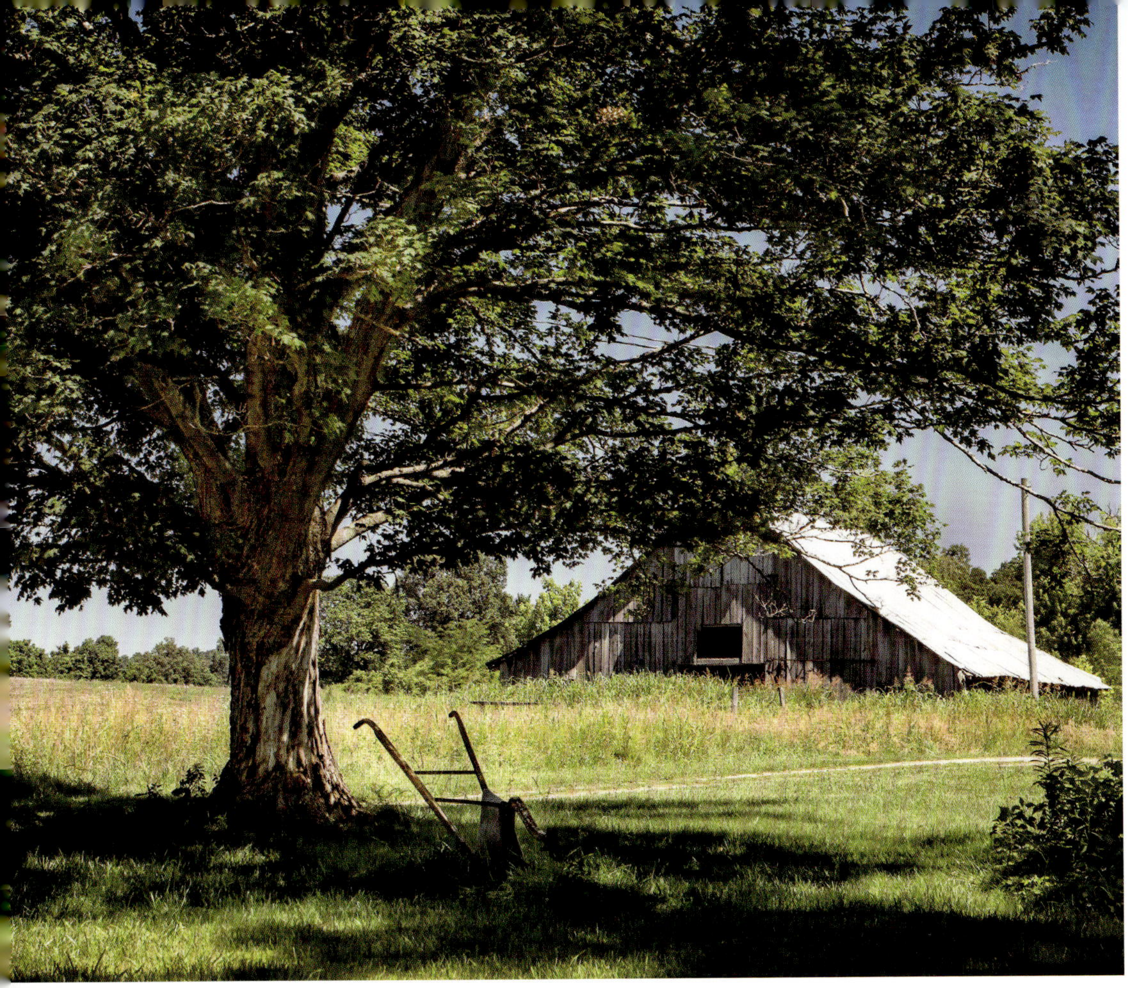

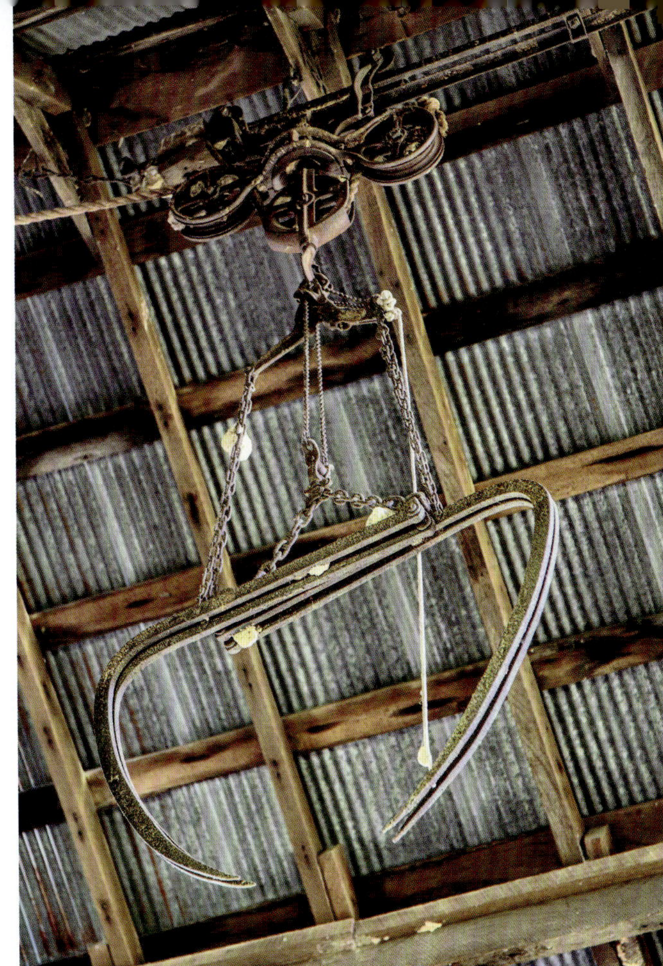

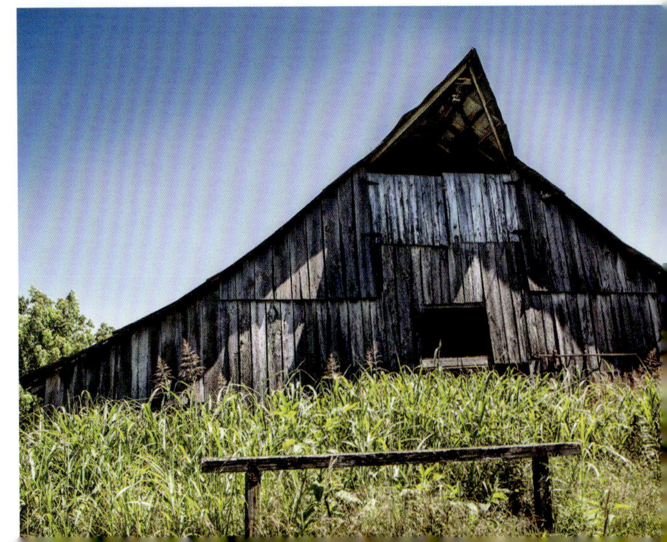

Lockheart Ridge Farm has been in the Lockheart family for several generations. This stock barn with a prominent hay hood and hayfork was constructed in the mid-twentieth century.

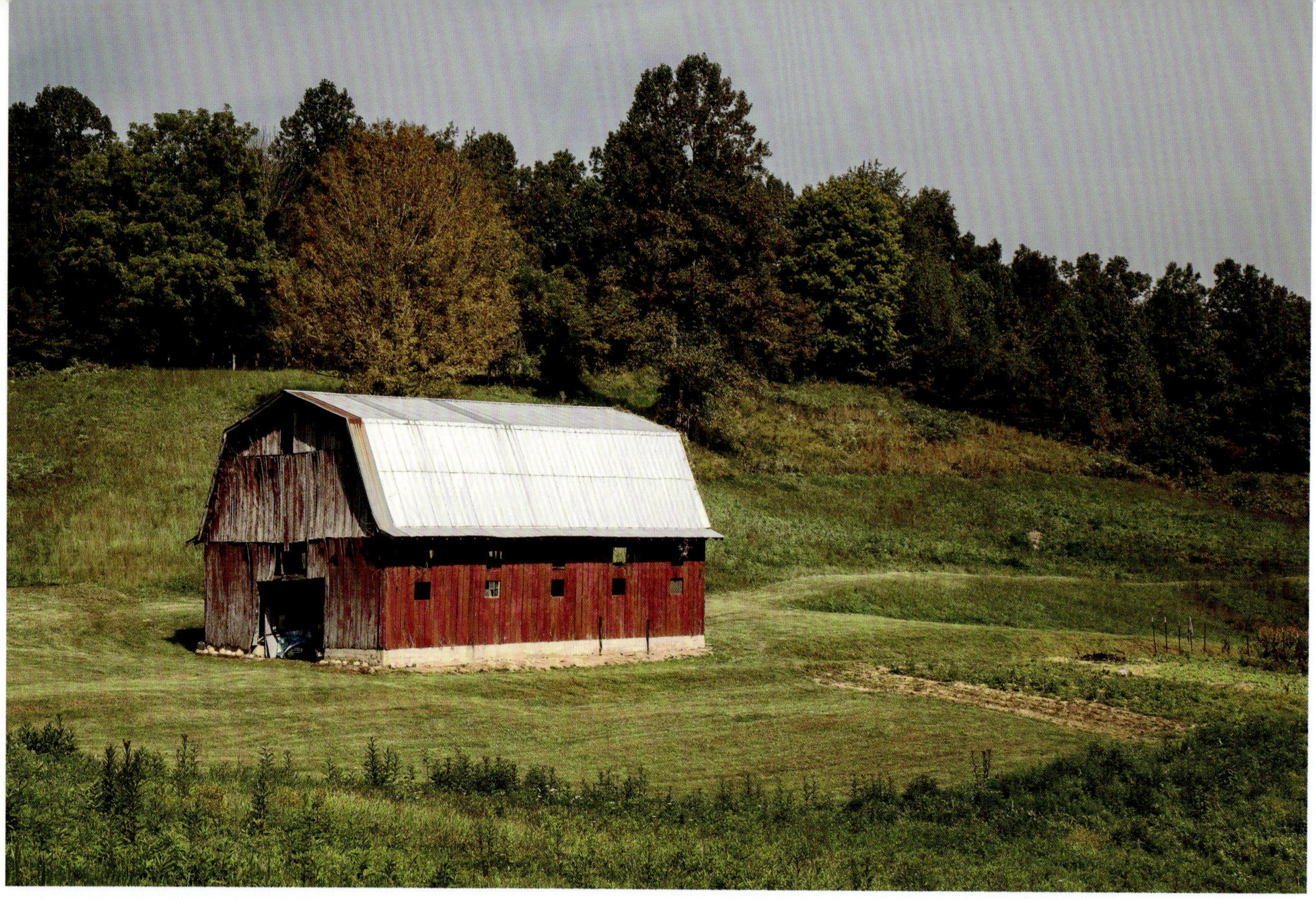

Lincoln County.

Ottenheim, a Swiss settlement near Stanford, had many structures obviously influenced by Swiss architecture and cultural traditions. This is one of the earliest barns in the old community, originally built with two stalls inside. It has been altered internally to accommodate tobacco and a variety of uses. There is an addition to the left side that was built later.

Lawrence County.

Down in a flat area of an otherwise hilly region, this gambrel-roofed barn has a stone foundation and high hayloft. A family garden grows nearby.

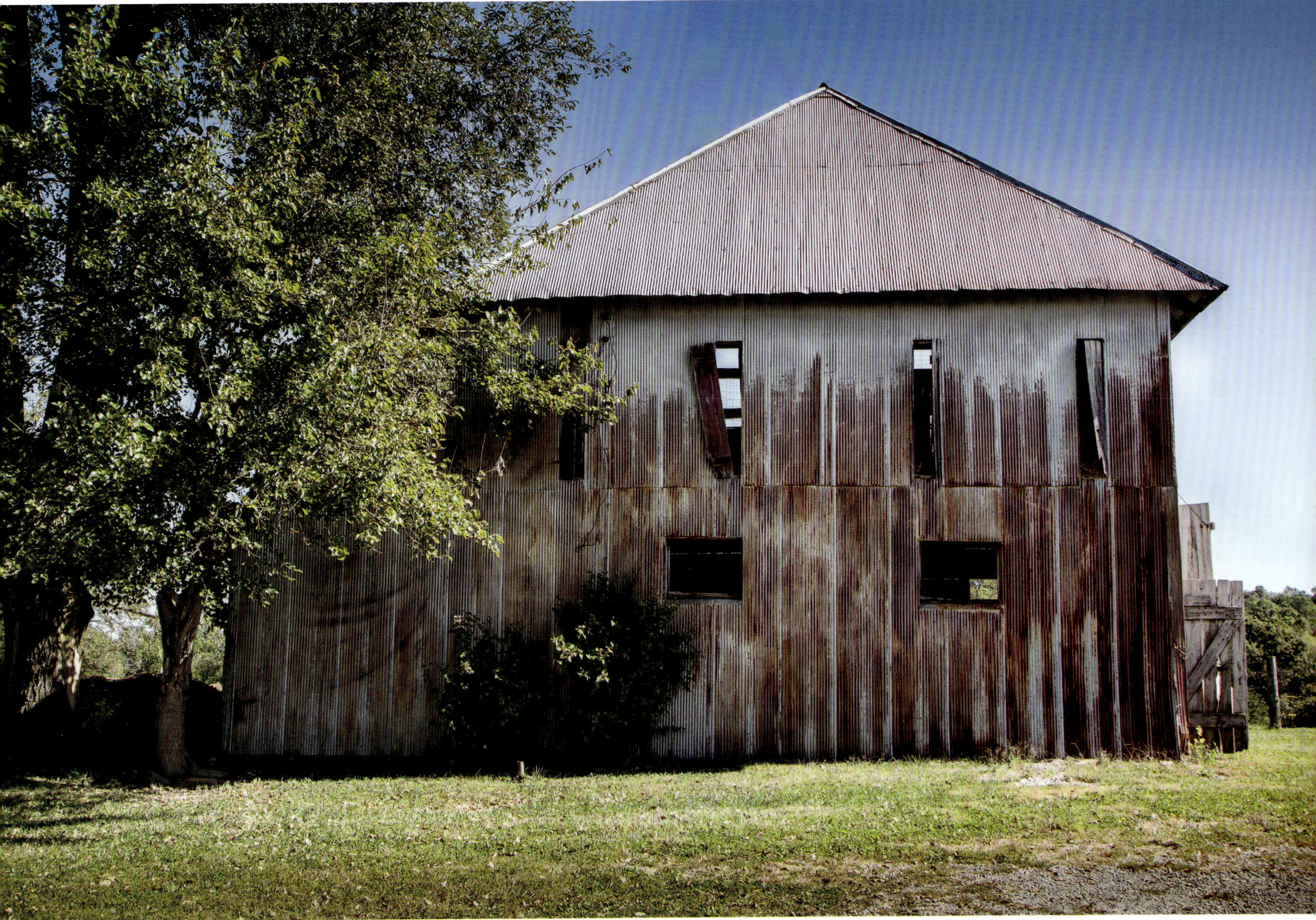

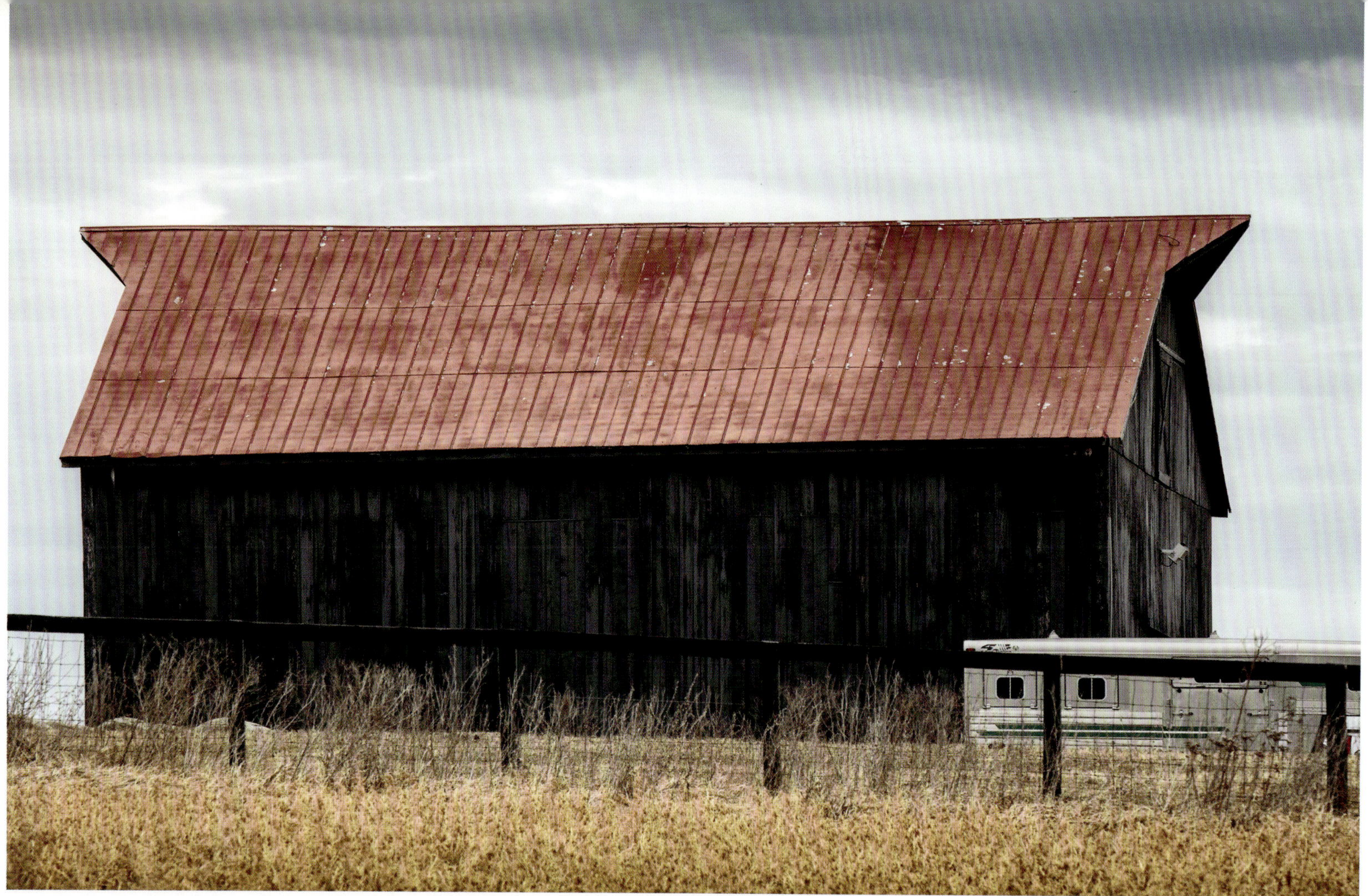

FACING PAGE

Bourbon County.

Because of the unusual design of this roof, this barn is known as a top hat barn; originally it would have been for tobacco curing and storage. The double hoods are for ventilation.

FACING PAGE

Lincoln County.

In the Swiss settlement of Ottenheim, John Papenhagen designed and built many of the buildings, including these three barns on the Spangler farm: a double-crib barn, a large dairy barn, and a general-purpose barn.

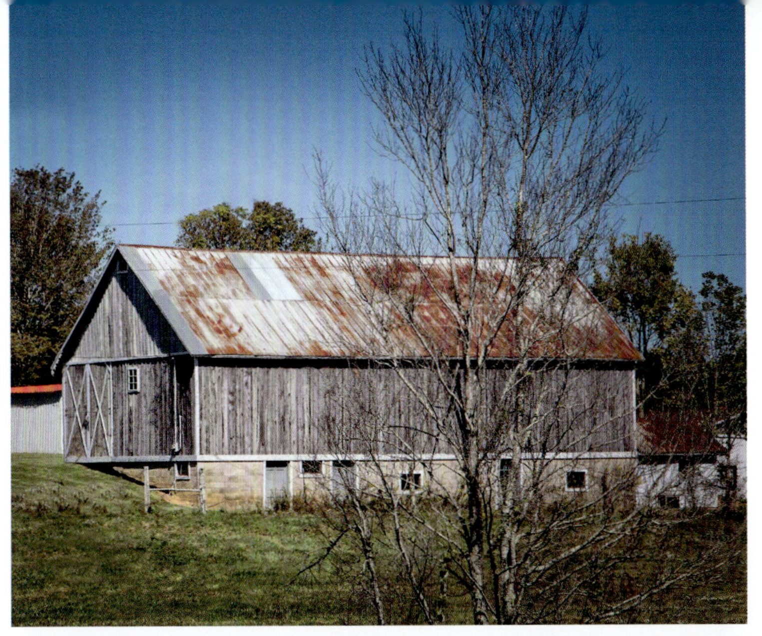
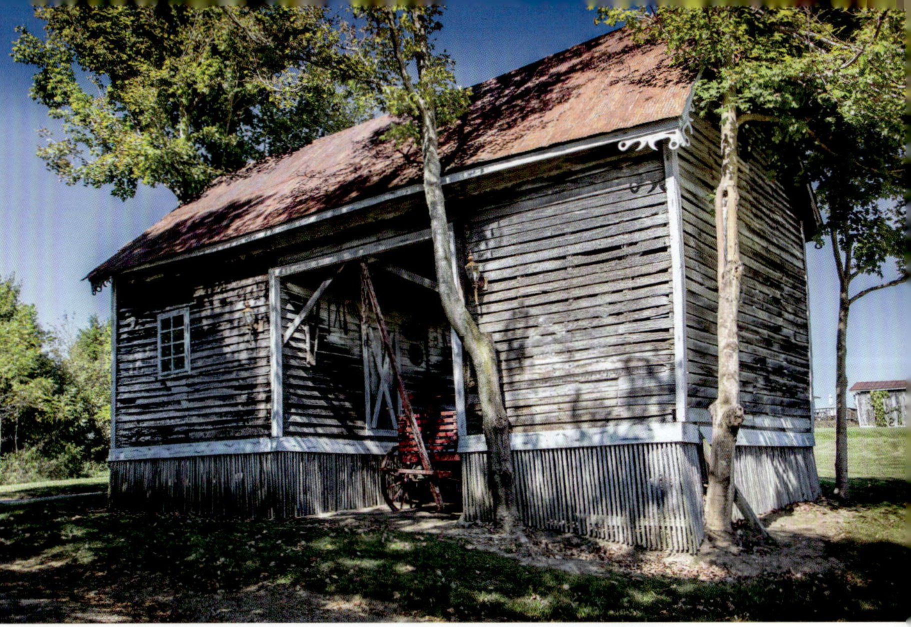

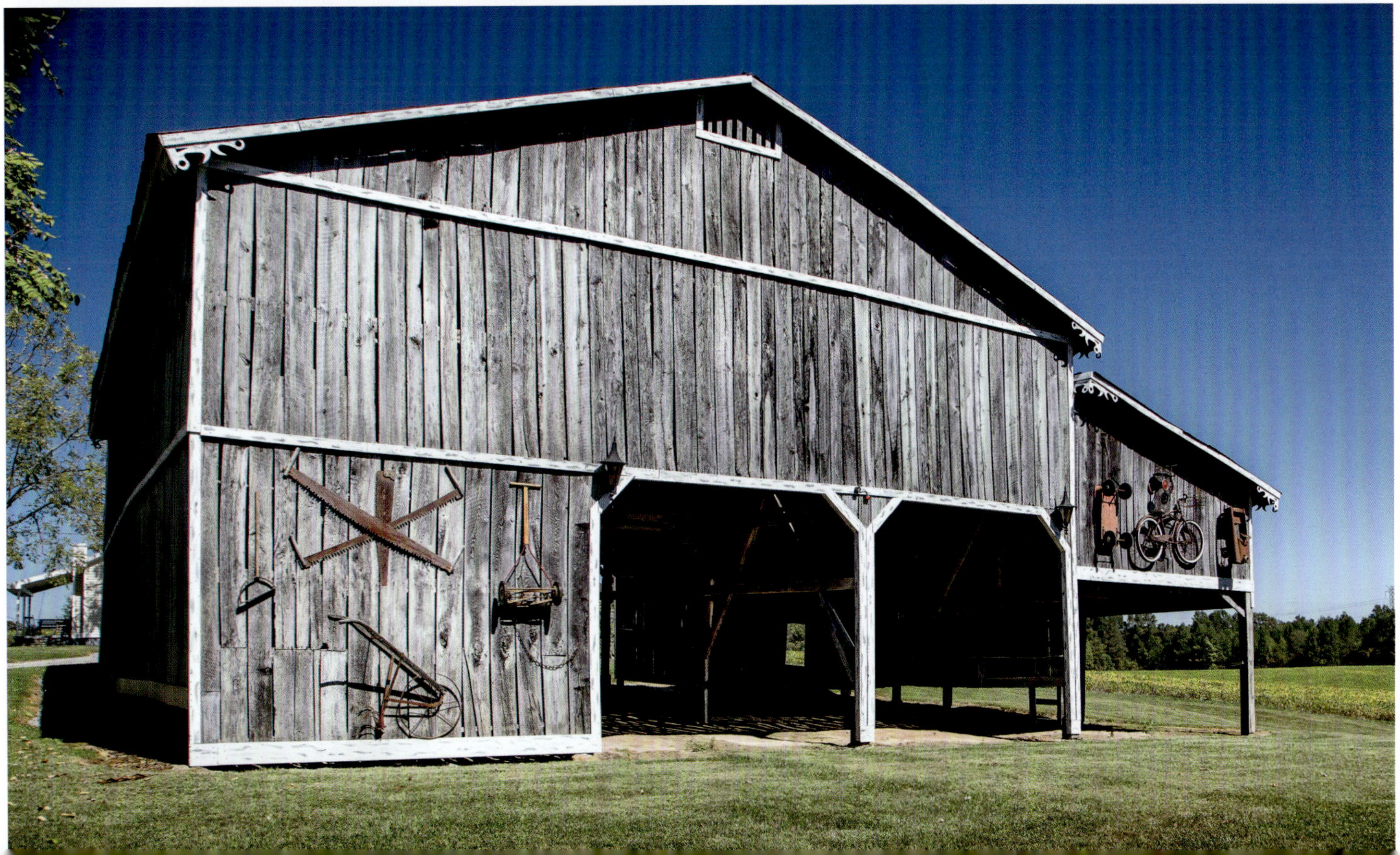

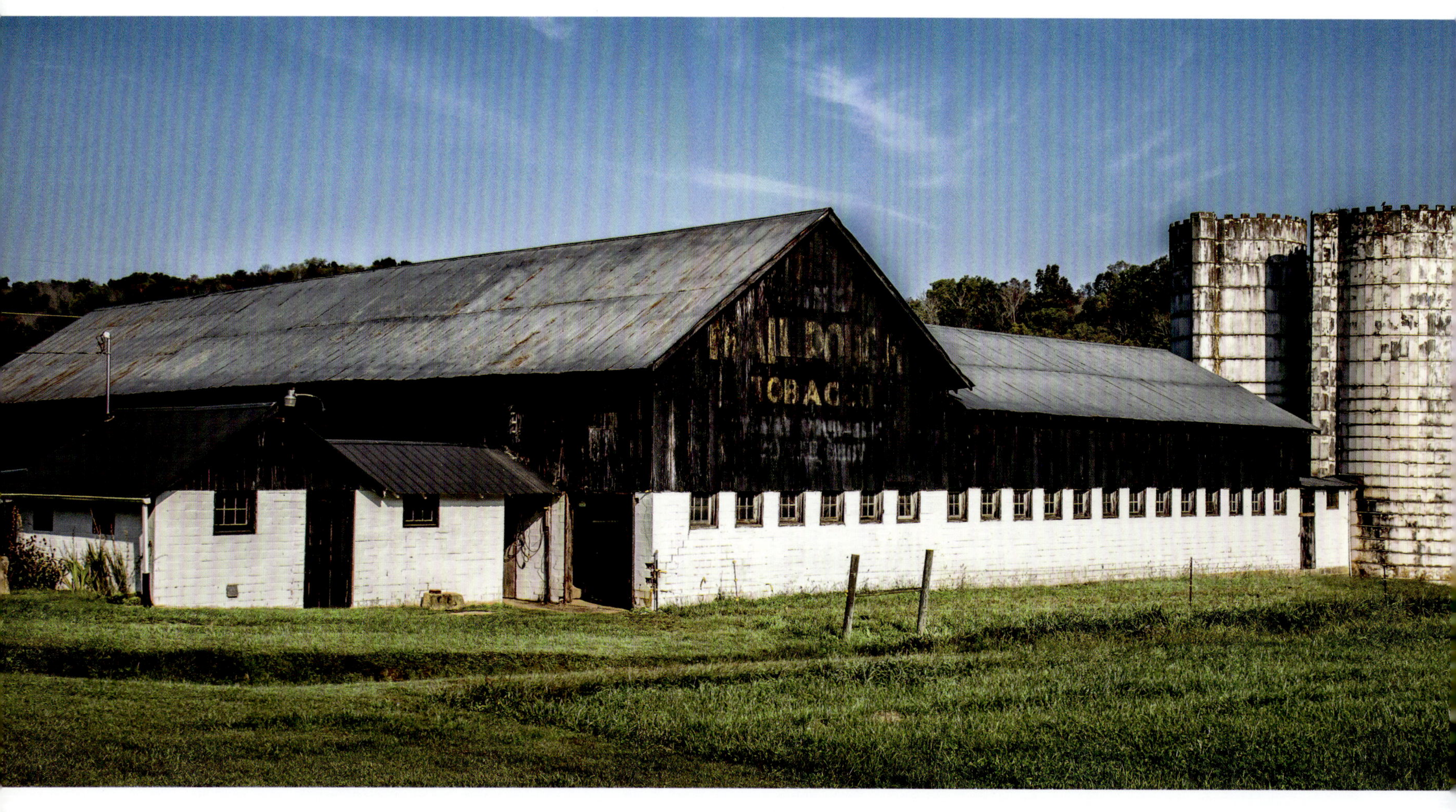

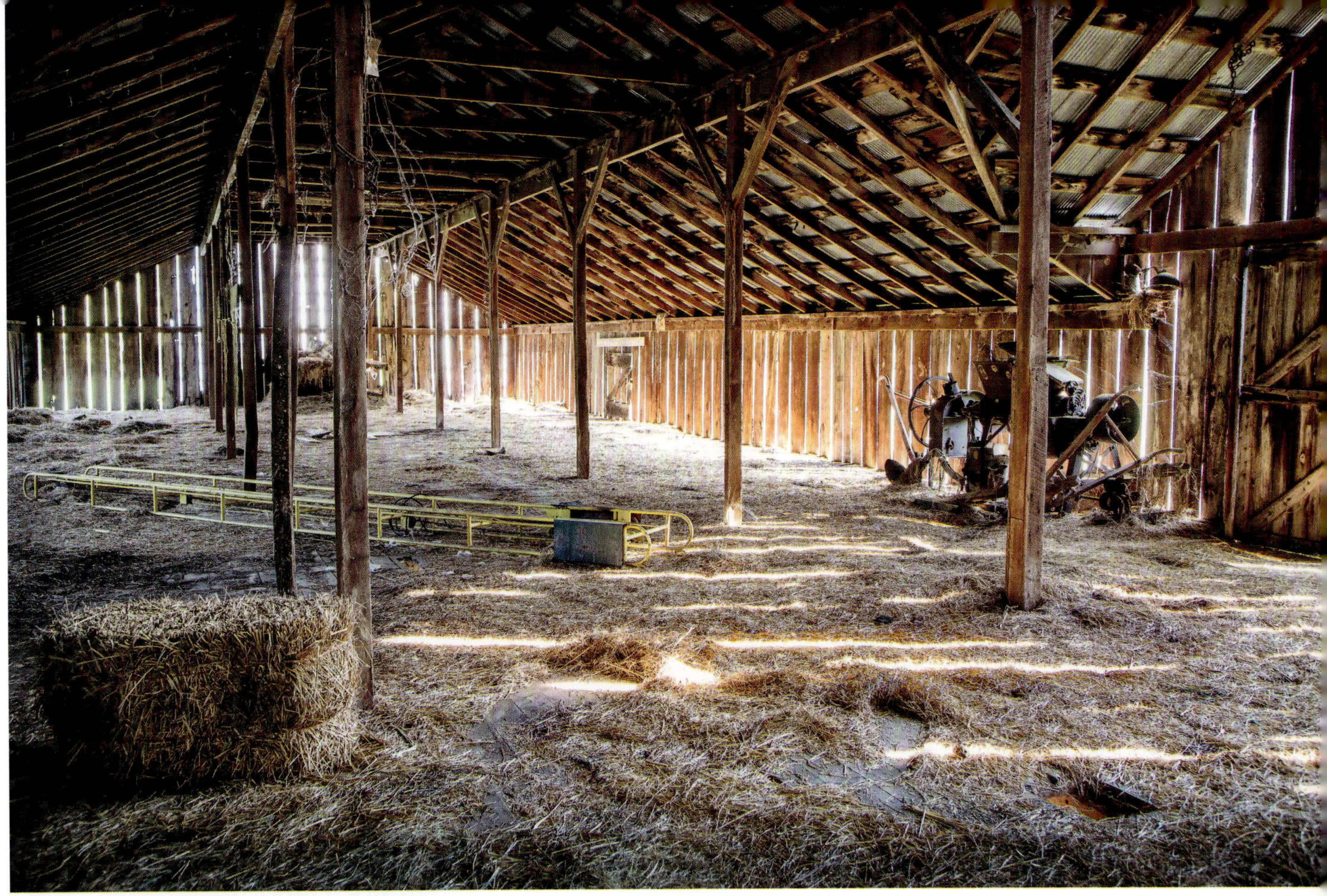

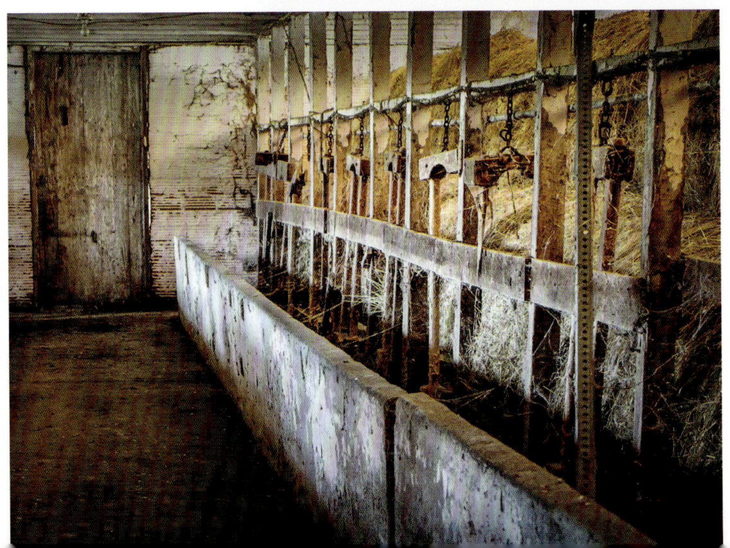

Lawrence County.

Dairy farming was the primary farming suited to northeastern Kentucky. The WB May dairy farm was one of the largest in the area in the mid-1900s. This stanchion barn design includes the typical block foundation, wooden stanchions, a block milk house, a block silo, and a wooden hayloft. Note that a faded Chew Mail Pouch Tobacco sign is also visible.

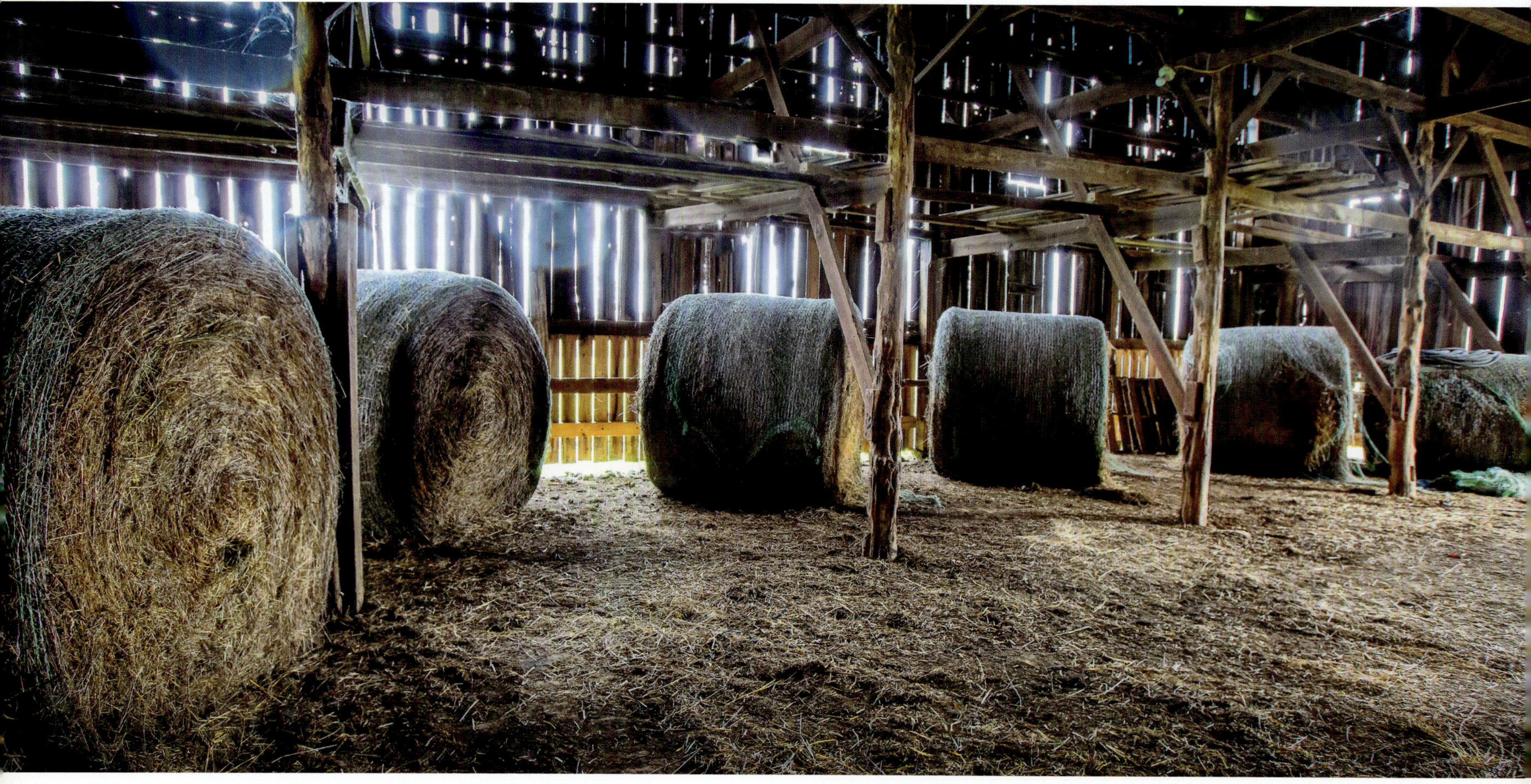

The Kubale homestead has been in that family for seven generations, with a large mule barn dating back to 1897. The scale barn and the crib barn are pictured here, along with the interior of one of two tobacco barns, now used to store hay.

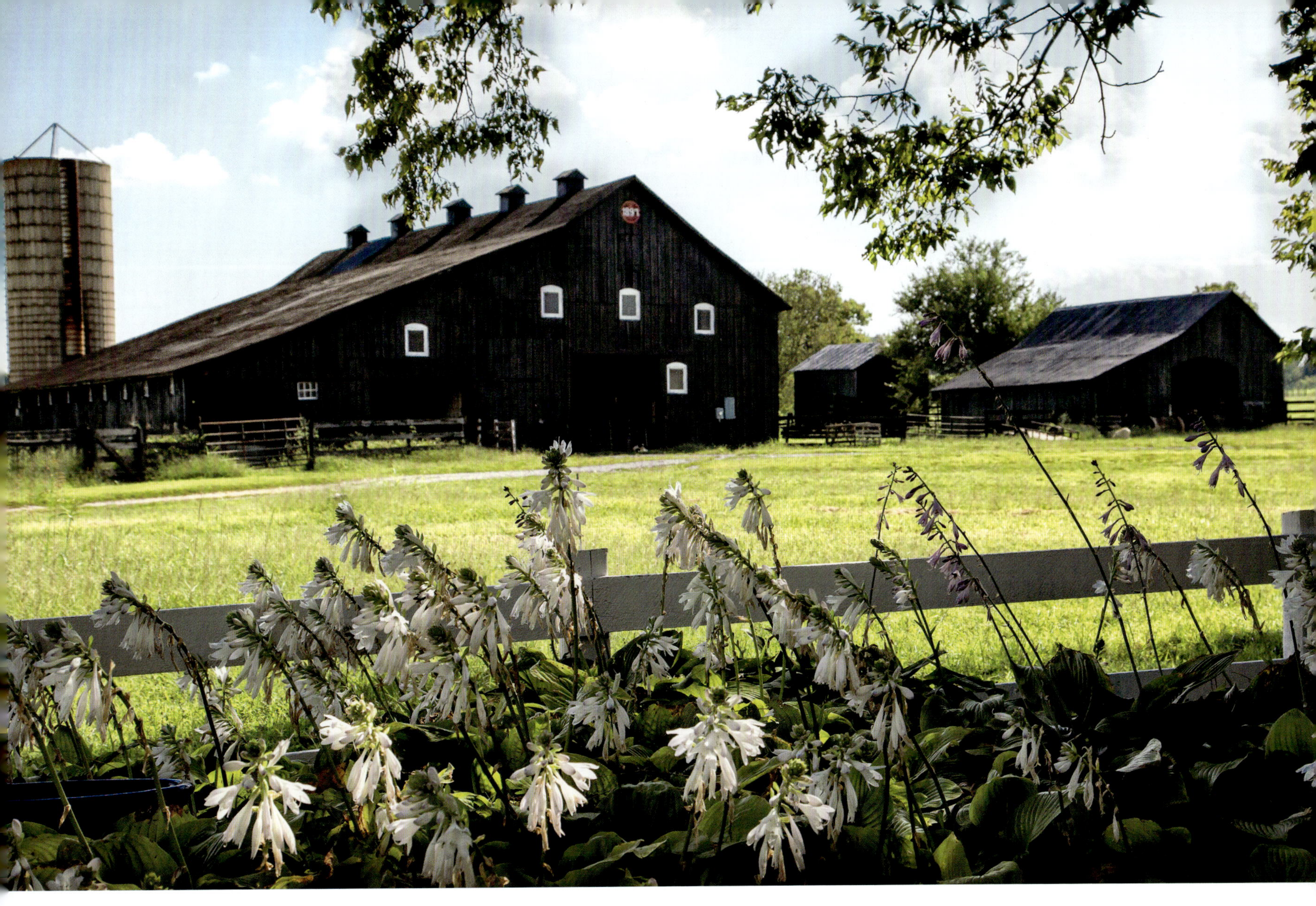

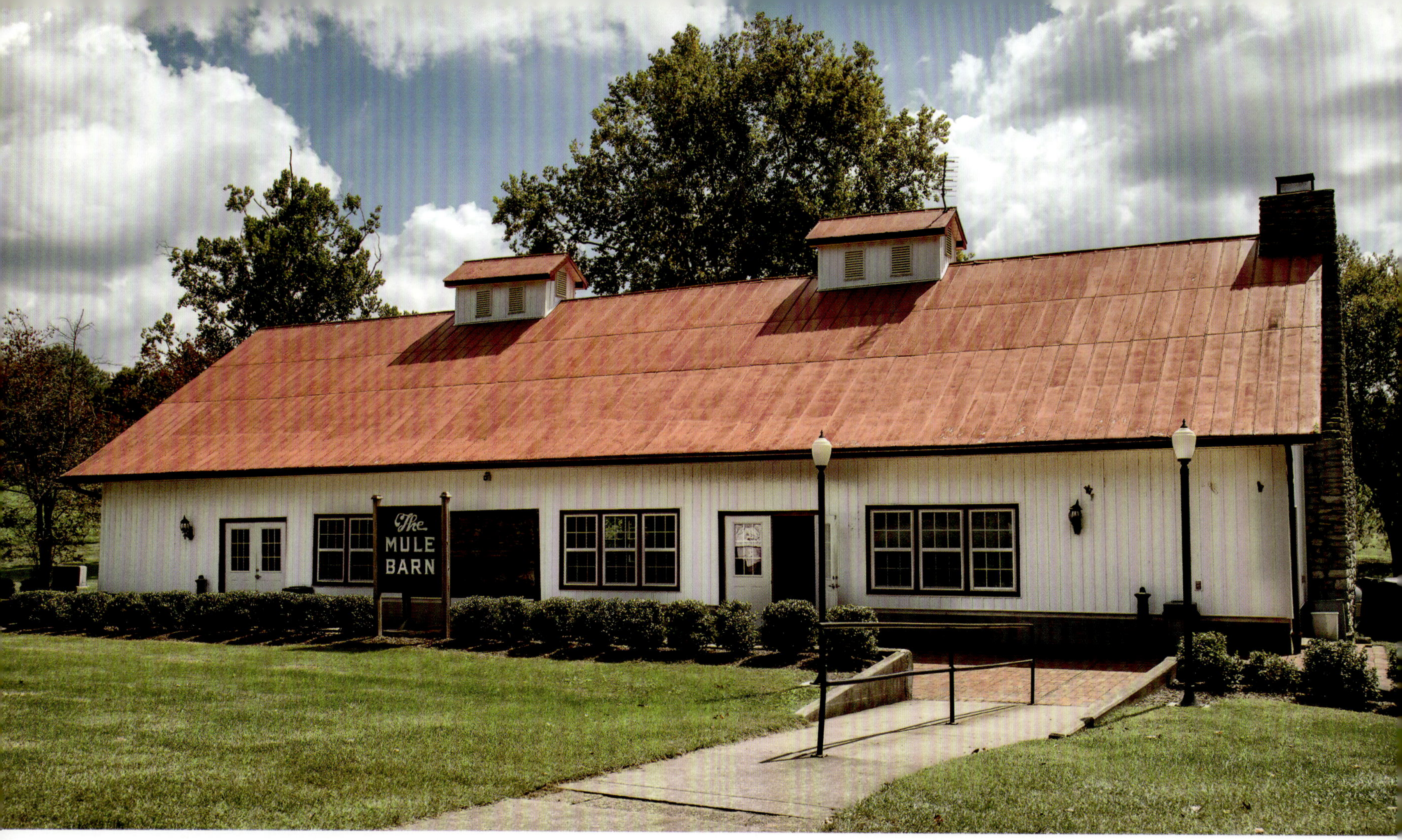

Madison County.

The Mule Barn at Eastern Kentucky University is now used for social functions associated with the university. It was once a true mule barn during the 1940s, located on farm property that belonged to the Hanger family.

FACING PAGE

Lawrence County.

This roadside barn appears to be an old dairy barn, with a low block foundation, primarily wooden construction, a block milking shed, and a gambrel roof. Note the covered well in the foreground.

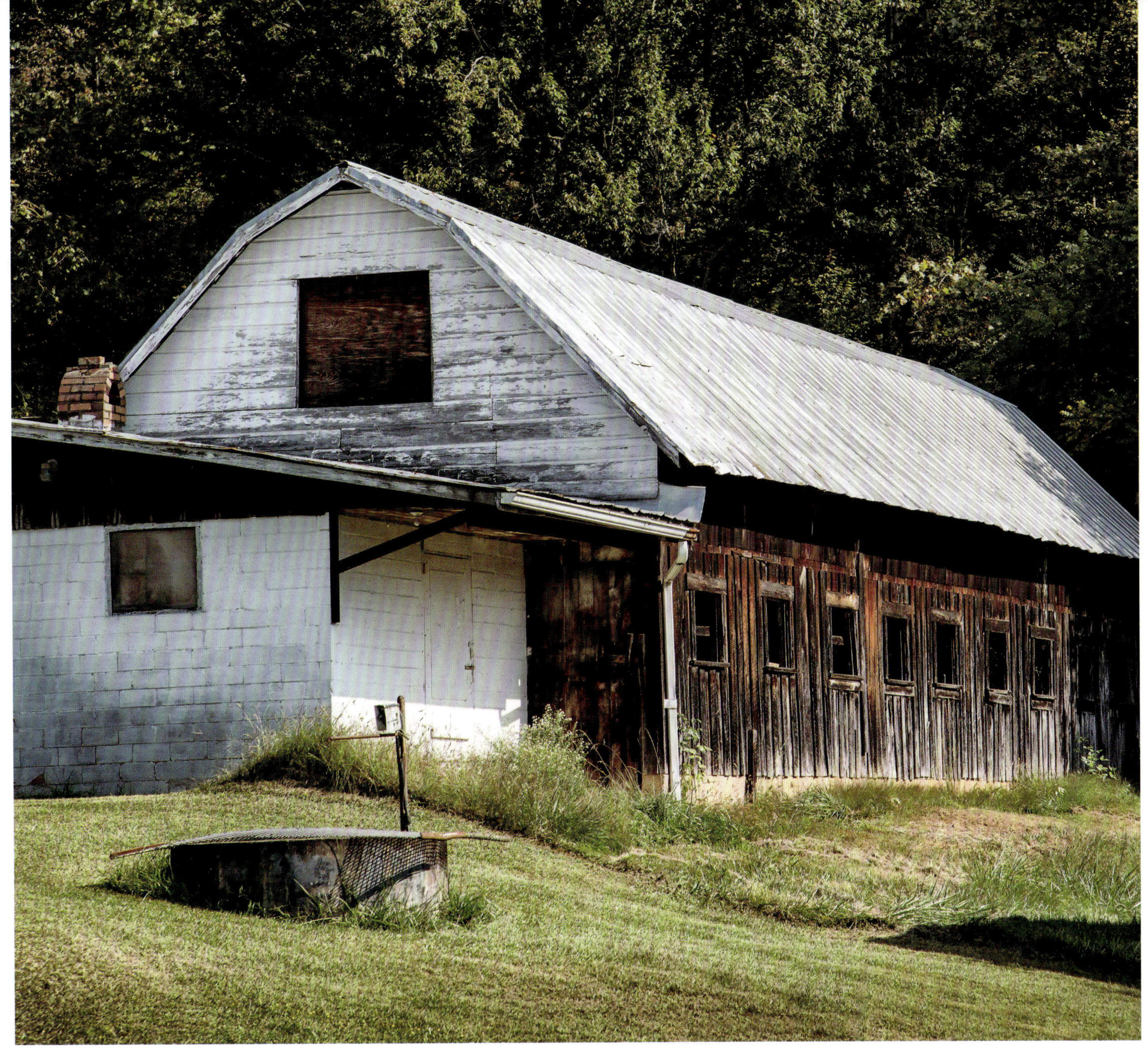

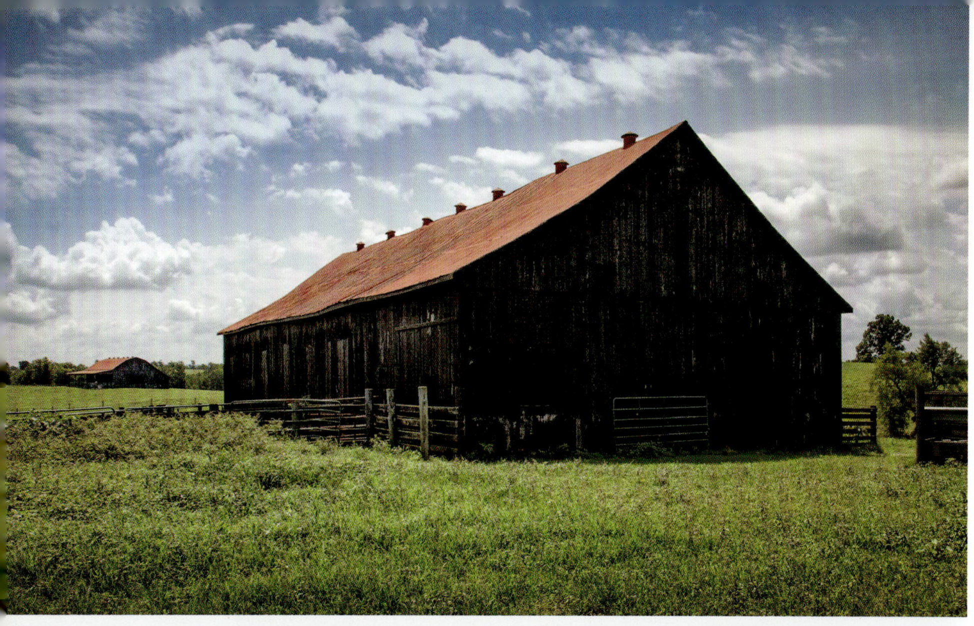

Madison County.

The William Malcom Miller Farm, circa the mid-nineteenth century, is part of a family farm originally established around 1784 by W. M. Miller's grandfather. The barn in these images is a single log structure on a stone foundation surrounded by a more recent wooden barn, built circa 1820–1840. Note the notched construction and various openings in the log building. Hayracks nailed to old logs serve livestock.

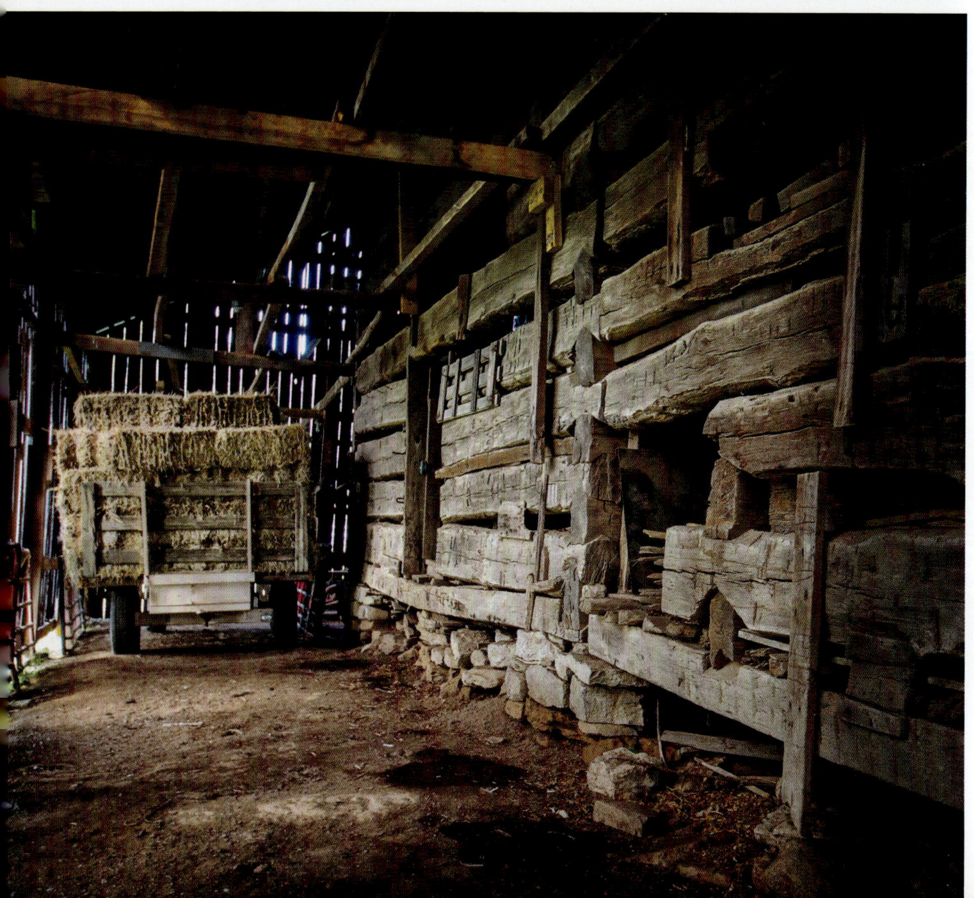

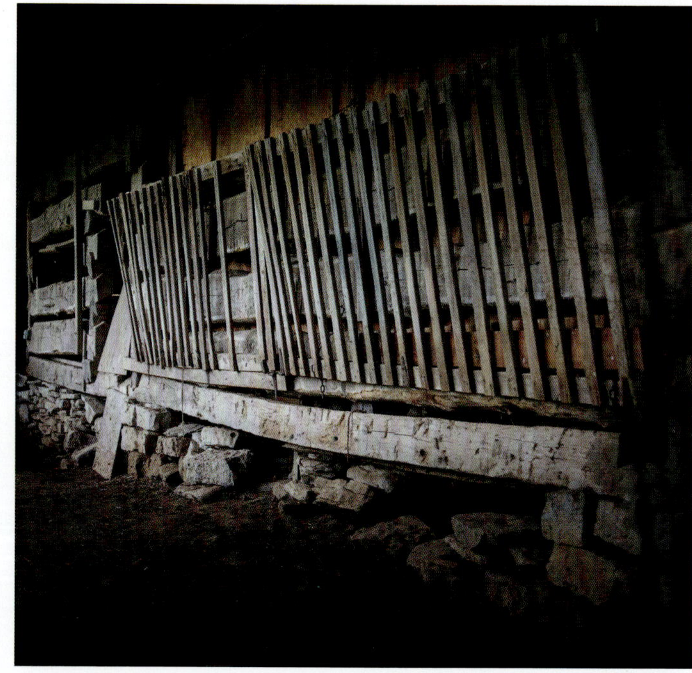

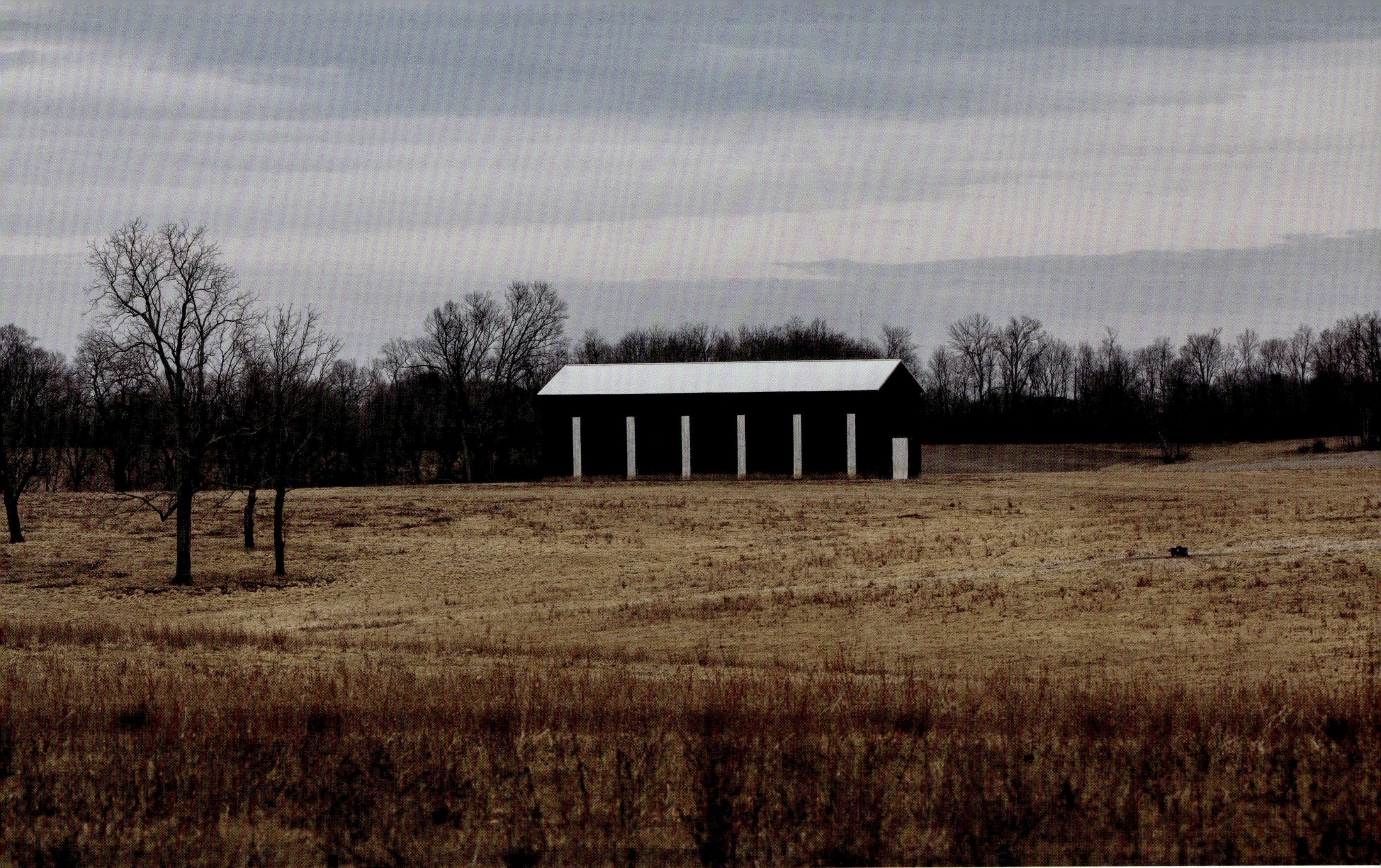

Bourbon County.

A tobacco barn anchoring a field is an iconic view in rural central
Kentucky. This barn has been given a tuxedo look, in black and white.

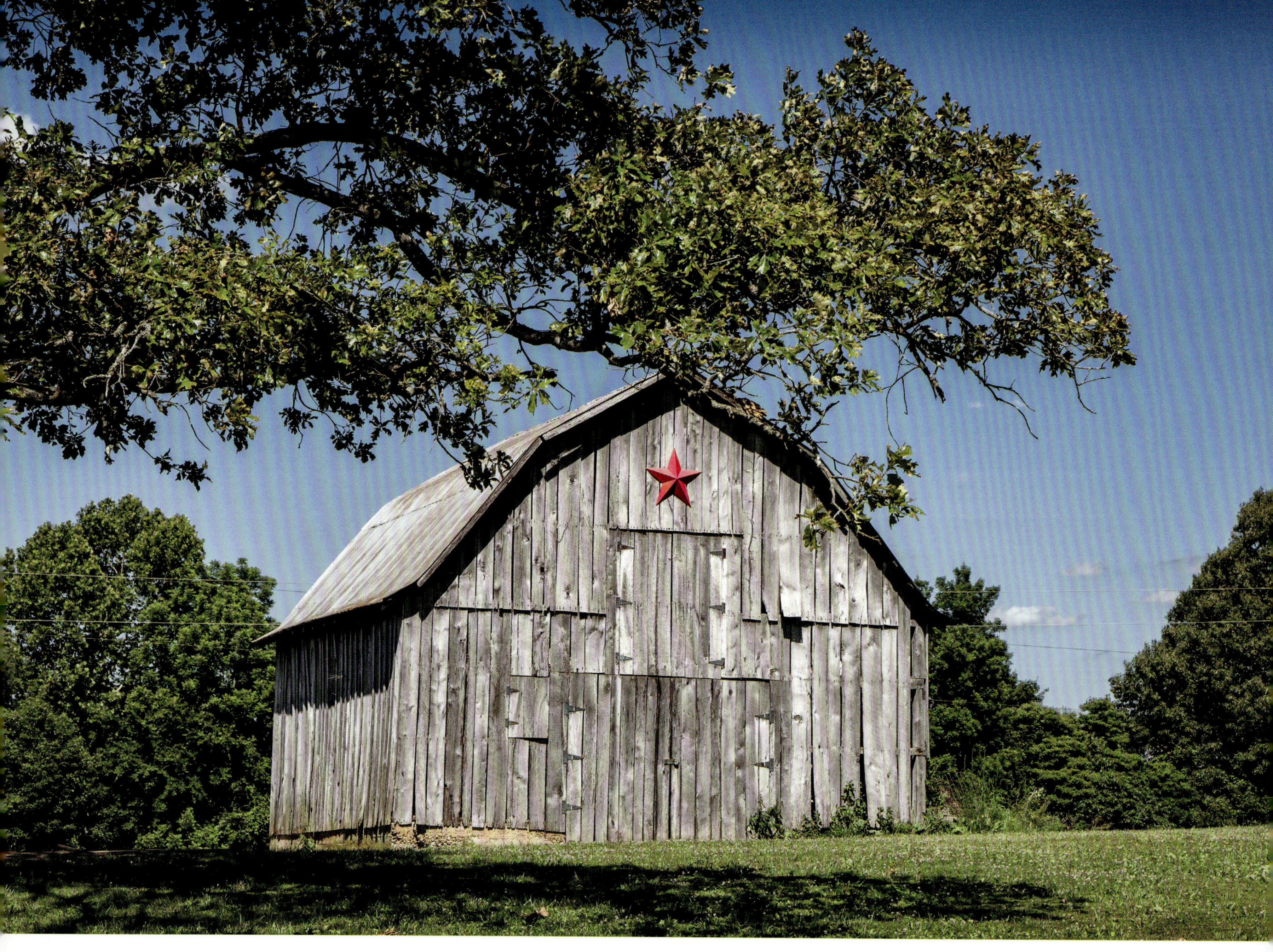

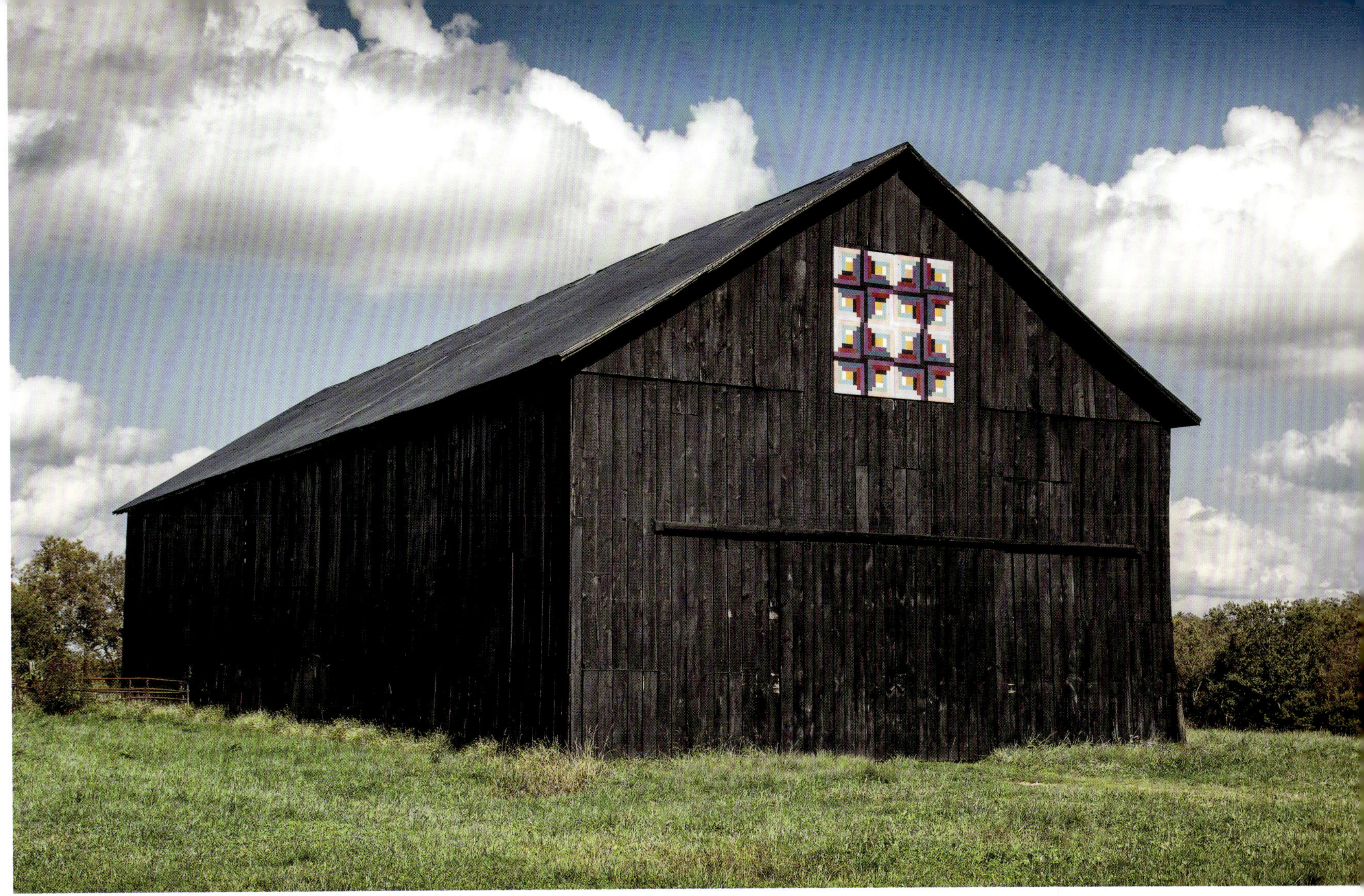

Lyon County.

The red star on this stone foundation timber barn
follows an old tradition of decorating one's barn.

Woodford County.

The most common color for the central Kentucky burley
tobacco barn is black, making it instantly recognizable.

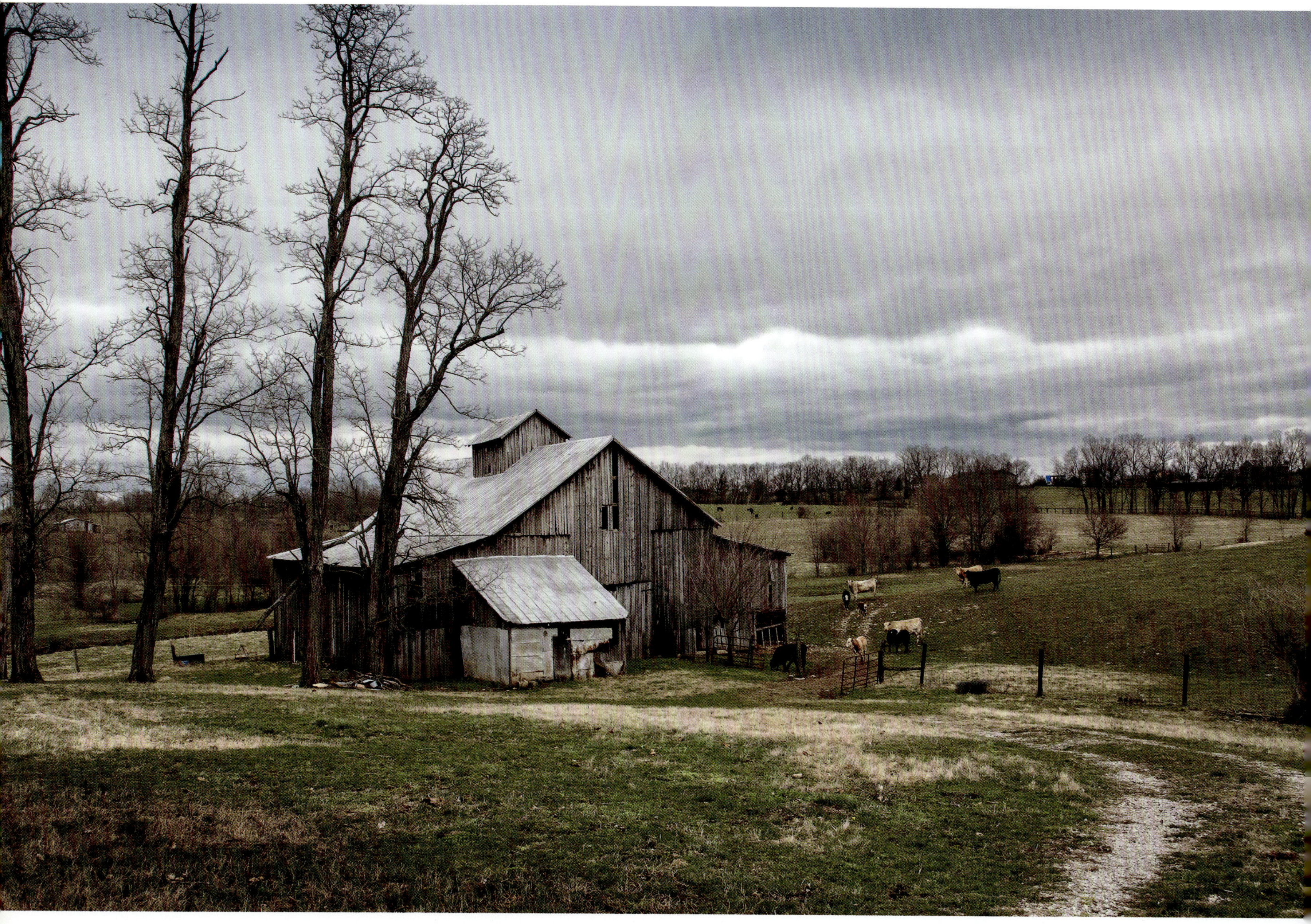

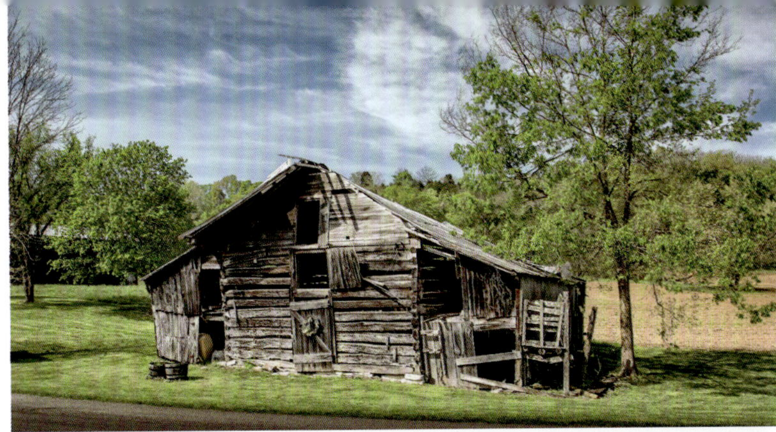

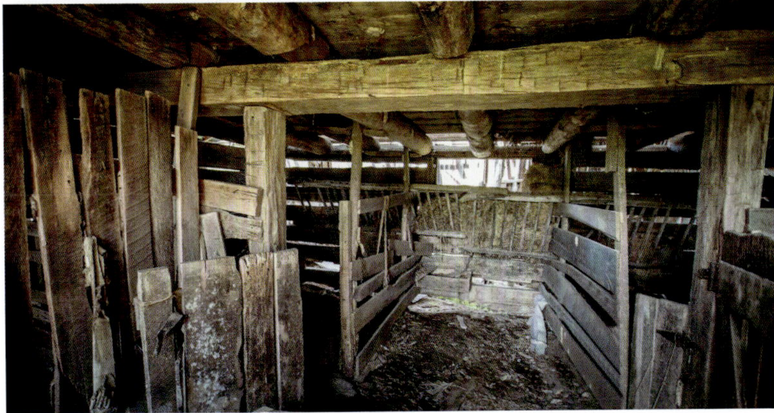

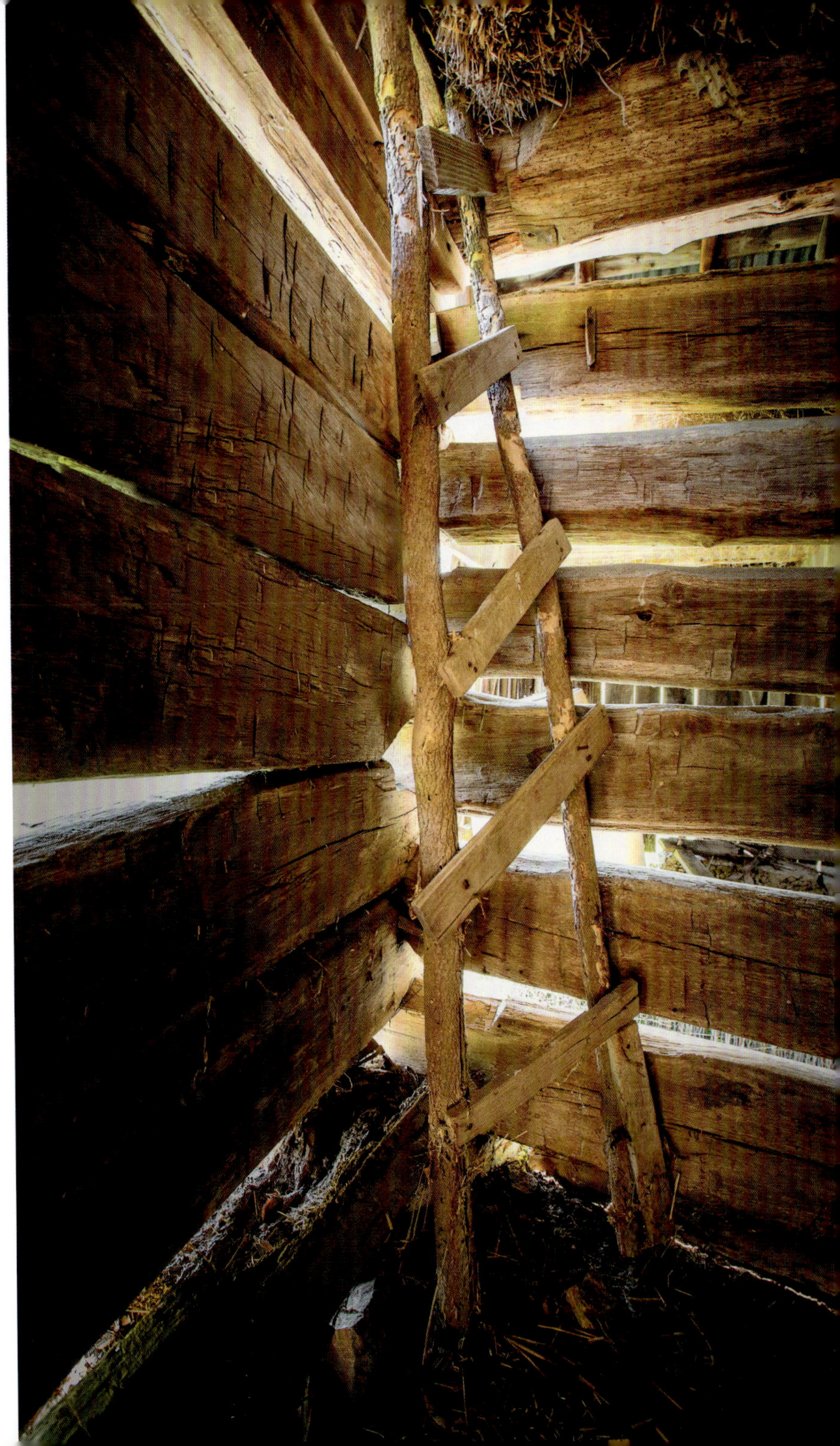

Franklin County.

This barn is a nice example of a nineteenth-century
V-notched single-crib log barn with frame extensions.

FACING PAGE

Bourbon County.

Often a barn's age is hard to determine. Mr. Chanslor,
current owner of this barn, dates his barn according to how
long he's owned the property—over fifty years.

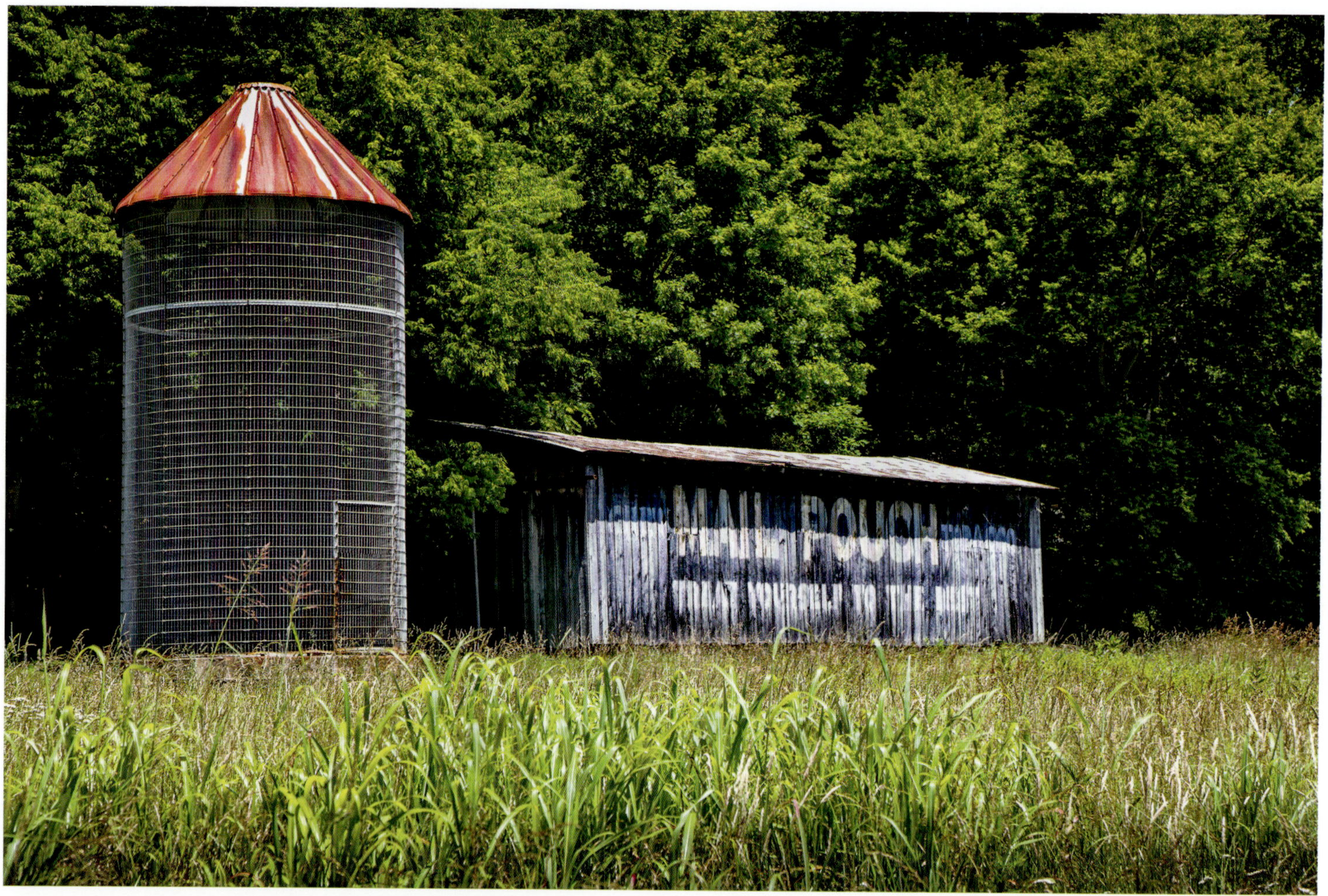

Crittenden County.

A Mail Pouch barn typically has white and yellow lettering, with a variety of combinations in those colors. The yellow Mail Pouch stands out on this silo barn.

Elliott County.

The lettering on this Mail Pouch barn is duplicated on the other side. The Ten Commandments were tacked to the door. Note the stripping shed with windows.

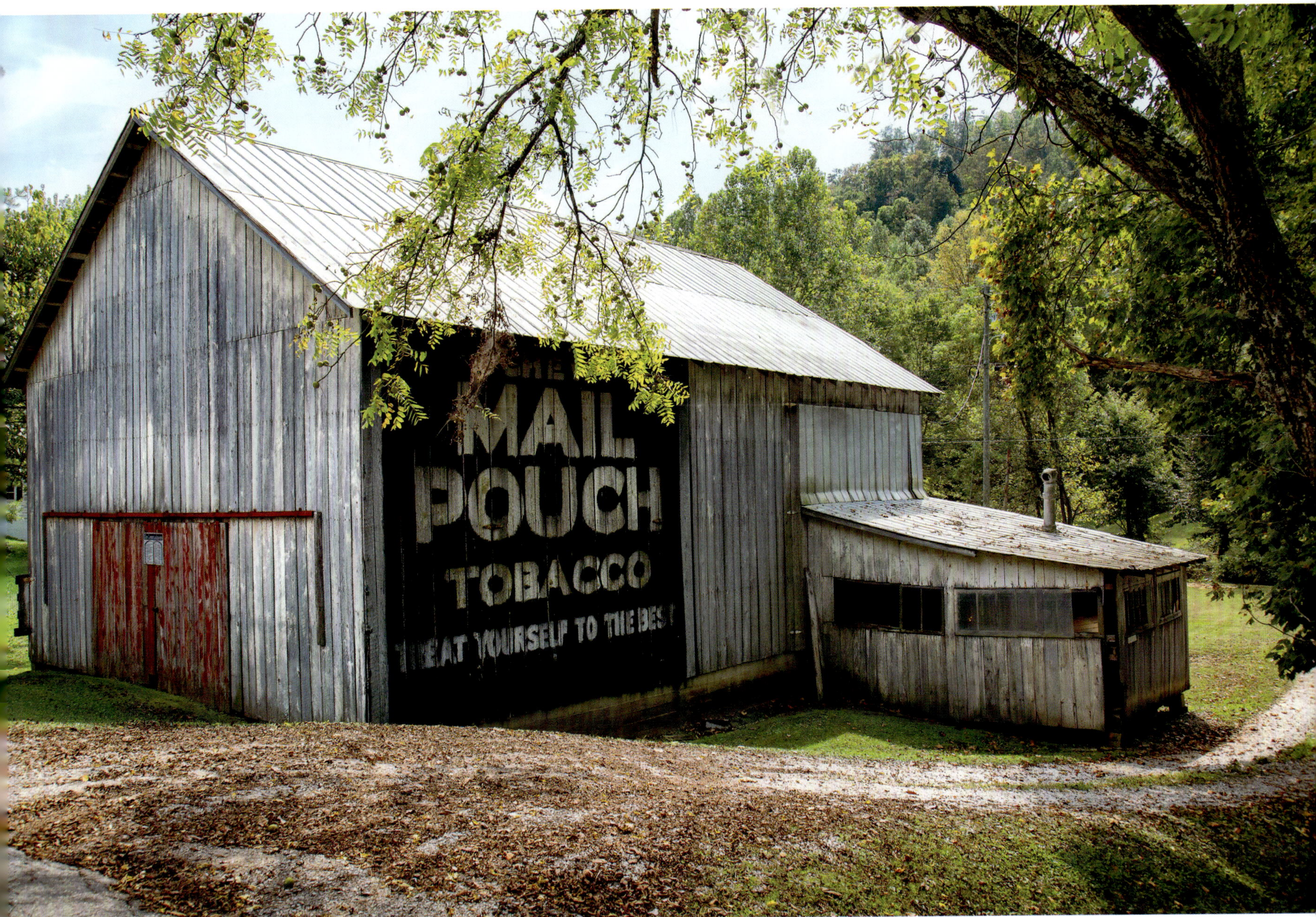
THE
MAIL
POUCH
TOBACCO
TREAT YOURSELF TO THE BEST

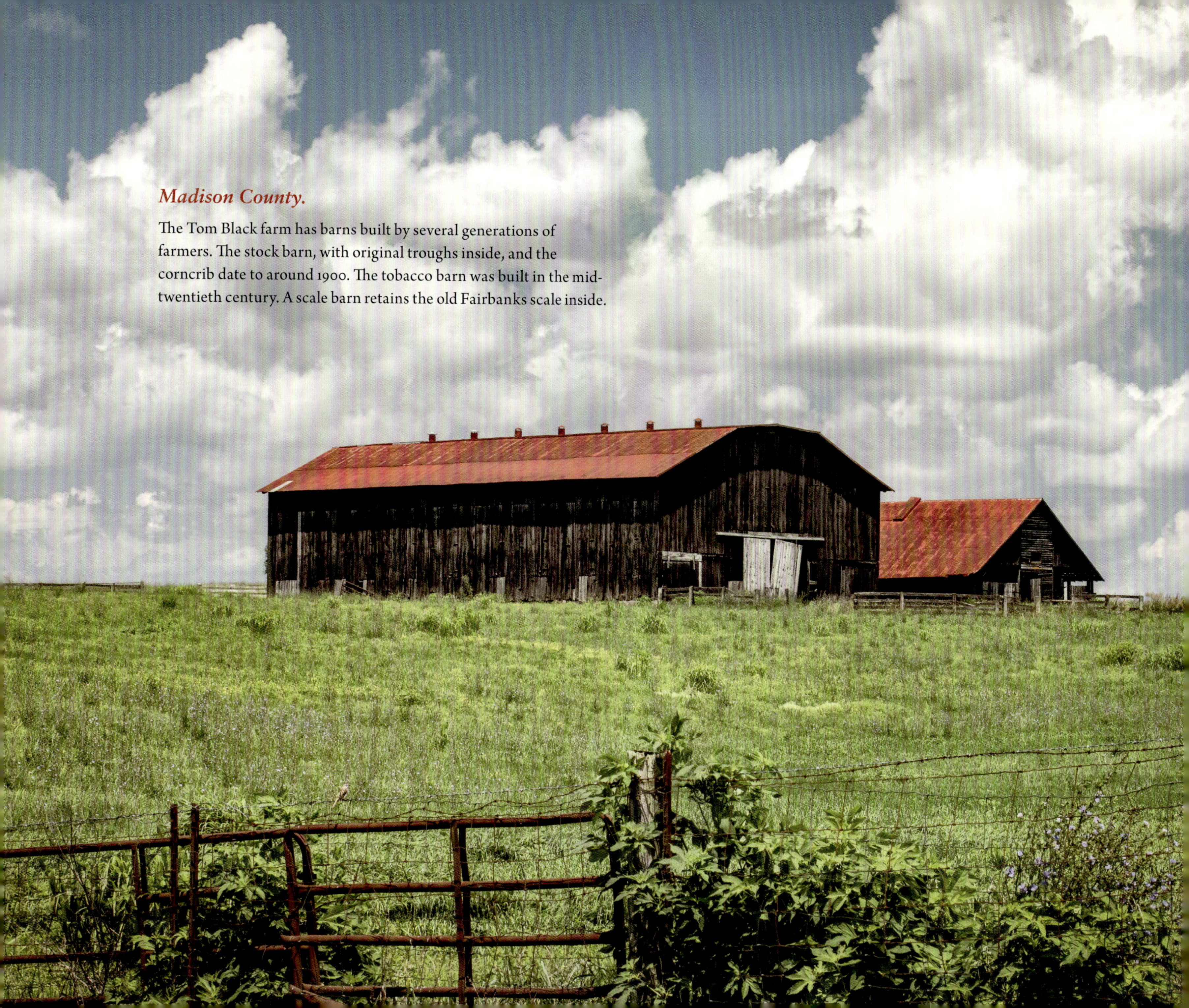

Madison County.

The Tom Black farm has barns built by several generations of farmers. The stock barn, with original troughs inside, and the corncrib date to around 1900. The tobacco barn was built in the mid-twentieth century. A scale barn retains the old Fairbanks scale inside.

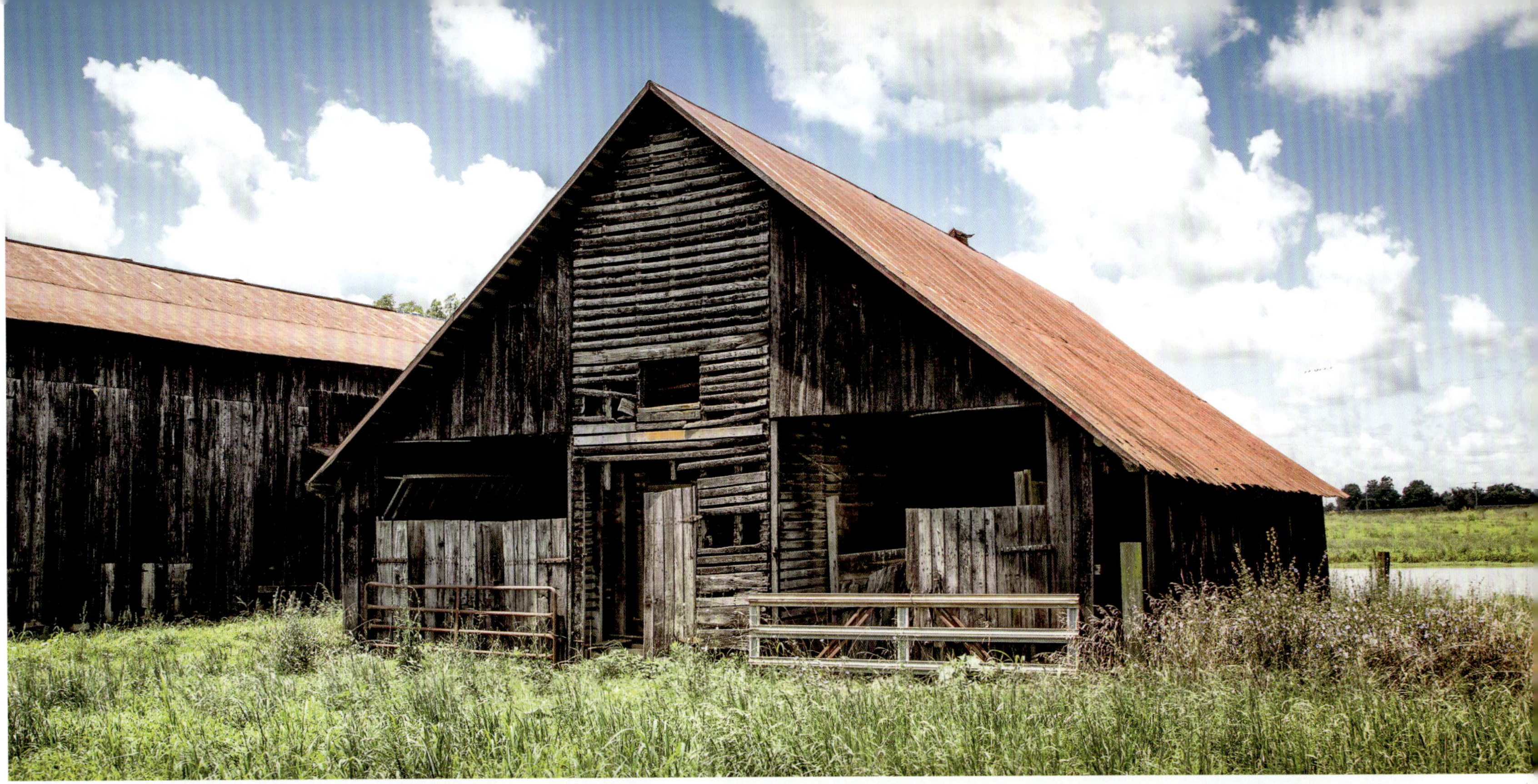

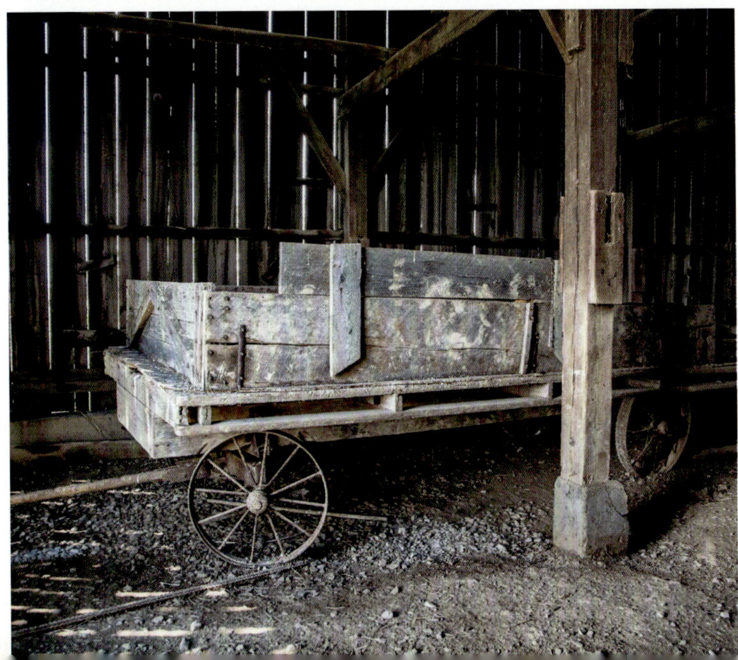
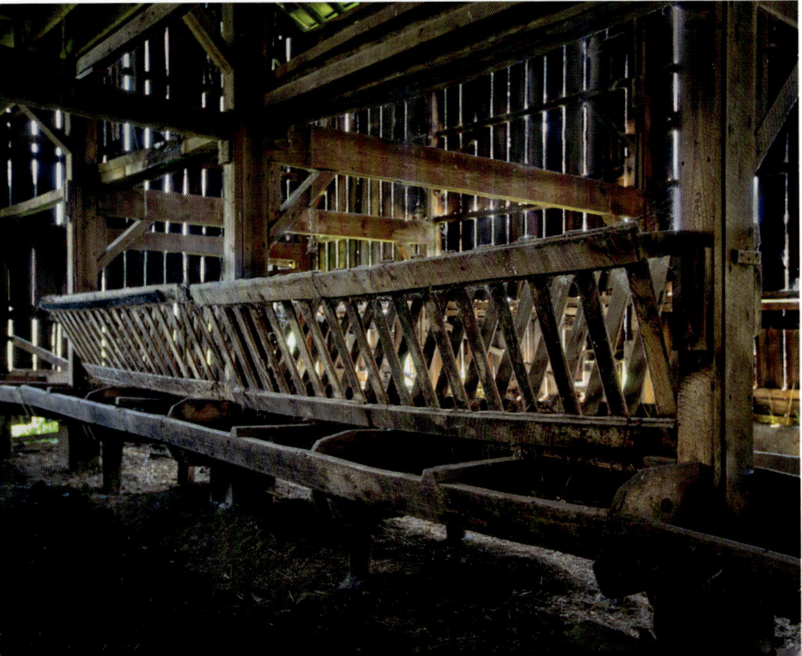

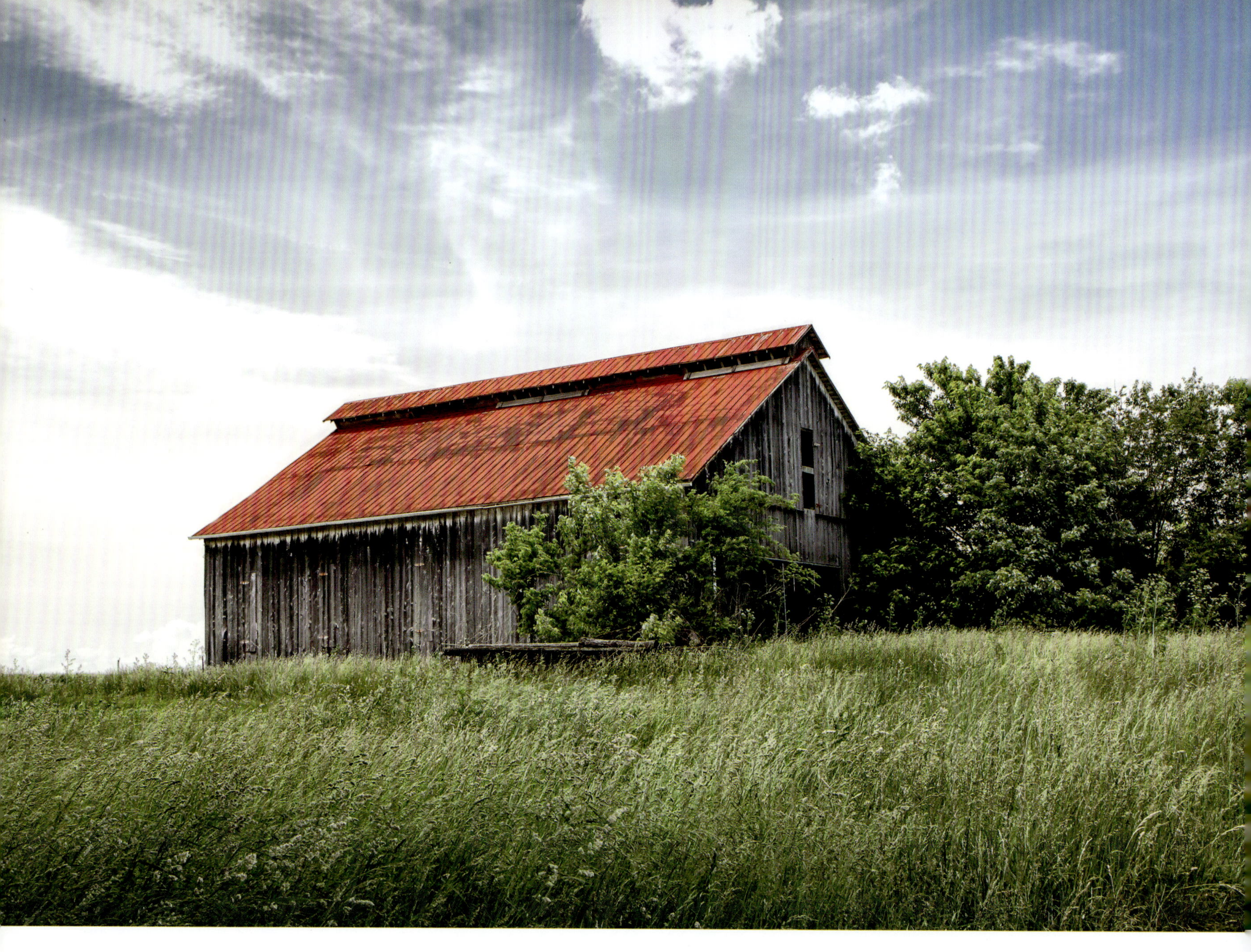

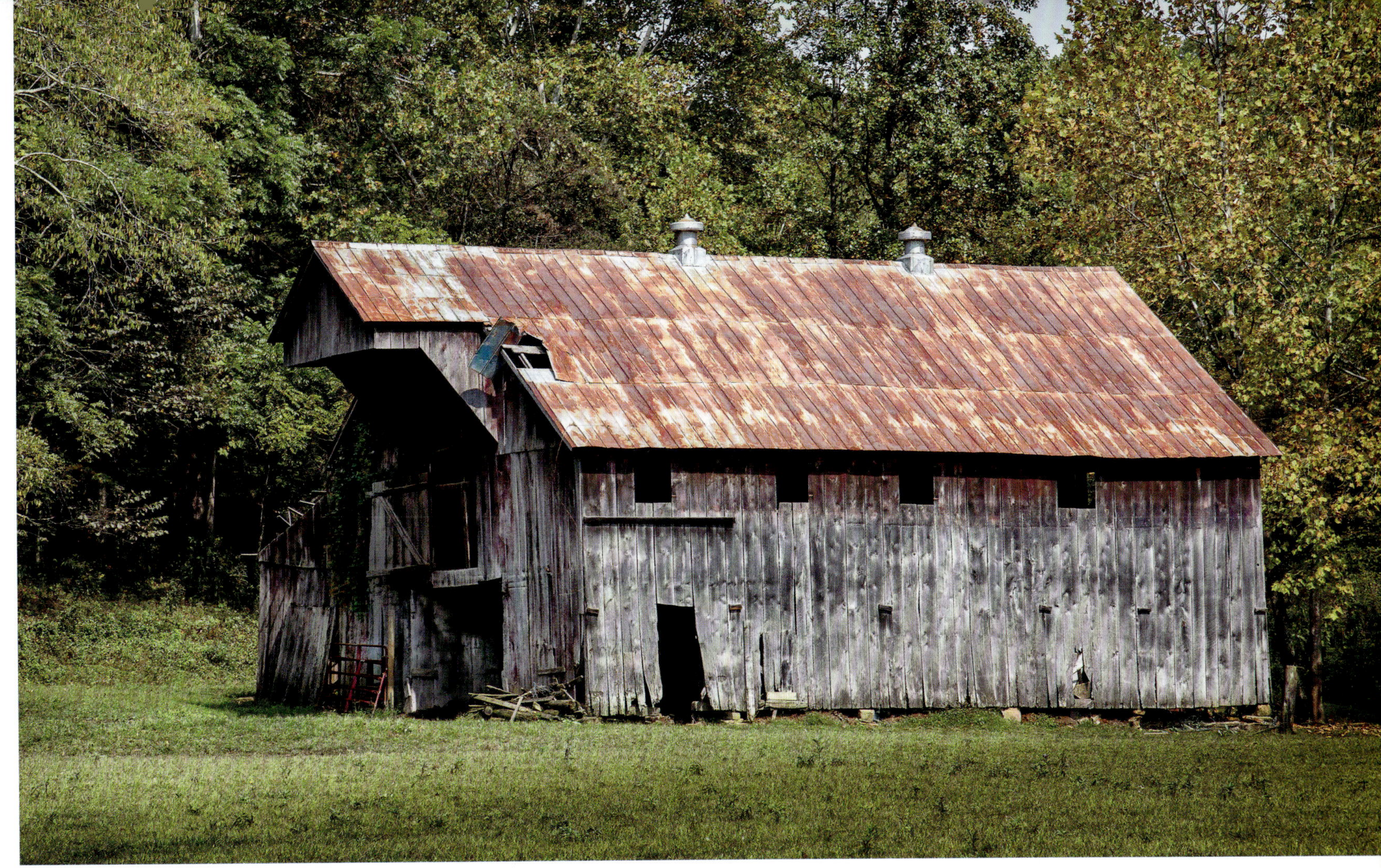

Mason County.

The venting for this tobacco barn, a continuous roofline ridge, is a variation on the usual individual vents placed at intervals along a roof line.

Lawrence County.

This old stock barn and stable is supported by stone pilings and has a very large hay hood. Nearby is a log corncrib.

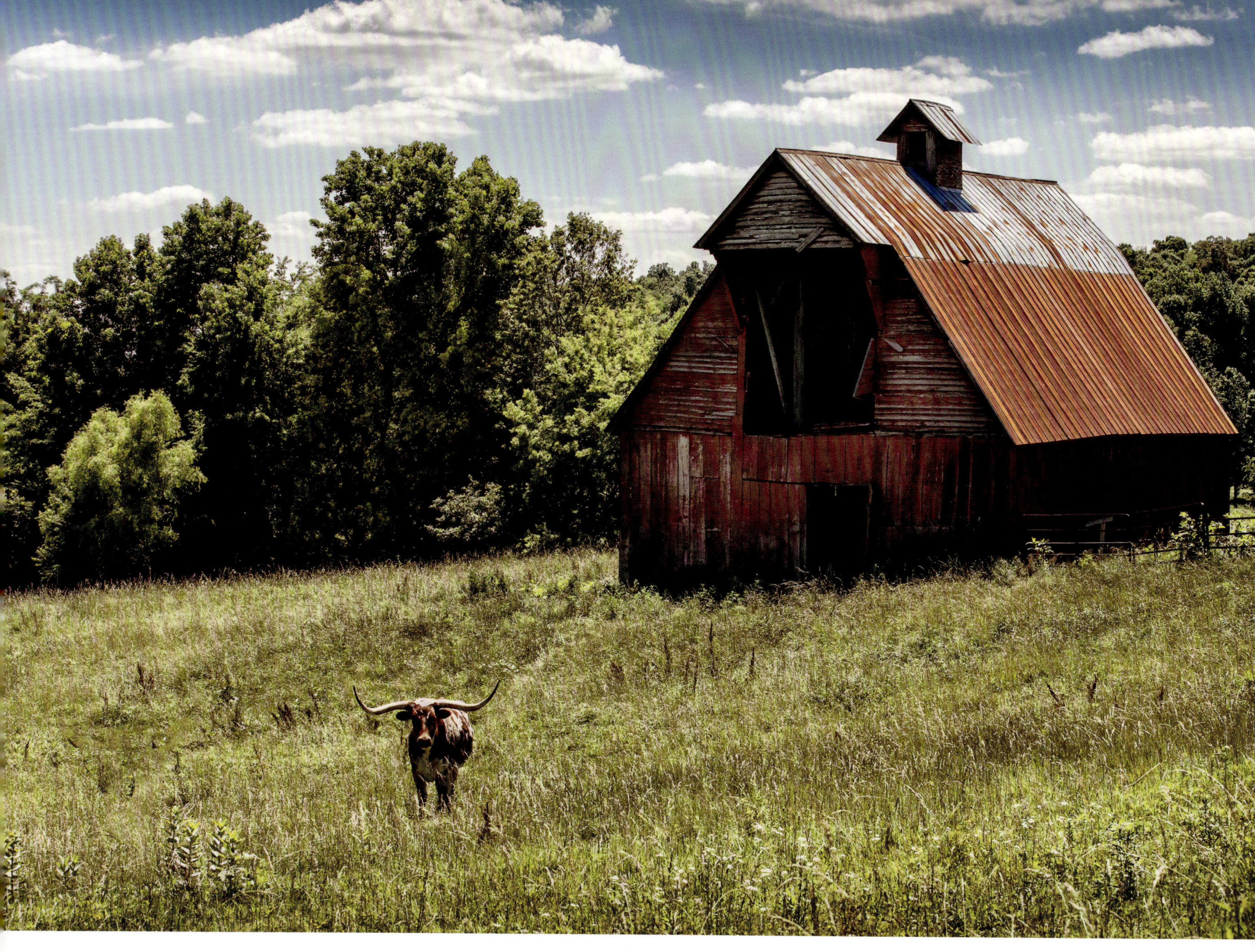

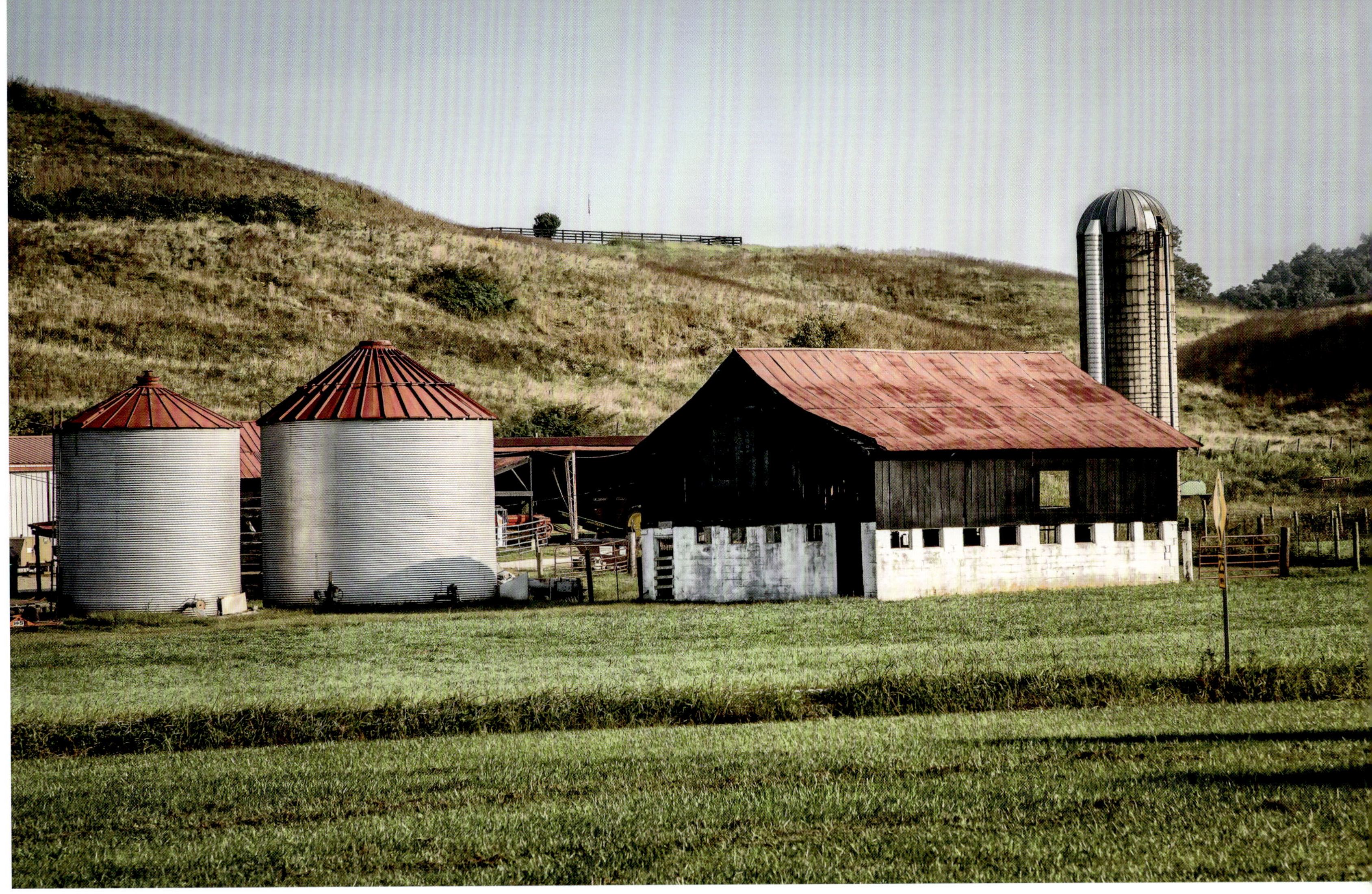

Caldwell County.

The timbers in this old stock barn are hand hewn. Note the variation in timber board direction—horizontal on the top and vertical on the bottom.

Boyd County.

Dairy was a leading industry in the hilly areas of northeastern Kentucky during the twentieth century. Only a few farms, such as this one, still operate.

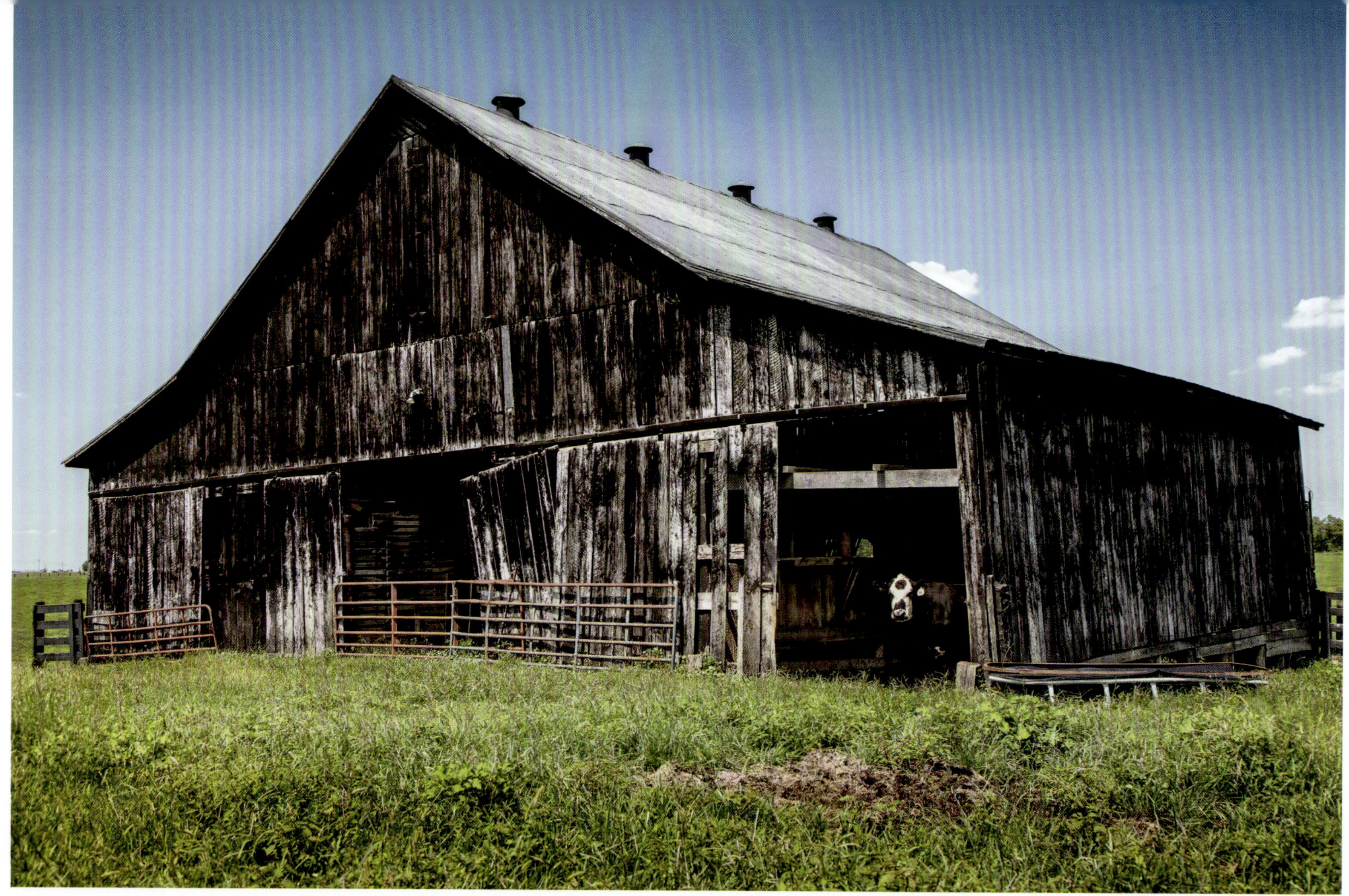

A bull looks out of this livestock barn on the old McWilliams
Revolutionary War land grant farm. The barn is multipurpose with
extensions for equipment, a corncrib inside, and hayracks for the animals.

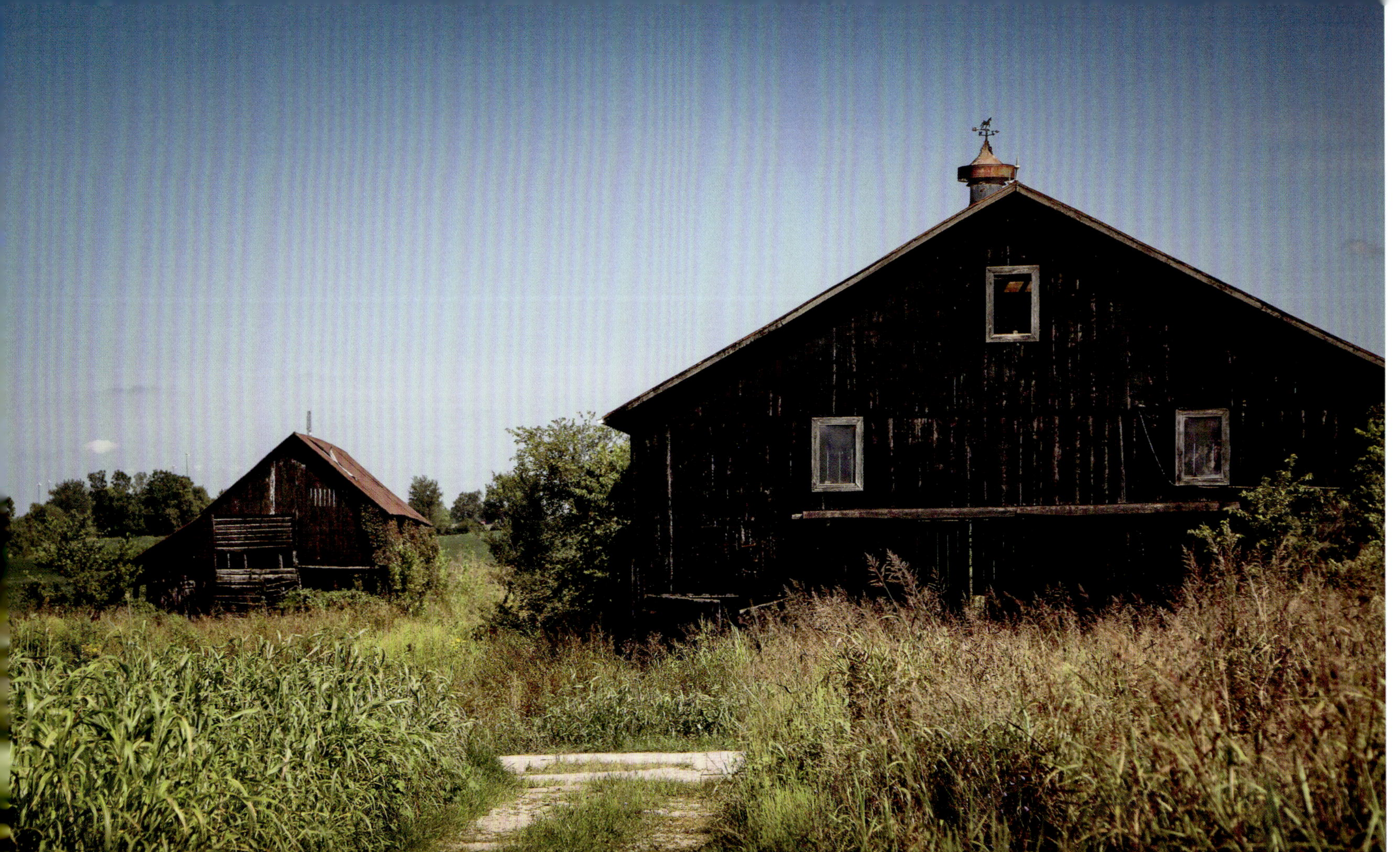

Madison County.

Once part of a vast land grant farm, these barns sit unused. In the foreground is a stable, while a drive-in corncrib barn is in the background.

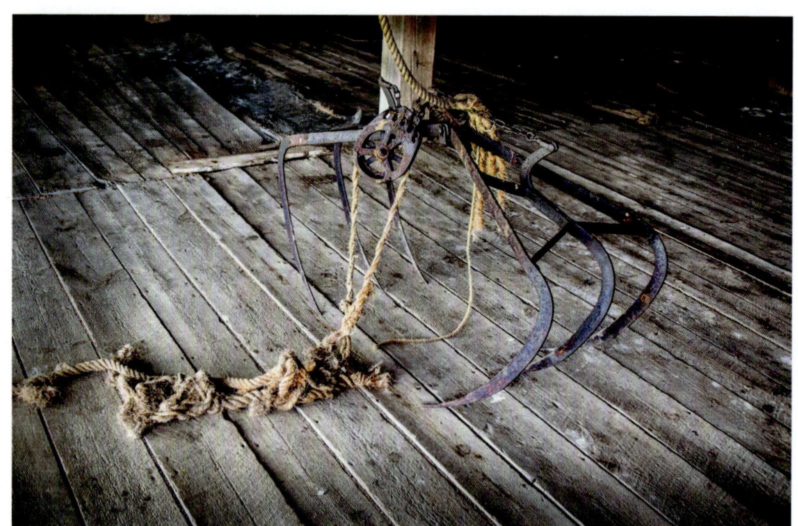

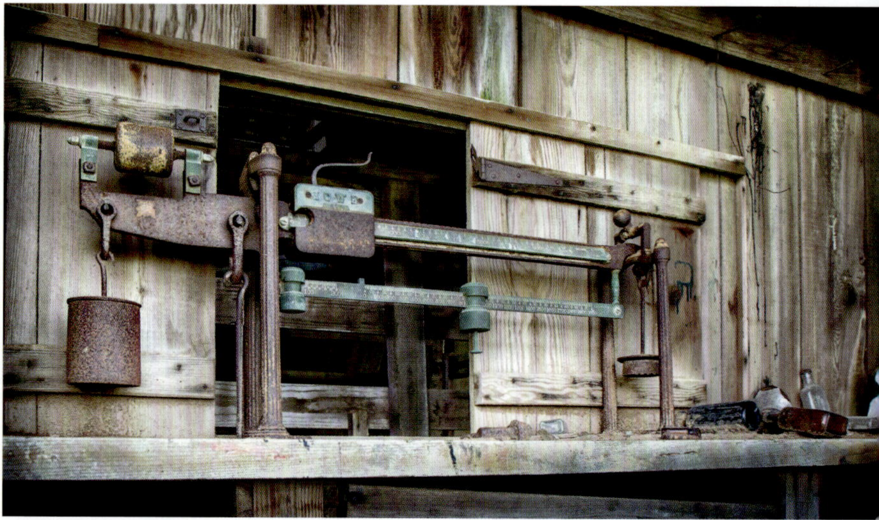

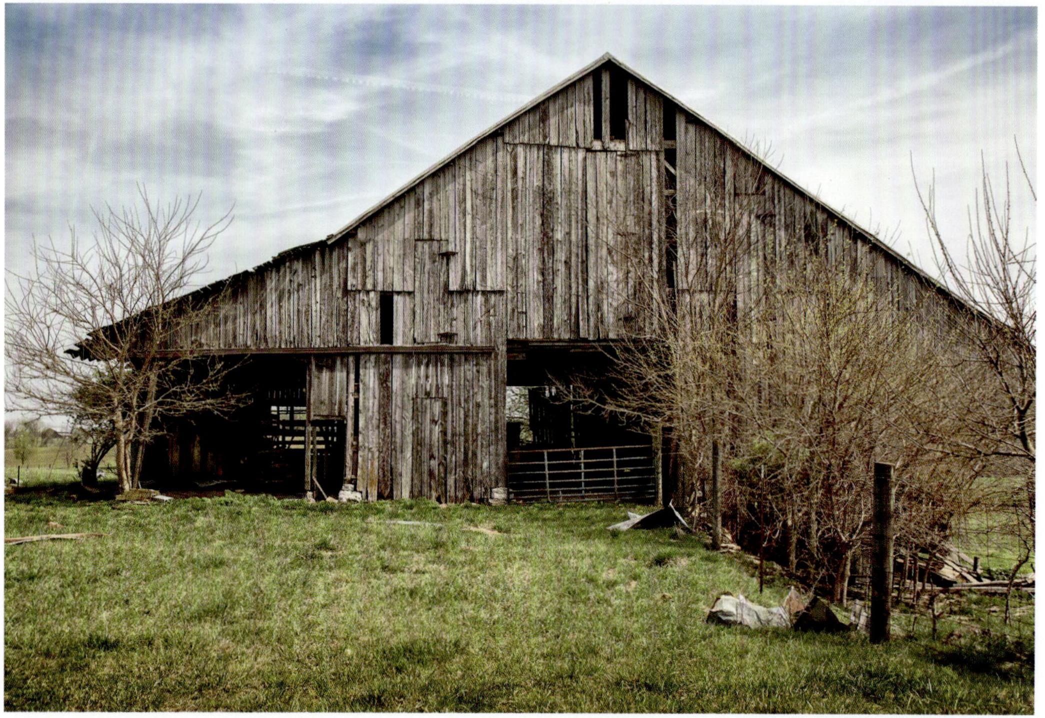

Montgomery County.

Inside the hayloft of this late nineteenth-century stock barn is a fine example of the hayfork. It stands about three feet tall when sitting on the floor like this. The barn also retains scales in the scale shed portion of the barn. A corncrib extends the length of the barn inside, and hayracks take up the lower level.

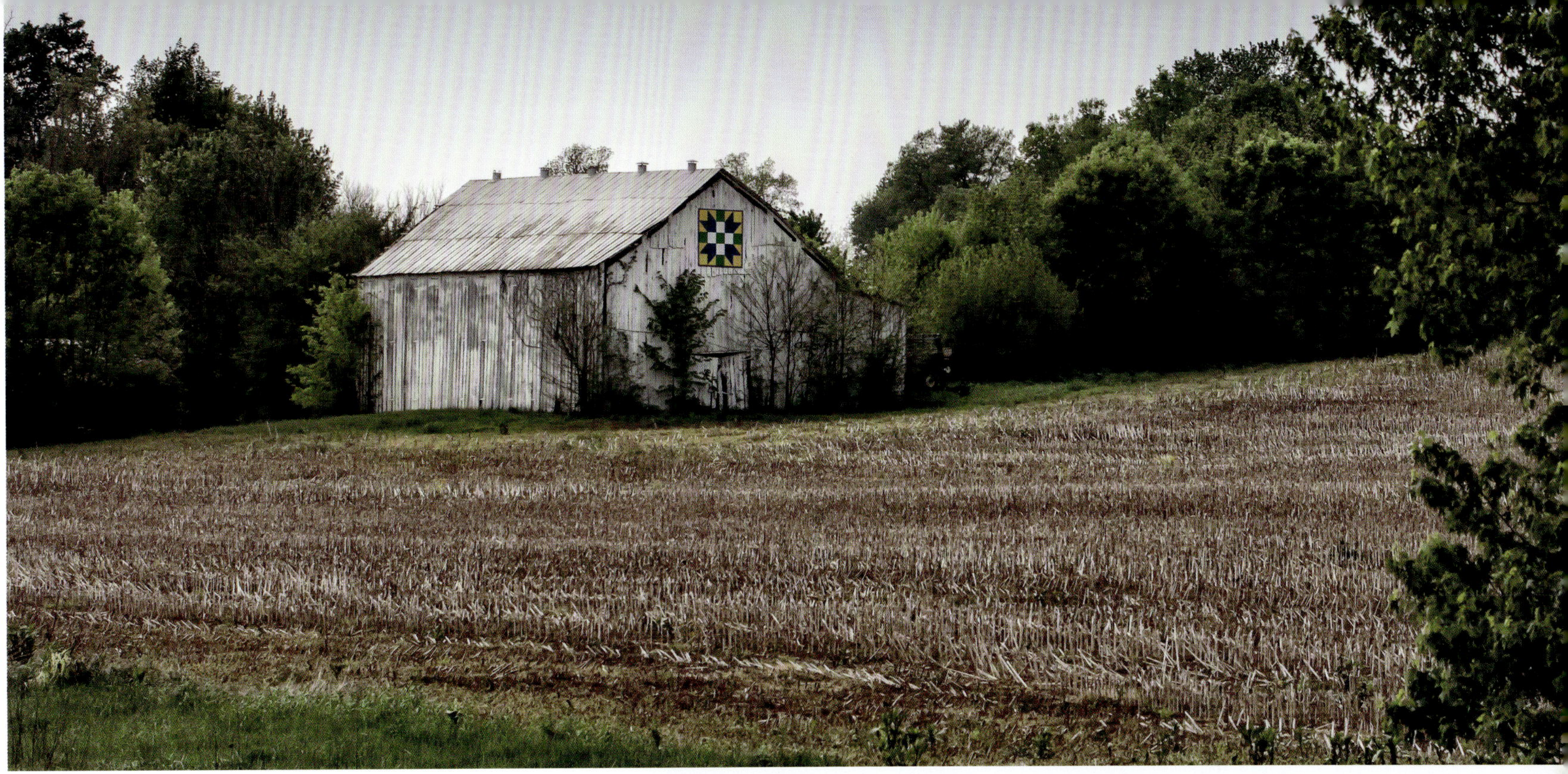

Woodford County.

Part of the Chandler family property, this quilt barn with a cross pattern quilt design sits in a harvested field, holding farm equipment.

Elliott County.

This classic quilt barn, with a wooden extension, is surrounded by grass rather than crops.

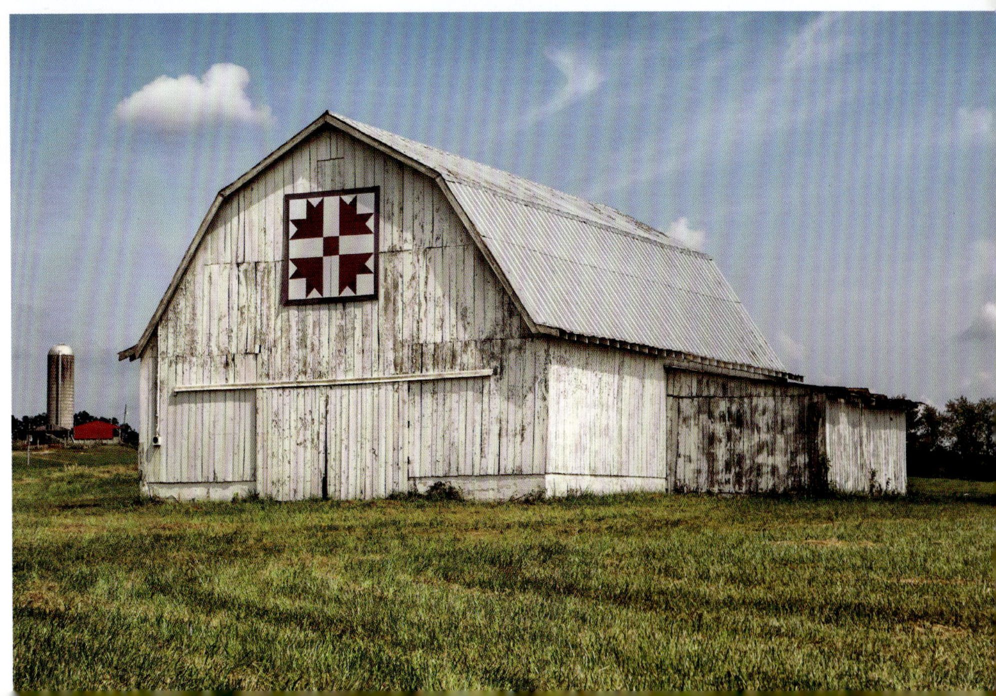

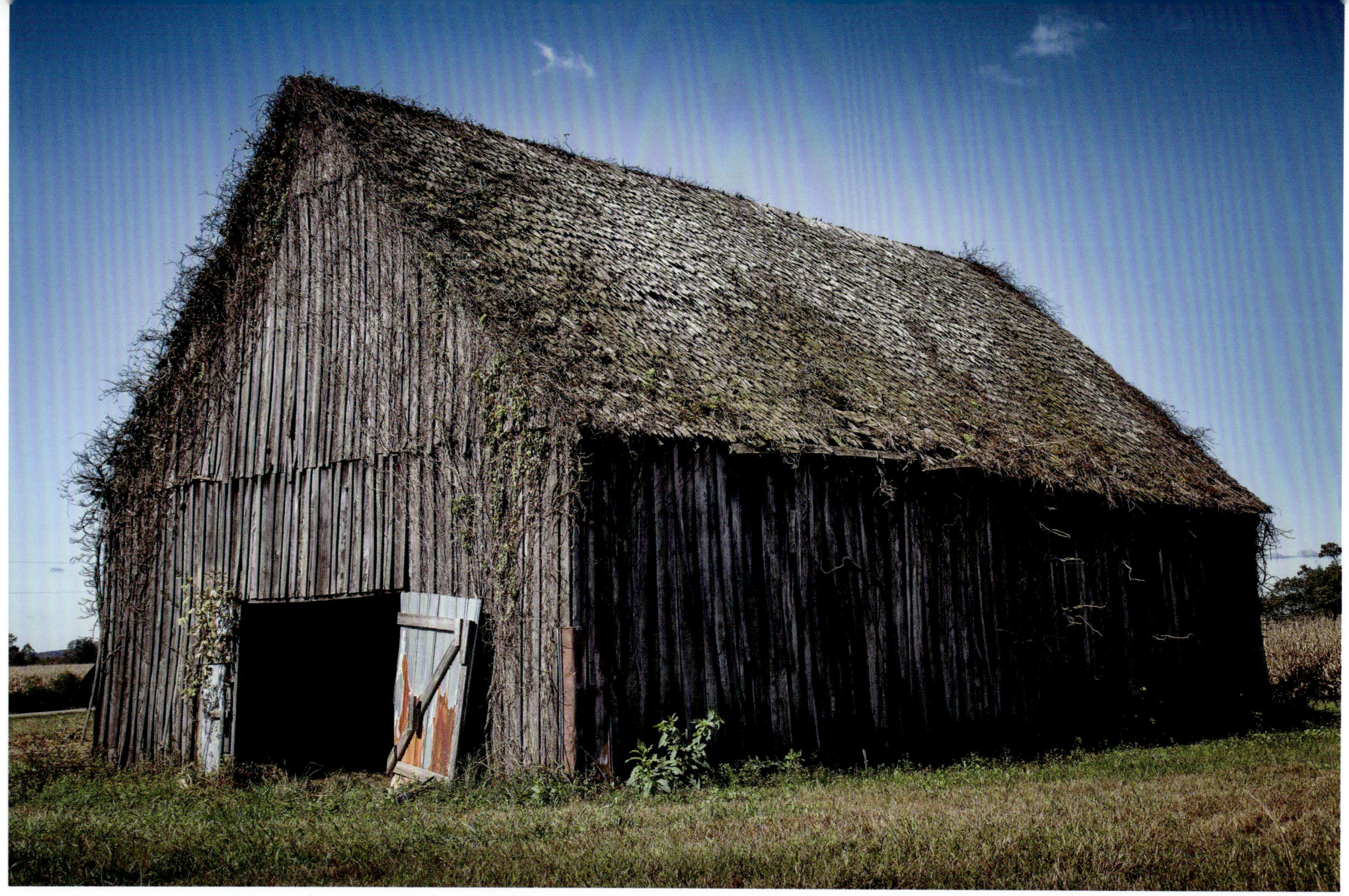

Christian County.

A beautiful and extremely rare sight in Kentucky, this barn has a thatched roof, reminiscent of Old World construction.

FACING PAGE

Mason County.

The owners of this white barn have embellished the architecture with painted circles around the square windows, creating a memorable impression.

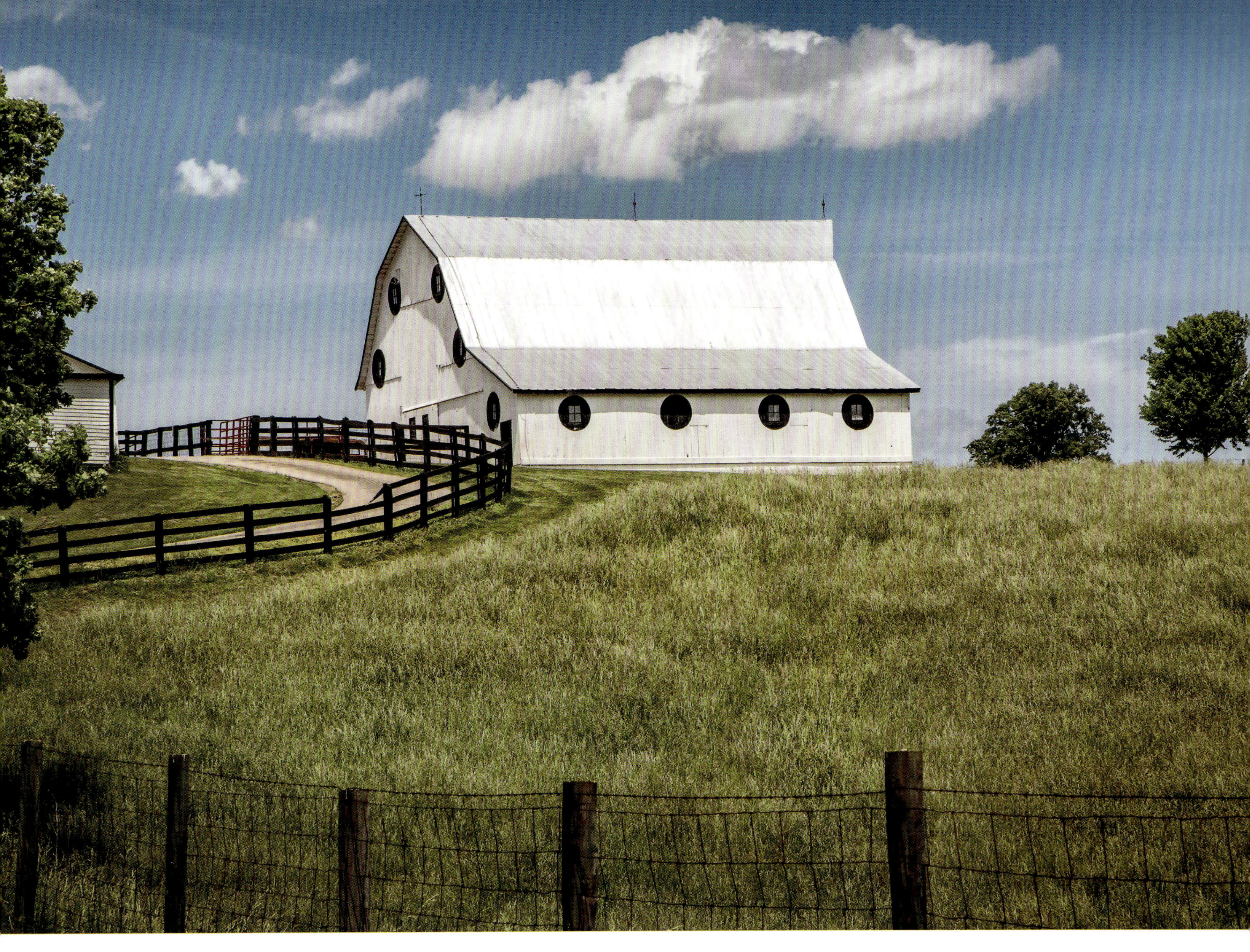

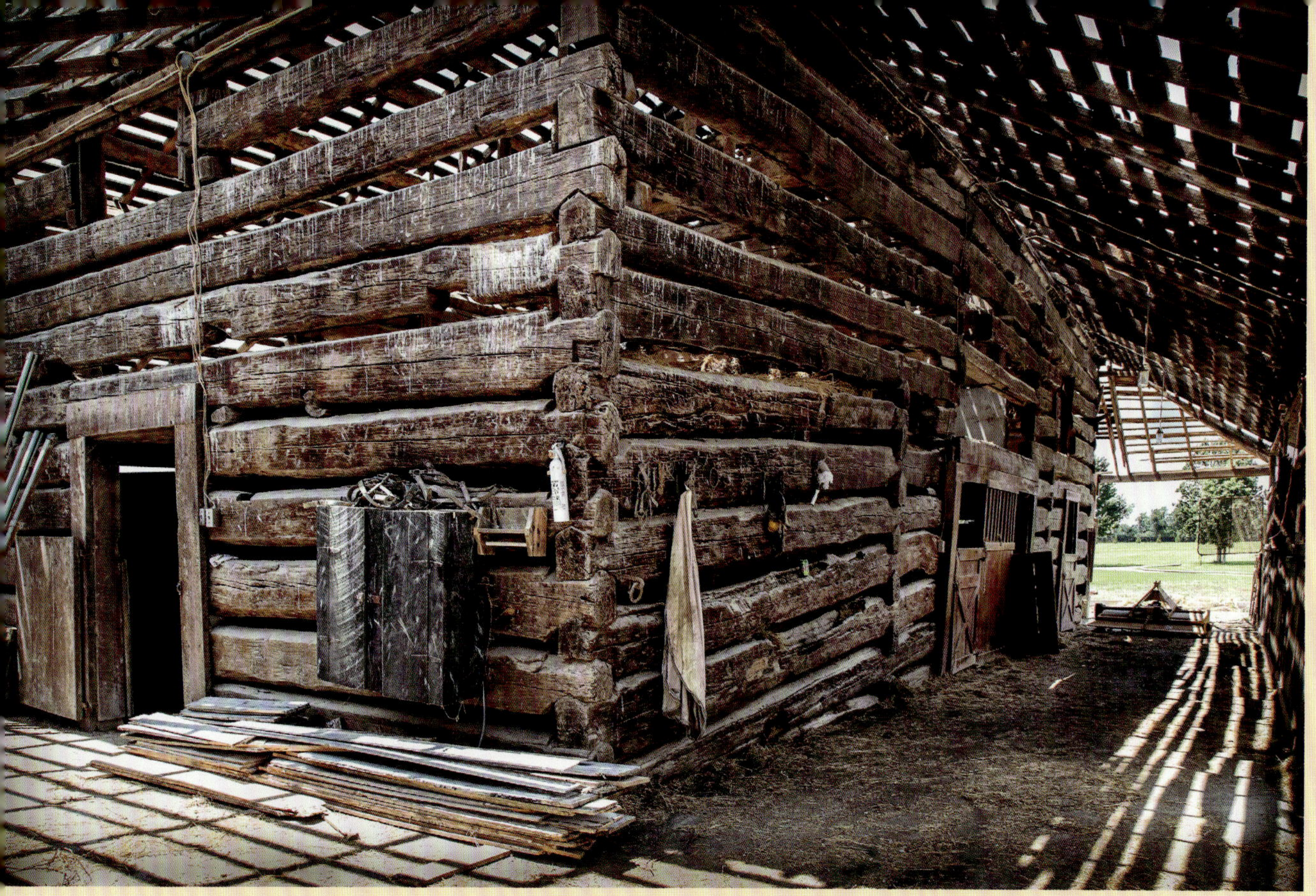

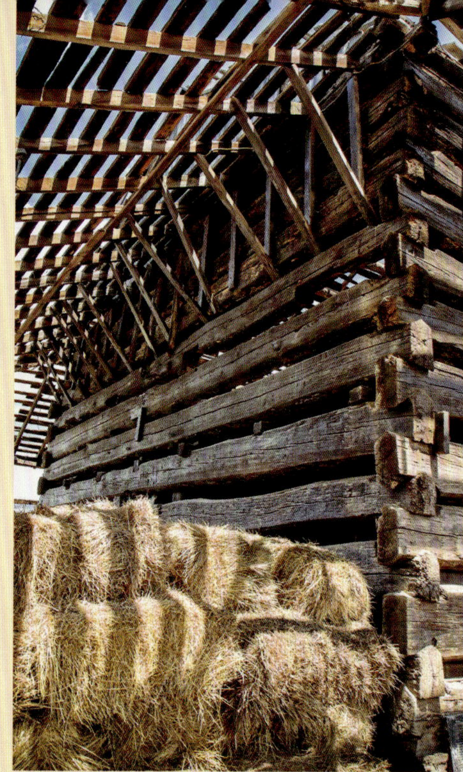

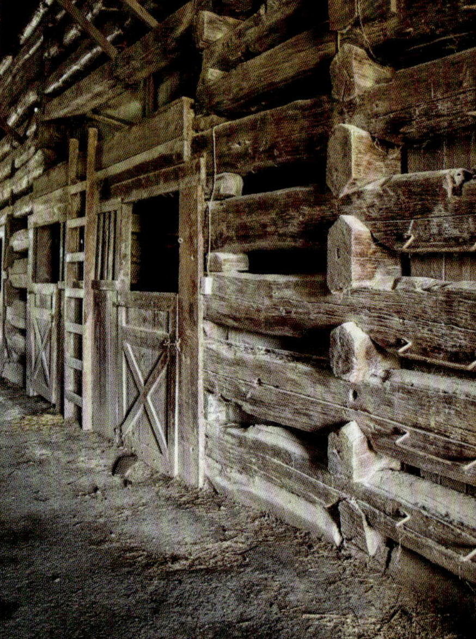

Fayette County.

When I first visited this two-story log stable, owner Mitch Potter was tearing down the surrounding timber barn for restoration. Inside, the V-notched logs, heavy timber doors on six stalls, and upstairs hayloft were easily visible. I returned later to see the beautifully redone barn, restored with no structural alteration to the historic interior.

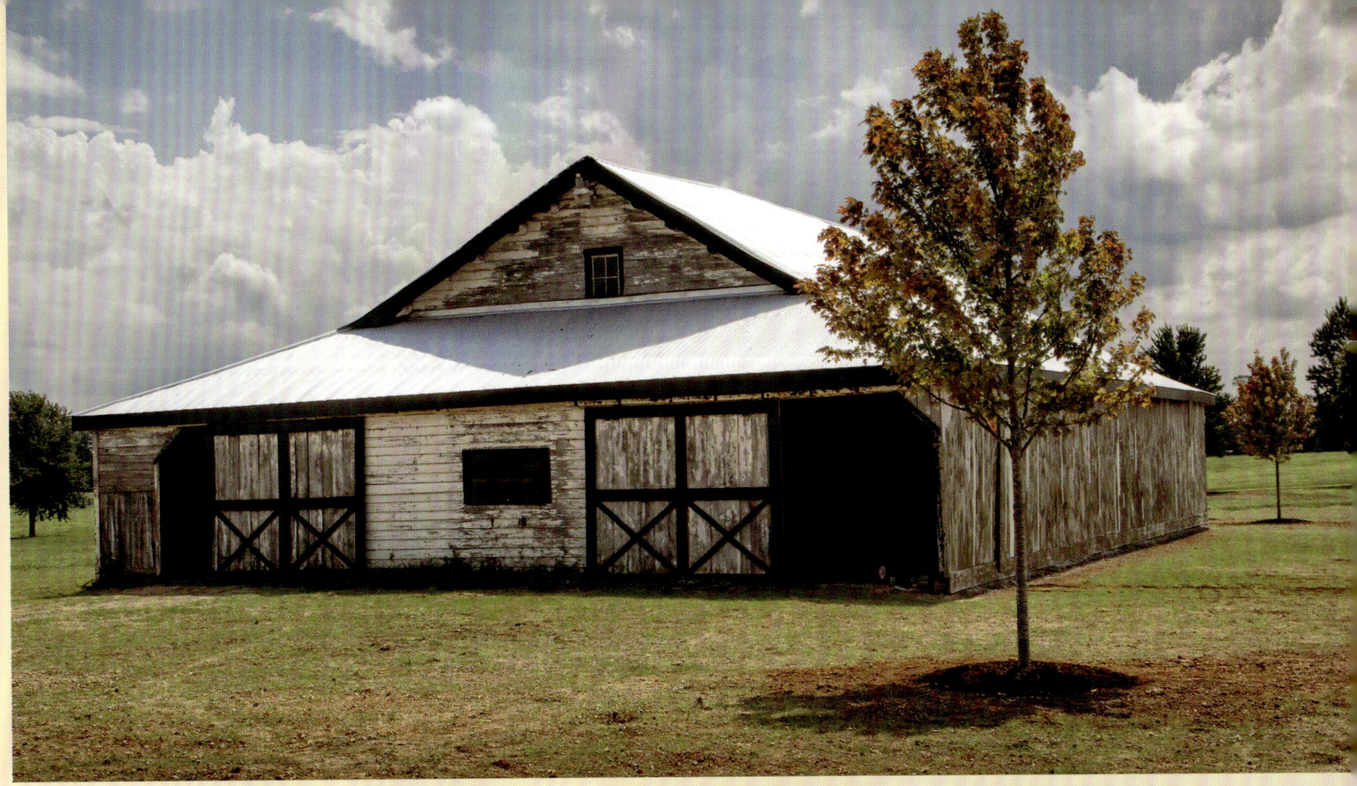

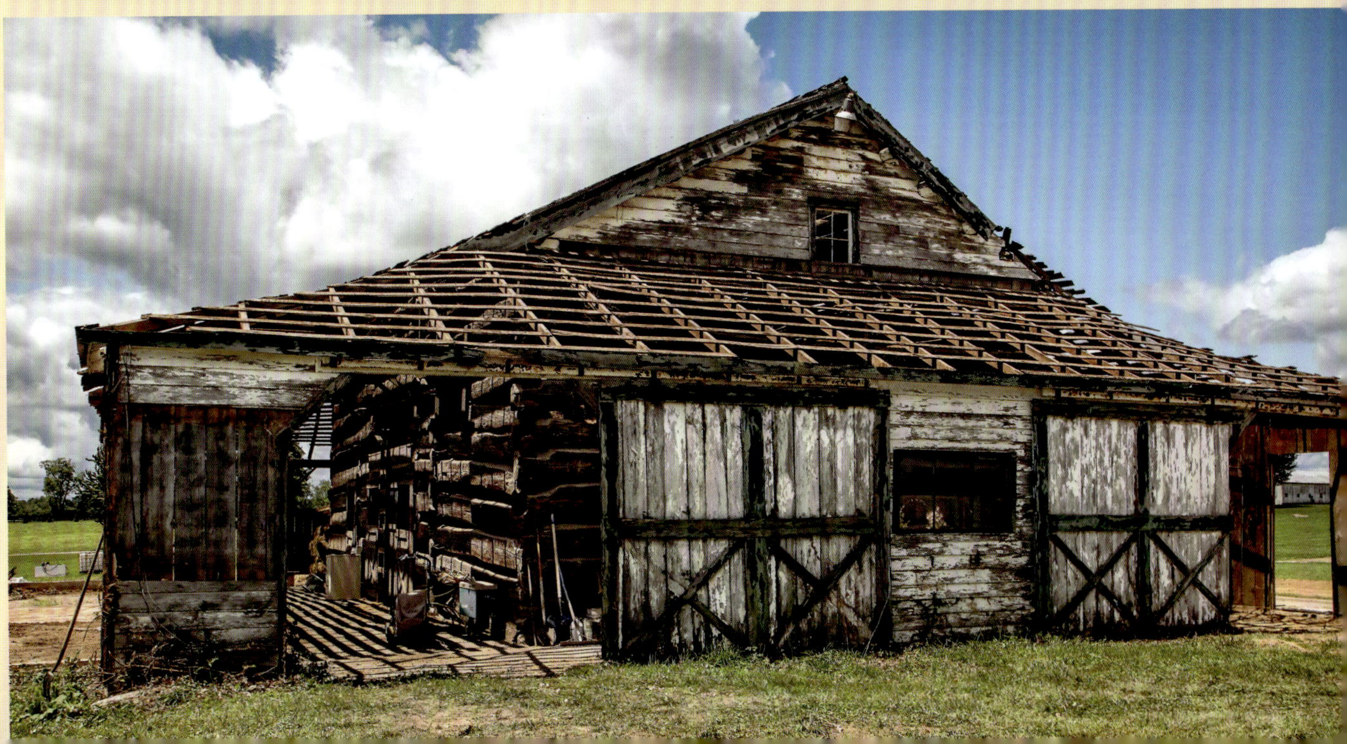

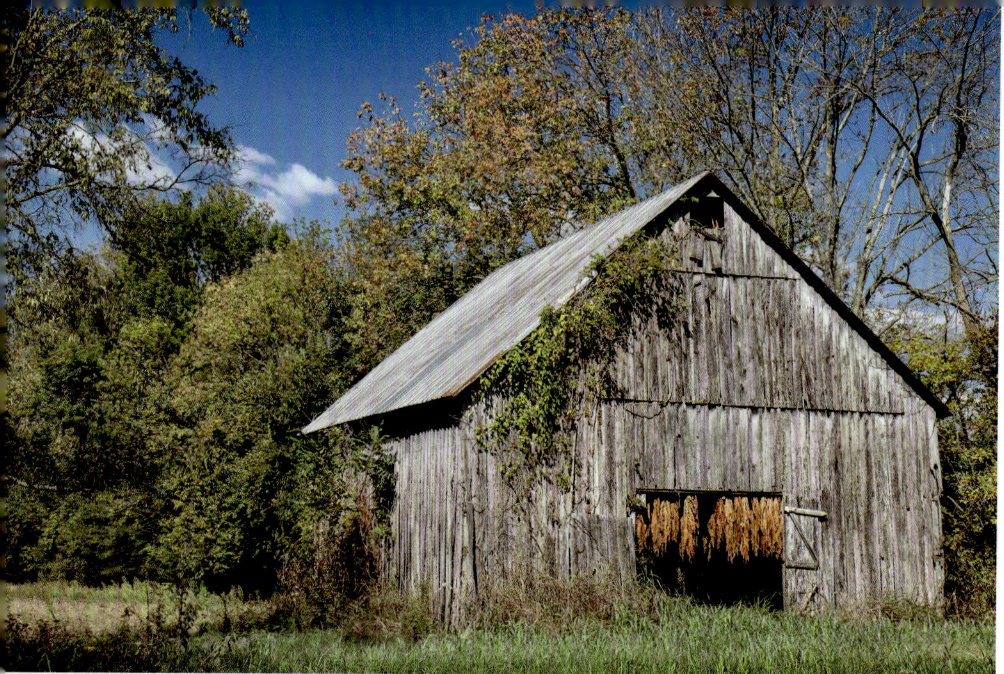

Christian County.

Although Western Kentucky is known for dark fired tobacco, there is still much burley tobacco being grown there, as seen hanging to dry in this old barn.

Todd County.

Barns like this field barn usually have many purposes. This structure, built to allow drive-ins, is used for cattle, tobacco, and storage.

FACING PAGE

Casey County.

Mules were a very profitable business in Kentucky in the nineteenth century, as they were cheaper than horses. The barn in this image is an early twentieth-century stock barn, continuing its function into the present.

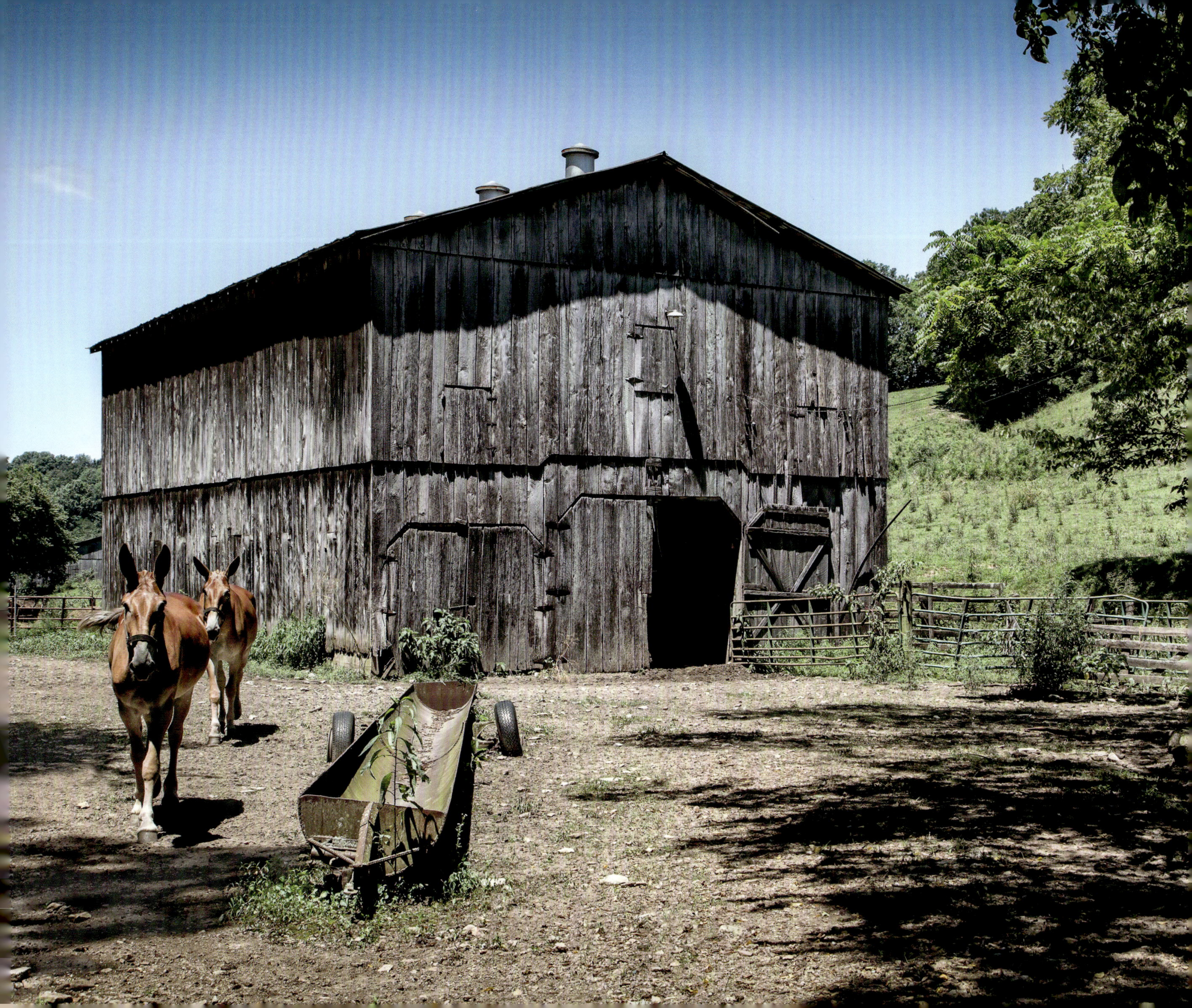

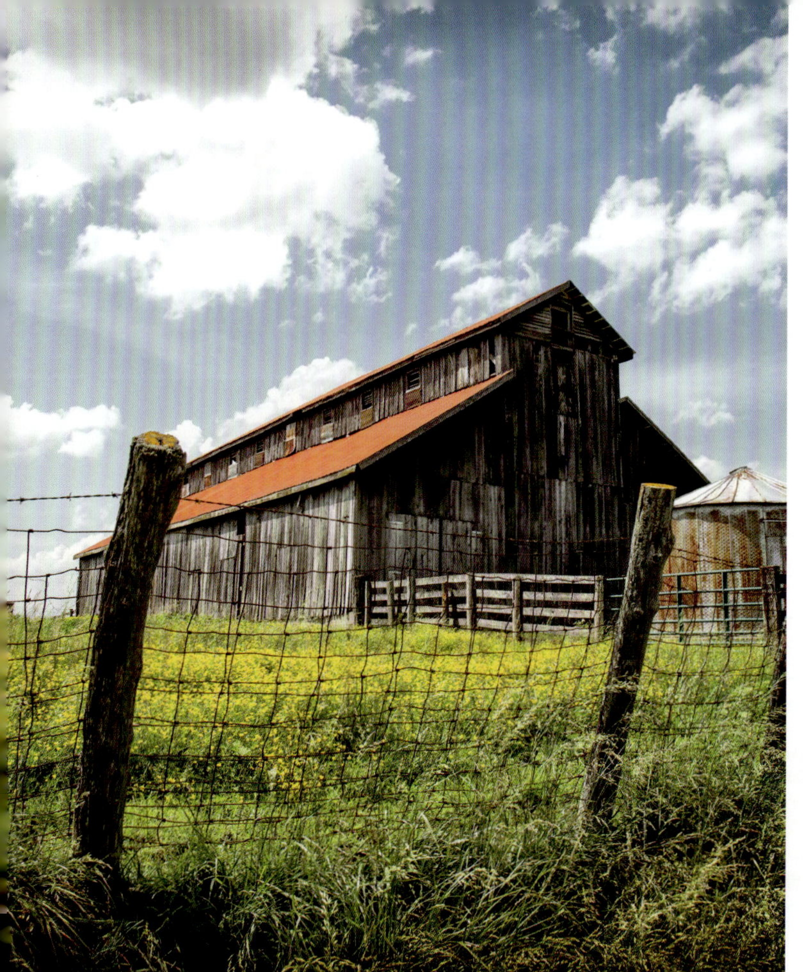

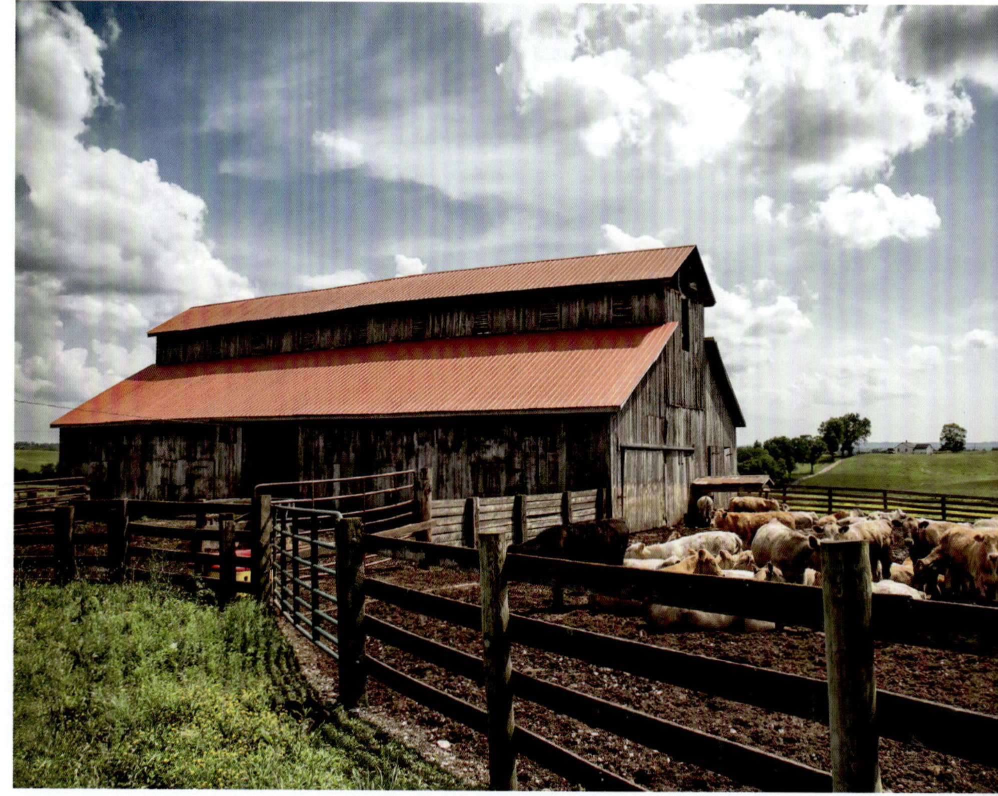

Garrard County.

This barn was once part of the Dalton farm. A house, which has been torn down, was built on that property around 1820 and had been listed on the National Register of Historic Places.

Ballard County.

The barn in this image sits on Eagle Rest Plantation, which was settled as a homestead circa 1845. In 1998, the farm was recognized as one of the oldest cattle ranches under continuous family ownership in the United States.

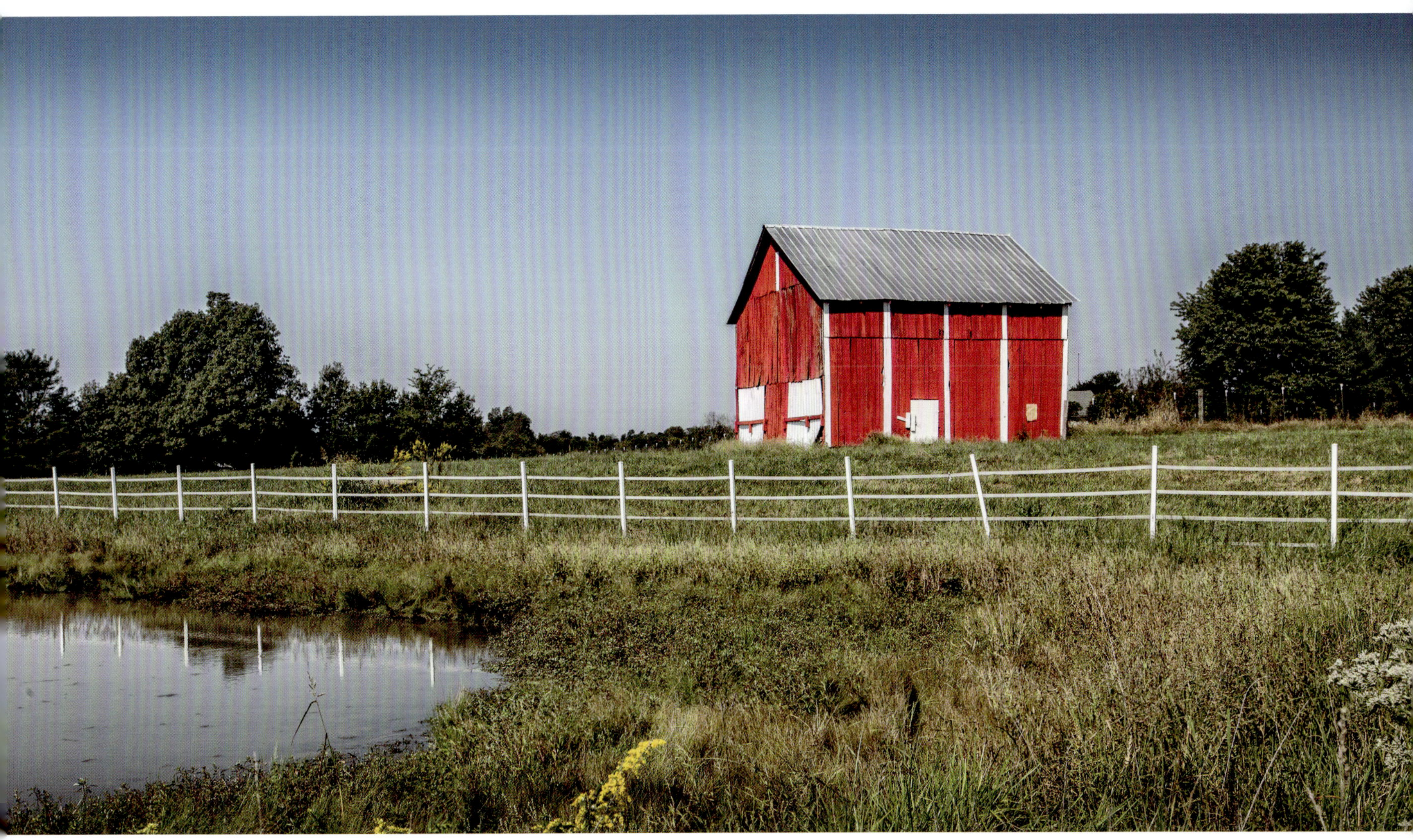

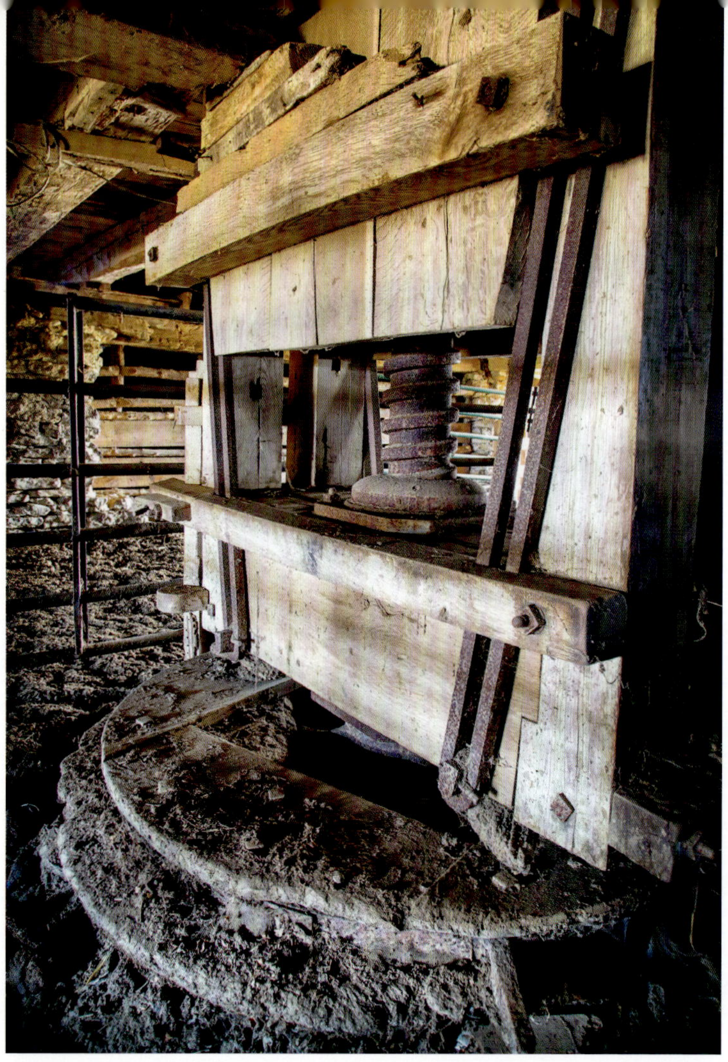

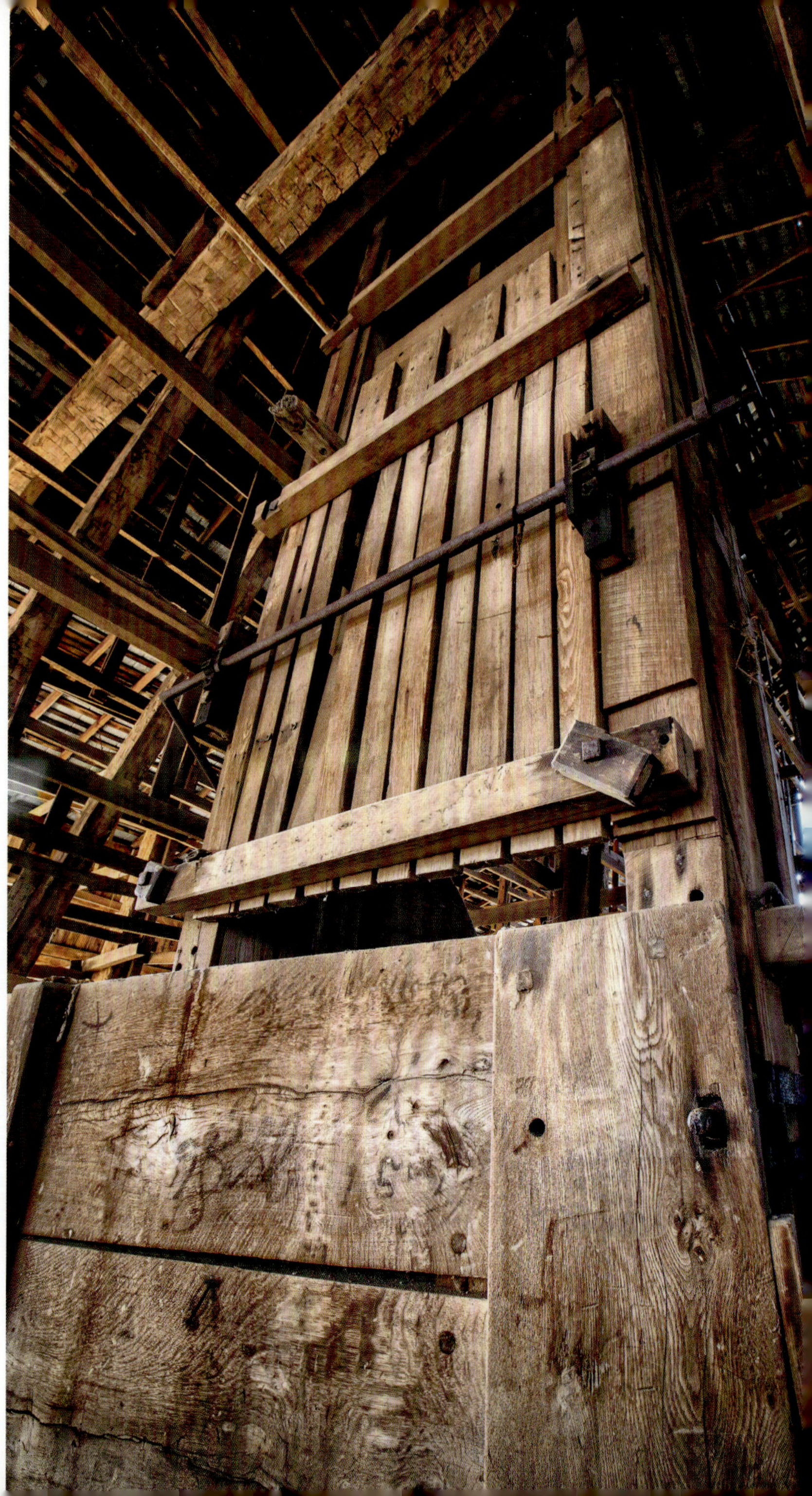

Gallatin County.

Possibly the only remaining hay press barn in Kentucky, this
barn was built circa 1873 by W. H. Gridley. Hay press barns
were located along the river, pressing hay into bales for easy
transport by boat down the river. The homestead remains
in the family, maintained by Gridley's grandchildren.

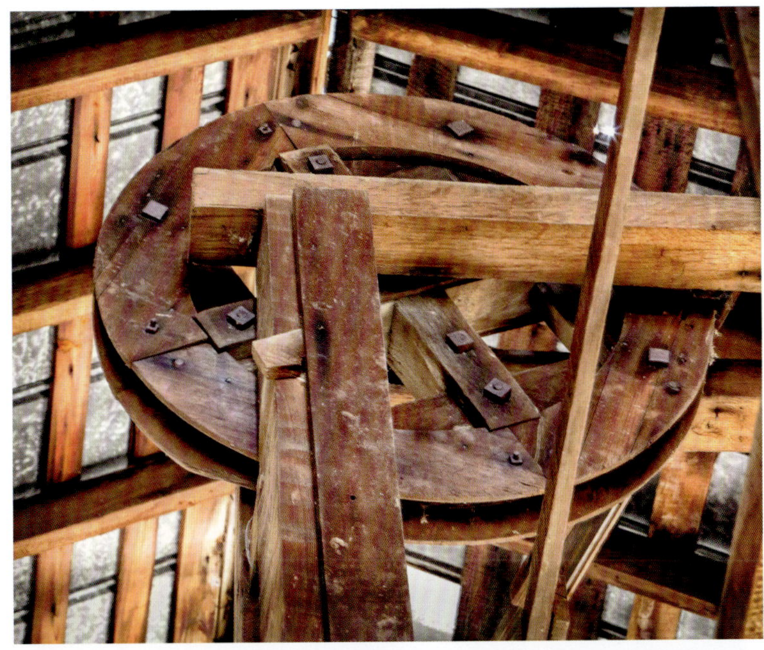

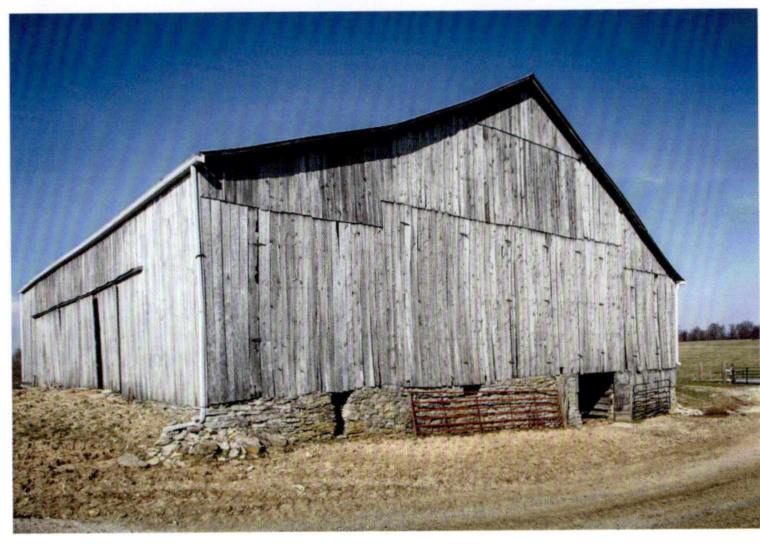

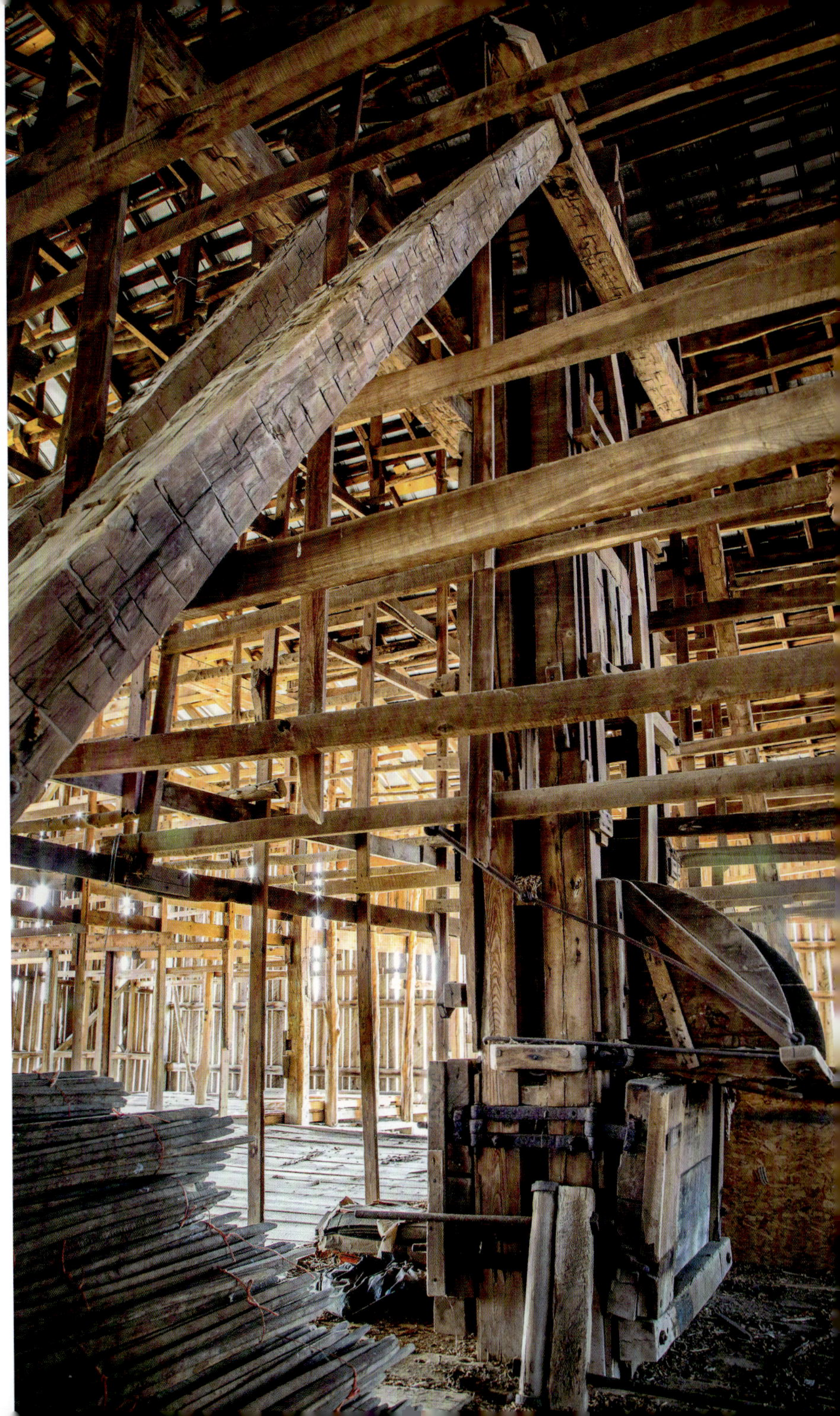

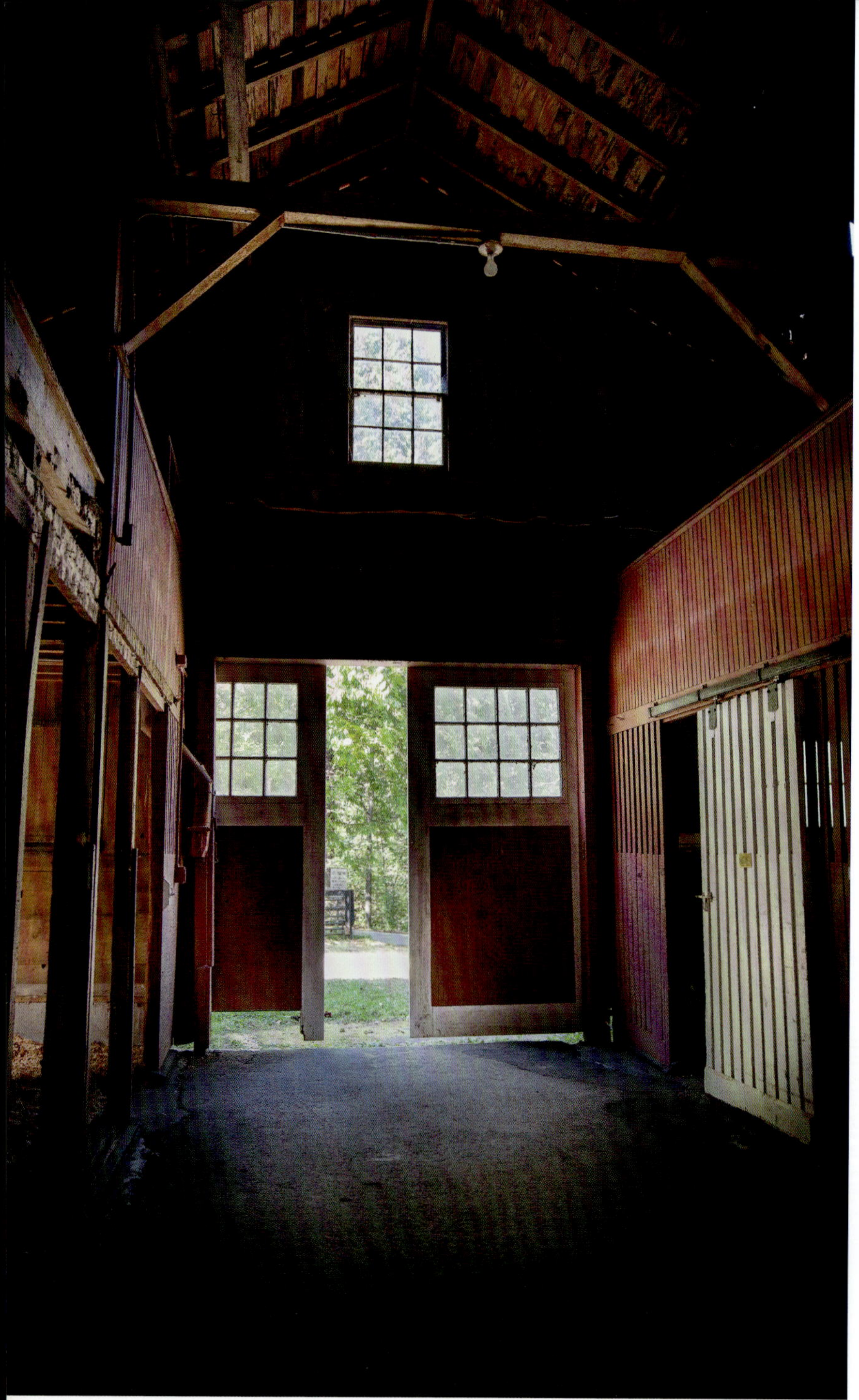

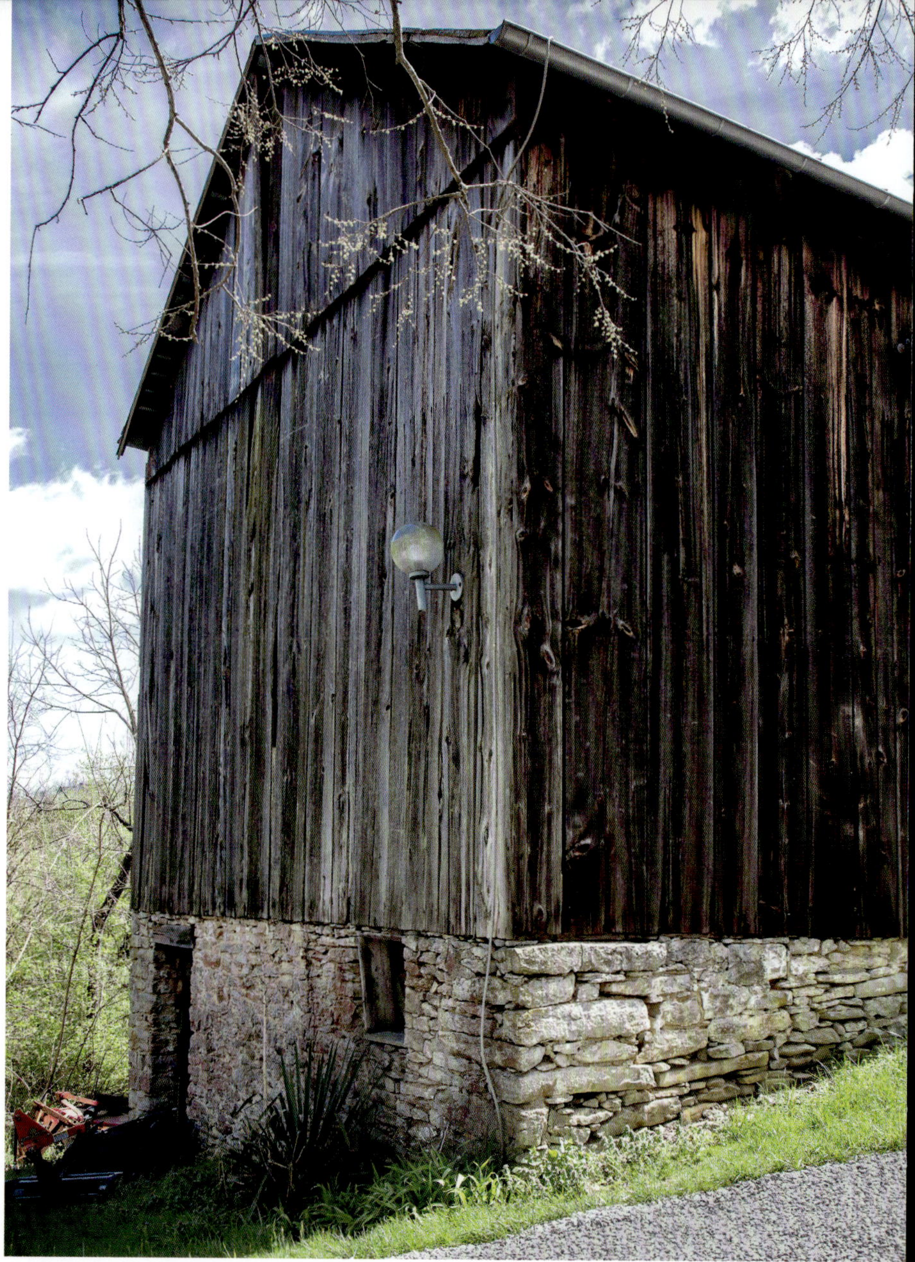

Casey County.

Nestled down on a hillside, this early twentieth-century stock barn stretches the length of the flatland created by the creek visible in the foreground.

FACING PAGE LEFT

Fayette County.

Grimes Mill was once a prominent mill as well as a distillery. In the early twentieth century, the Iroquois Hunt Club—who restored the mill building for their headquarters—built the stable shown here.

FACING PAGE RIGHT

Campbell County.

The Sandfoss house, built by German immigrants in the mid-nineteenth century, is only a few yards away from this bank barn. Both are listed on the National Register of Historic Places.

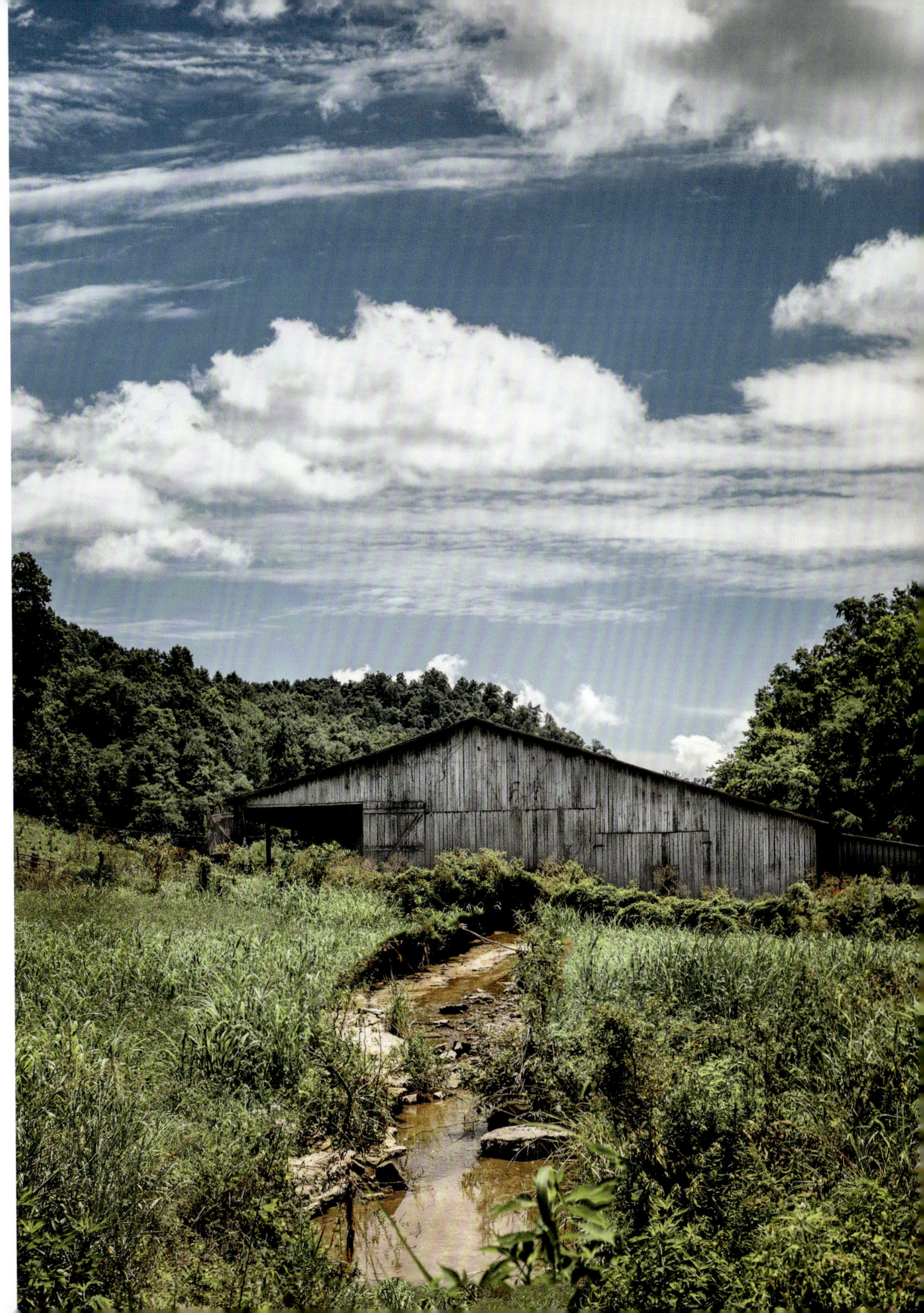

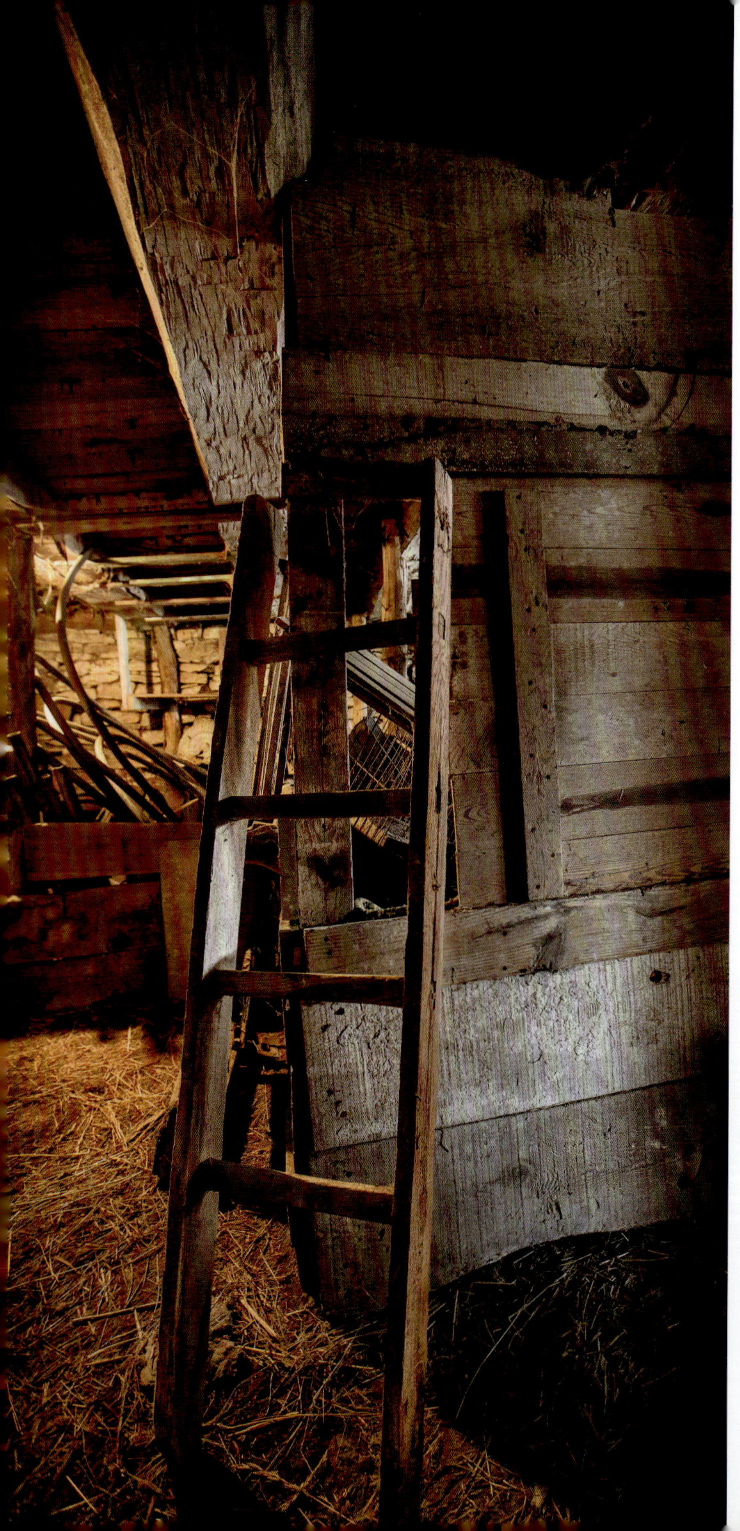

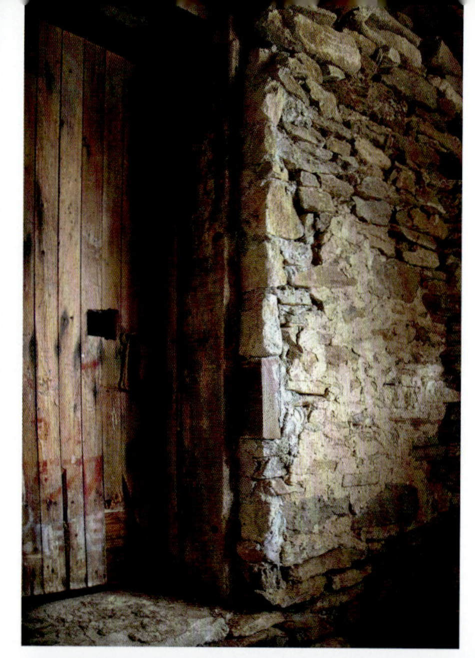
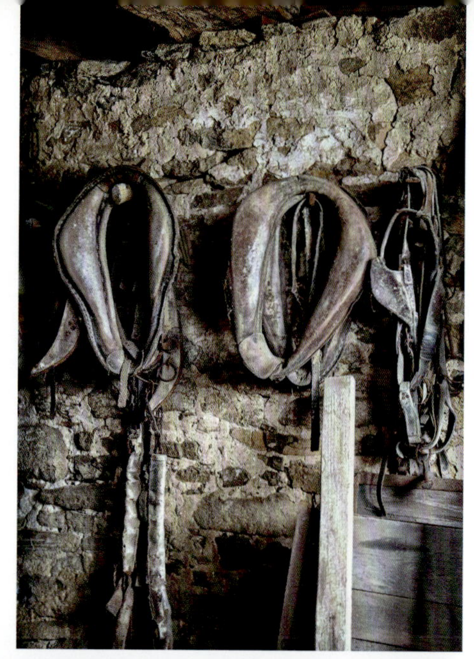
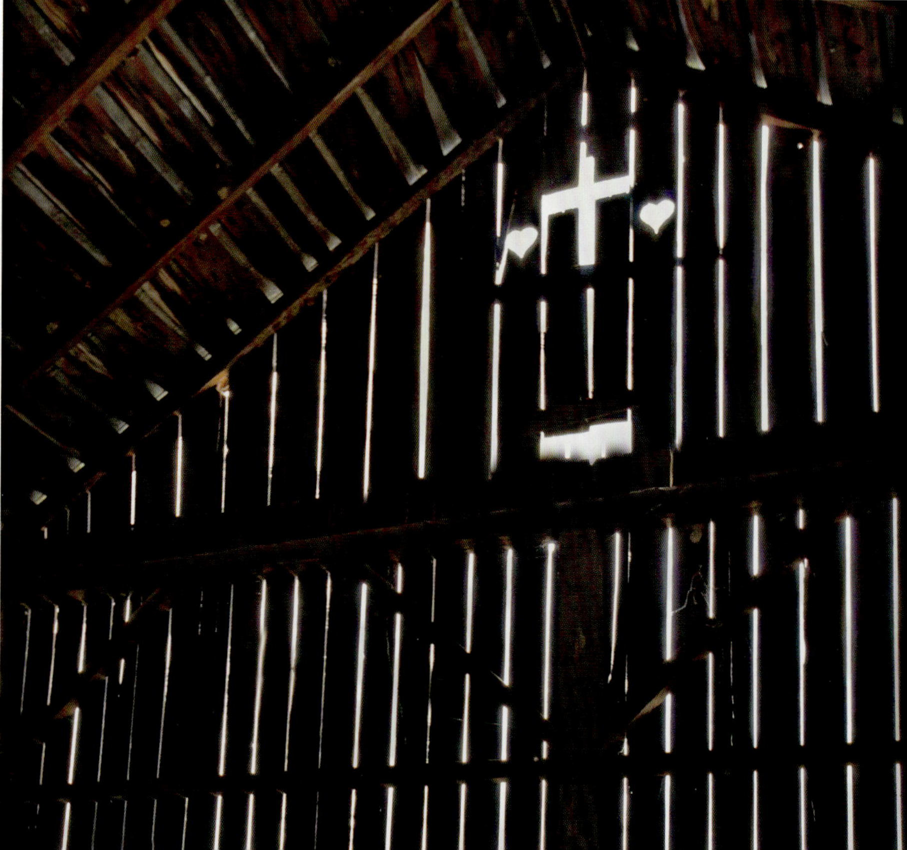

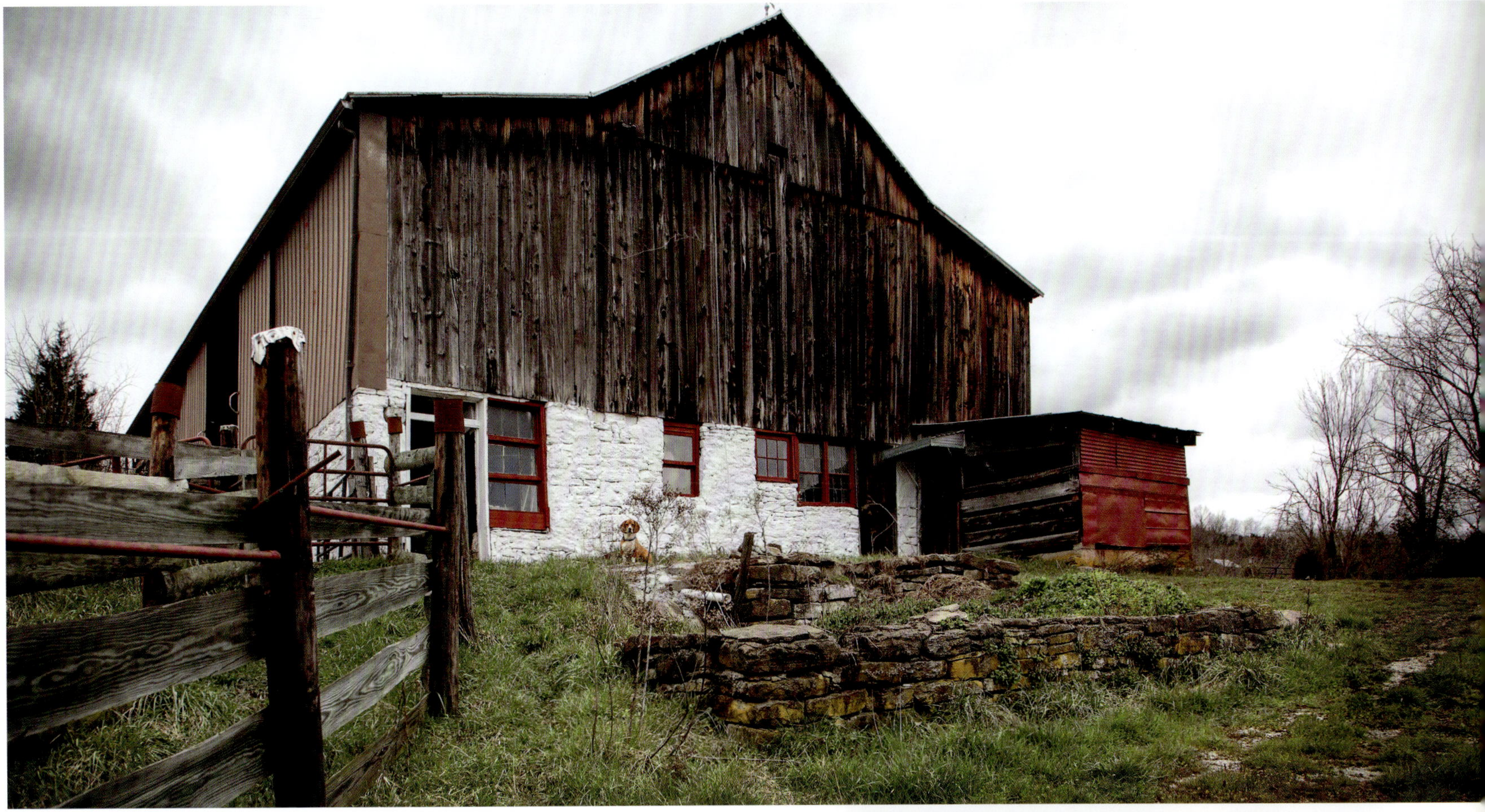

The German settlement of Camp Springs maintains many of the original dwellings and barns from the mid-nineteenth century. The motif of a cross with two hearts indicates that Andrew Ritter built the barn. Owned now by Whitey Heebs, the barn has stayed in continuous use as a stock barn.

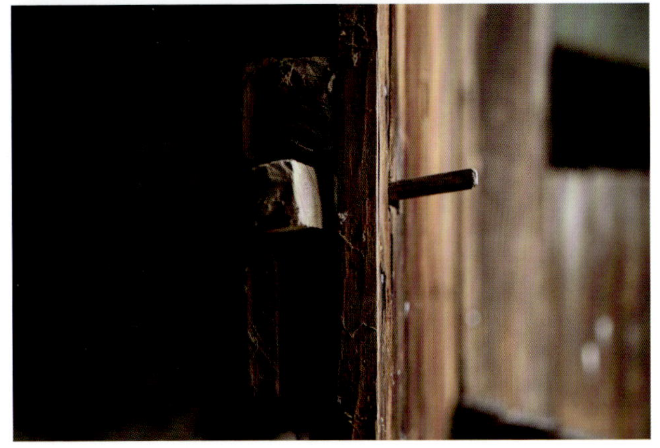

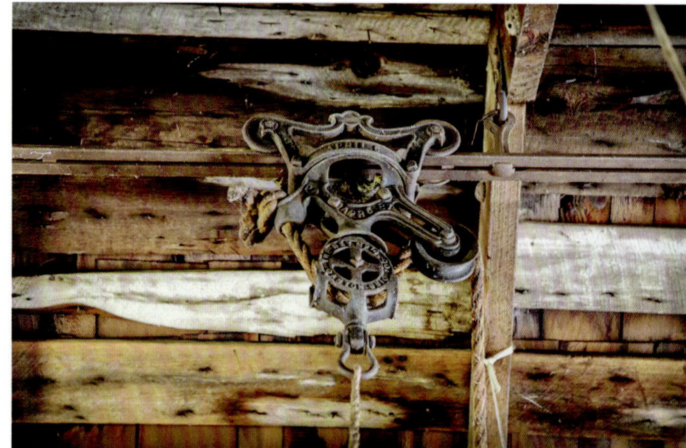

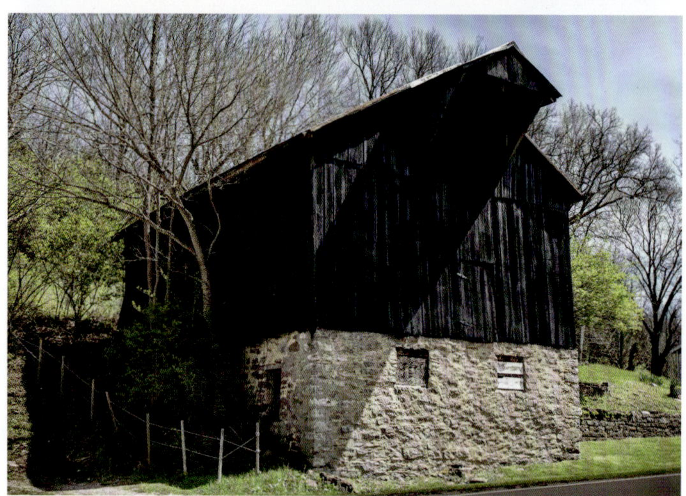

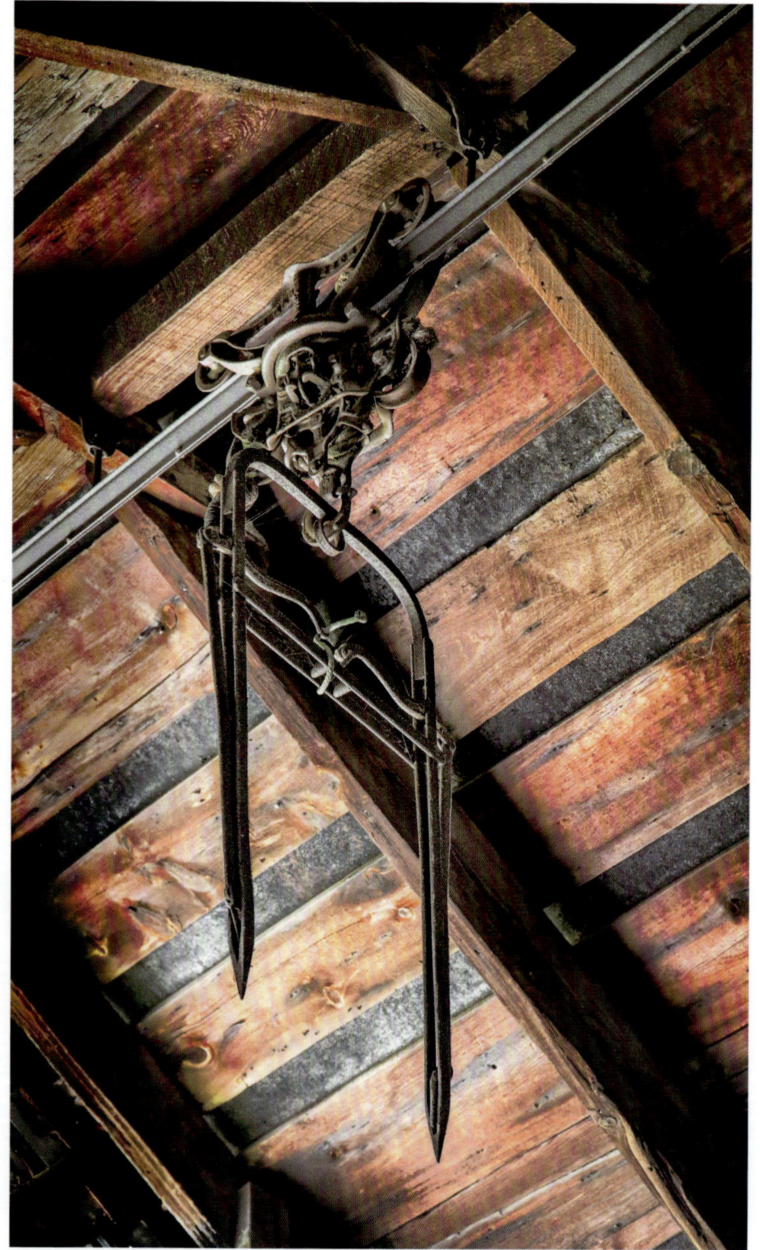

Campbell County.

The Reitman barn is another example of German architectural style from the mid-1800s preserved by the families of Camp Springs. Inside, an old hayfork and hay track are still in place. The fork was drawn along the track using a long rope, spearing the hay and lifting it into the loft from street level.

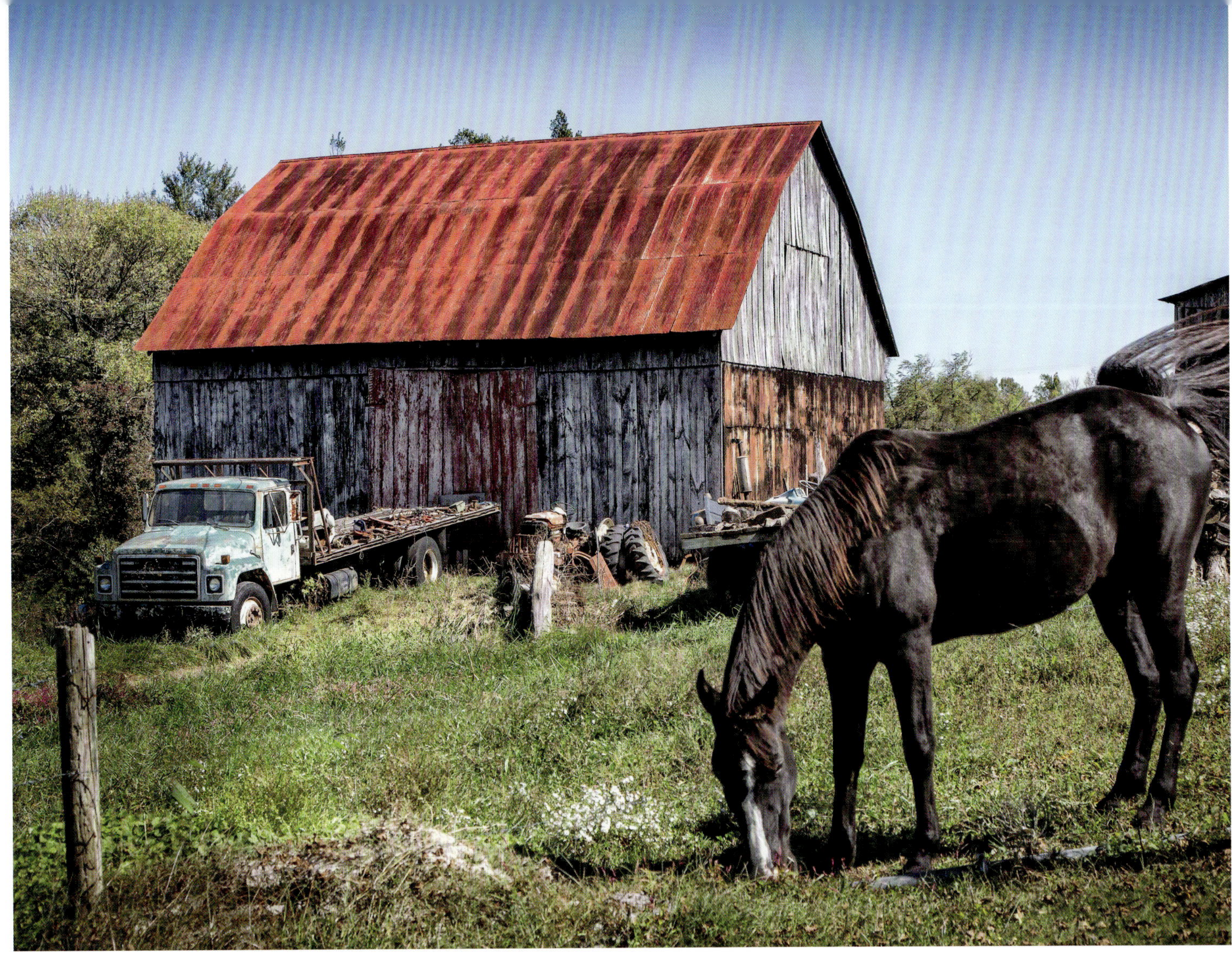

In the Swiss settlement area once known as Ottenheim, this
English barn with entry on the sides was built circa 1885.

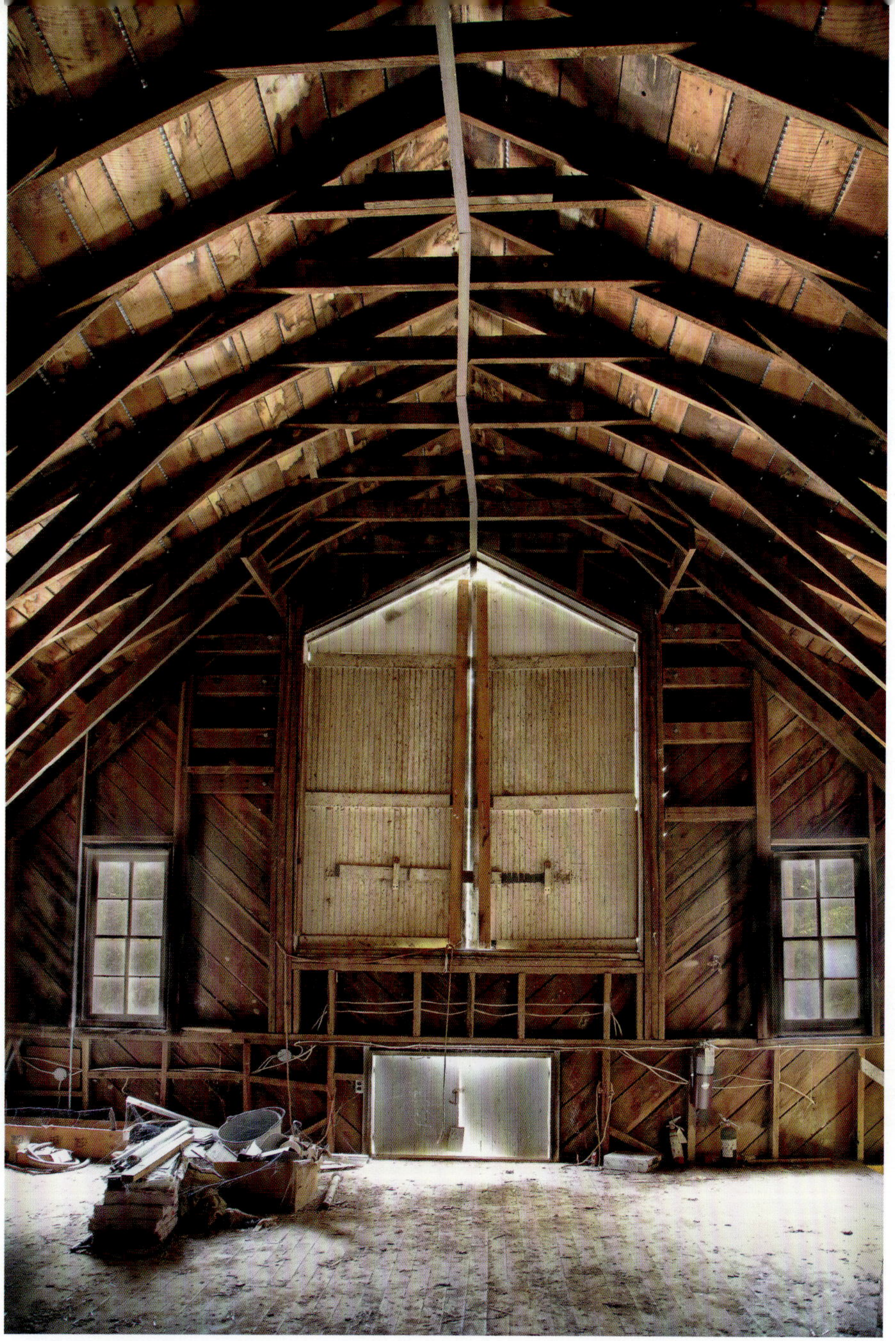

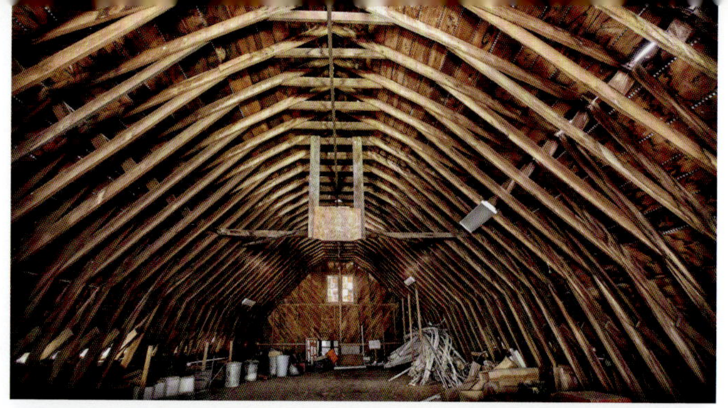

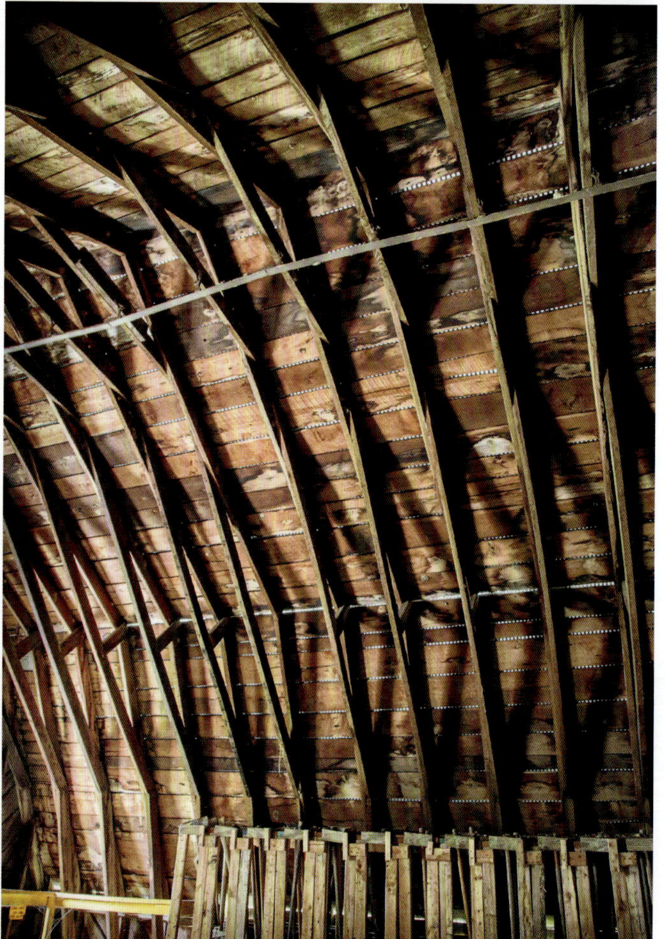

Clay County.

The Redbird Ranger Station includes on its property a structure that was once a horse barn, built circa 1927 by Oscar Bowling, who lived nearby at Big Creek. While the construction foreman did not specify whether he built the barn from plans, others associated with its construction made the assertion that the barn was inspired by the Sears, Roebuck and Company catalogue's pattern for the "Honor Bilt" barn.

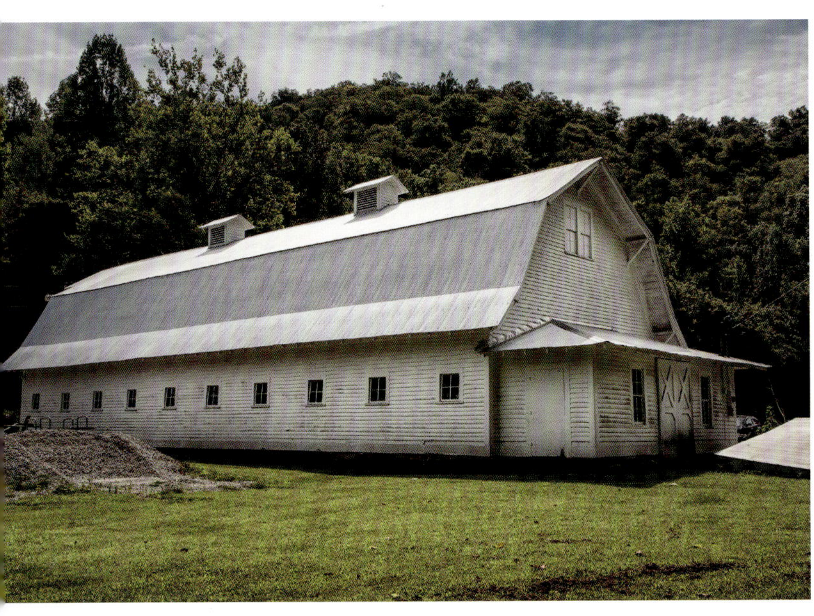

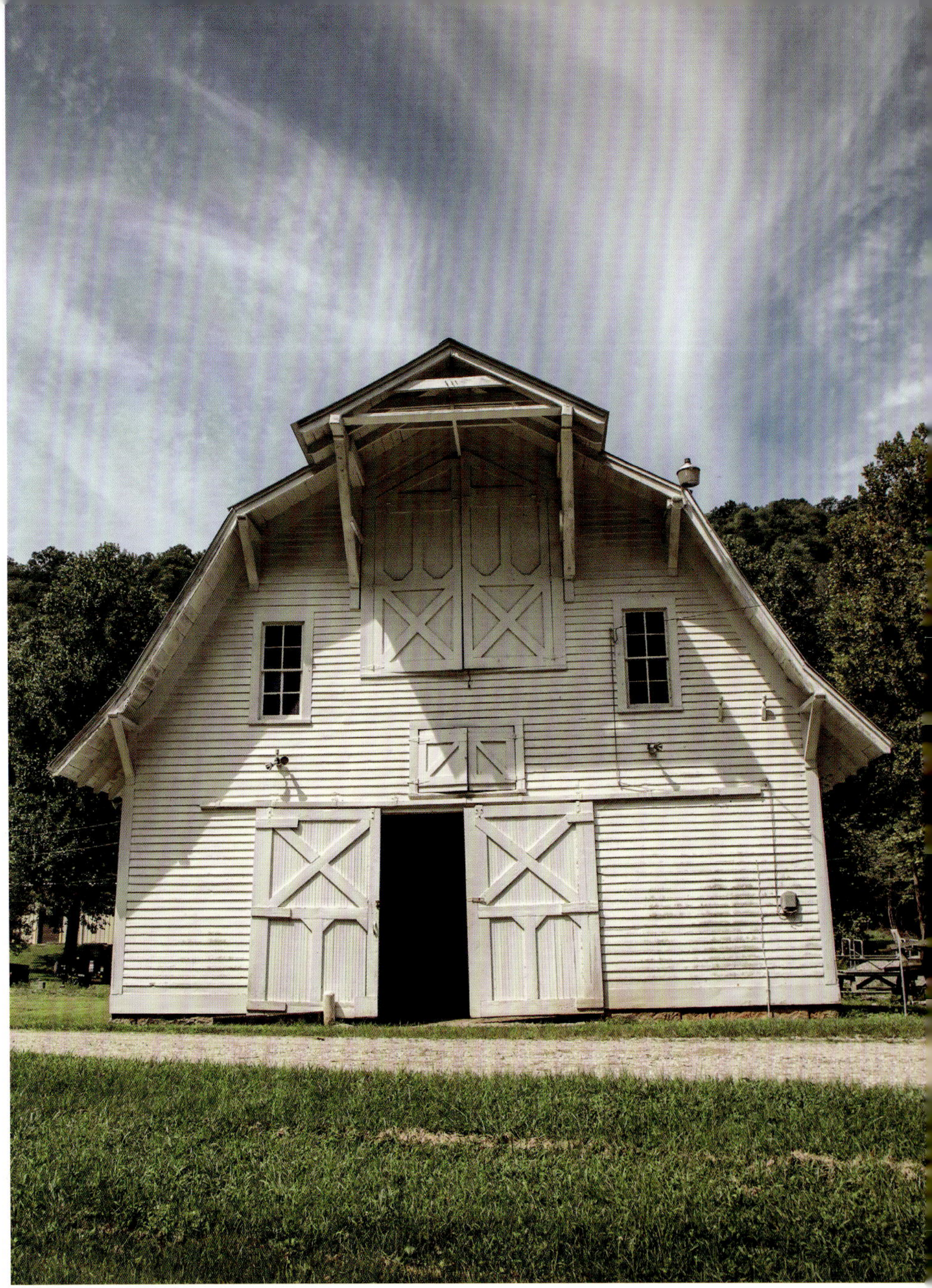

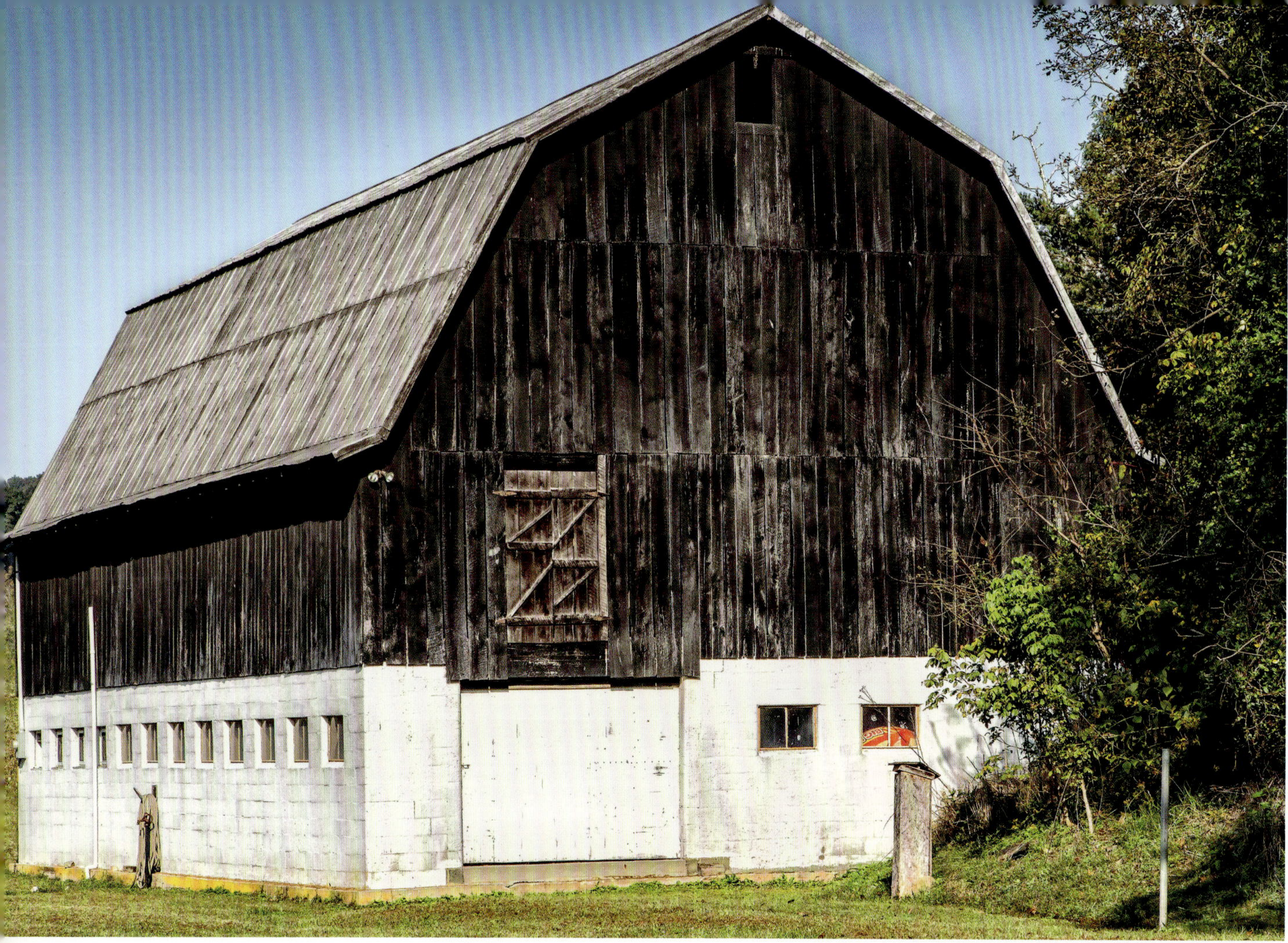

Woodford County.

Mintwood Farm was established in the mid-1800s and bought by J. L. Cleveland in 1917 to be used as a thoroughbred farm. This barn was built by the Clevelands in the early twentieth century and is used as a livestock barn.

Lawrence County.

The dairy barn in this image is designed with a block first level for easy and sanitary cleaning and a wooden hayloft above.

158

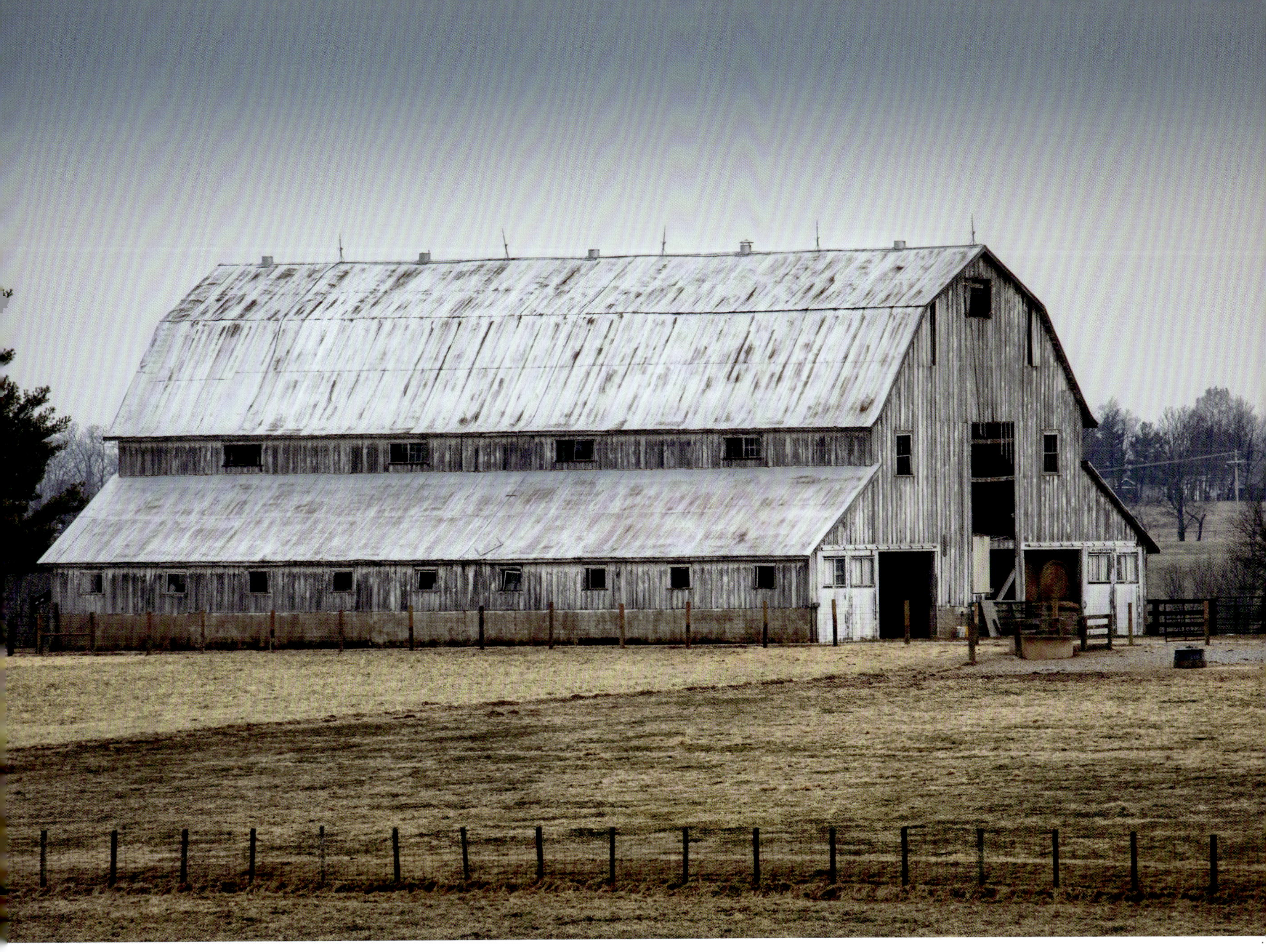

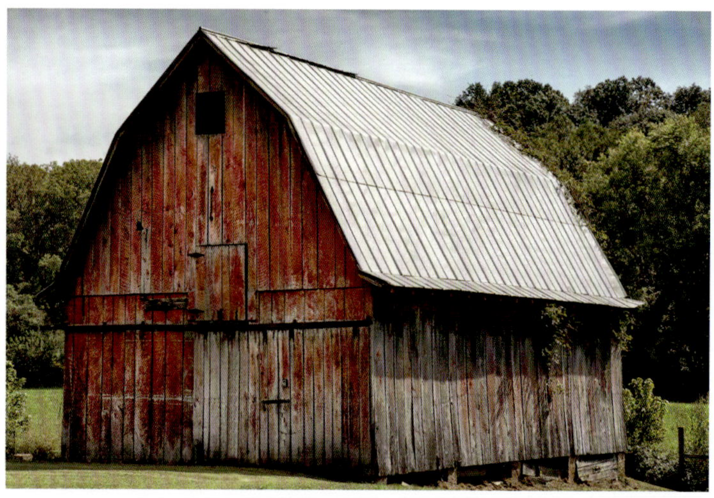

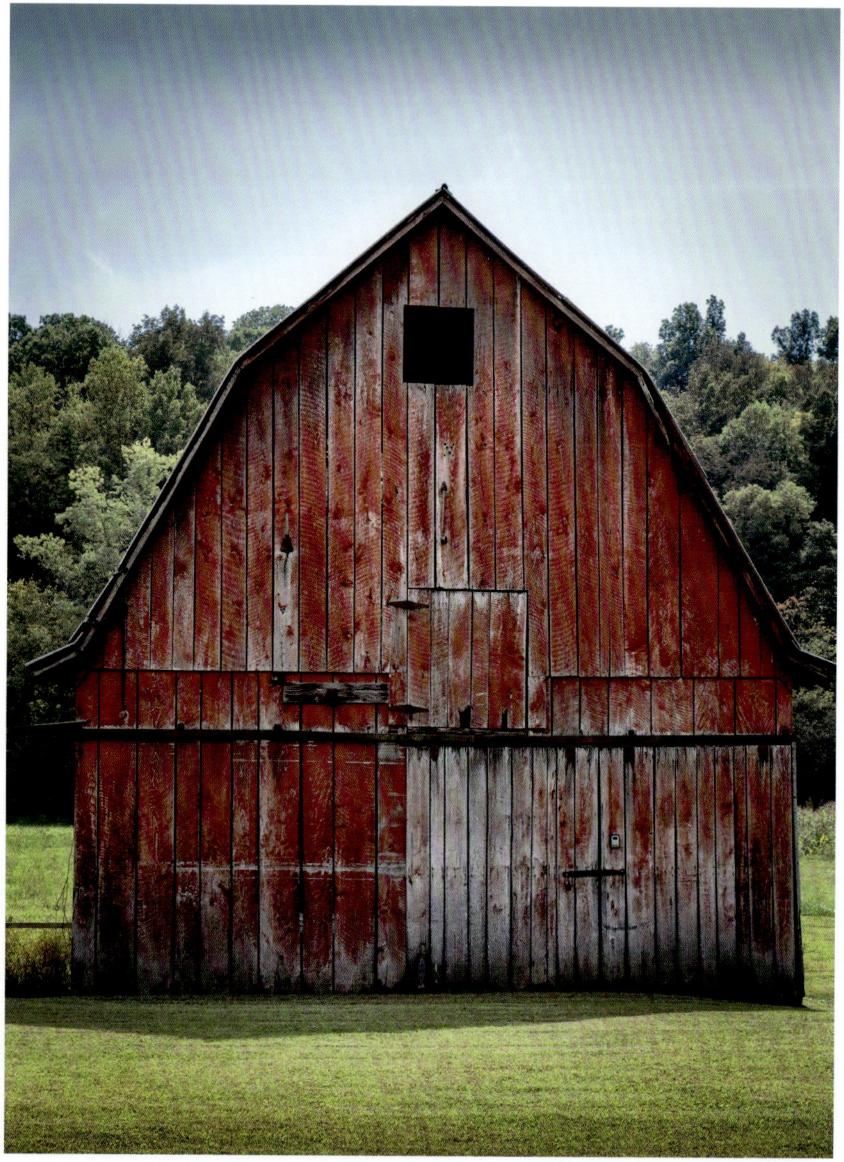

Lawrence County.

These roadside barns have a common design for the area, with a gambrel roof and off-center entrance at the gable end of the barn. The prevalence of the design suggests either the same barn builder or a purchased design.

FACING PAGE

Elliott County.

This double quilt barn is unique in its vernacular building method. Six sections of board in varying lengths compose the siding visible at this angle.

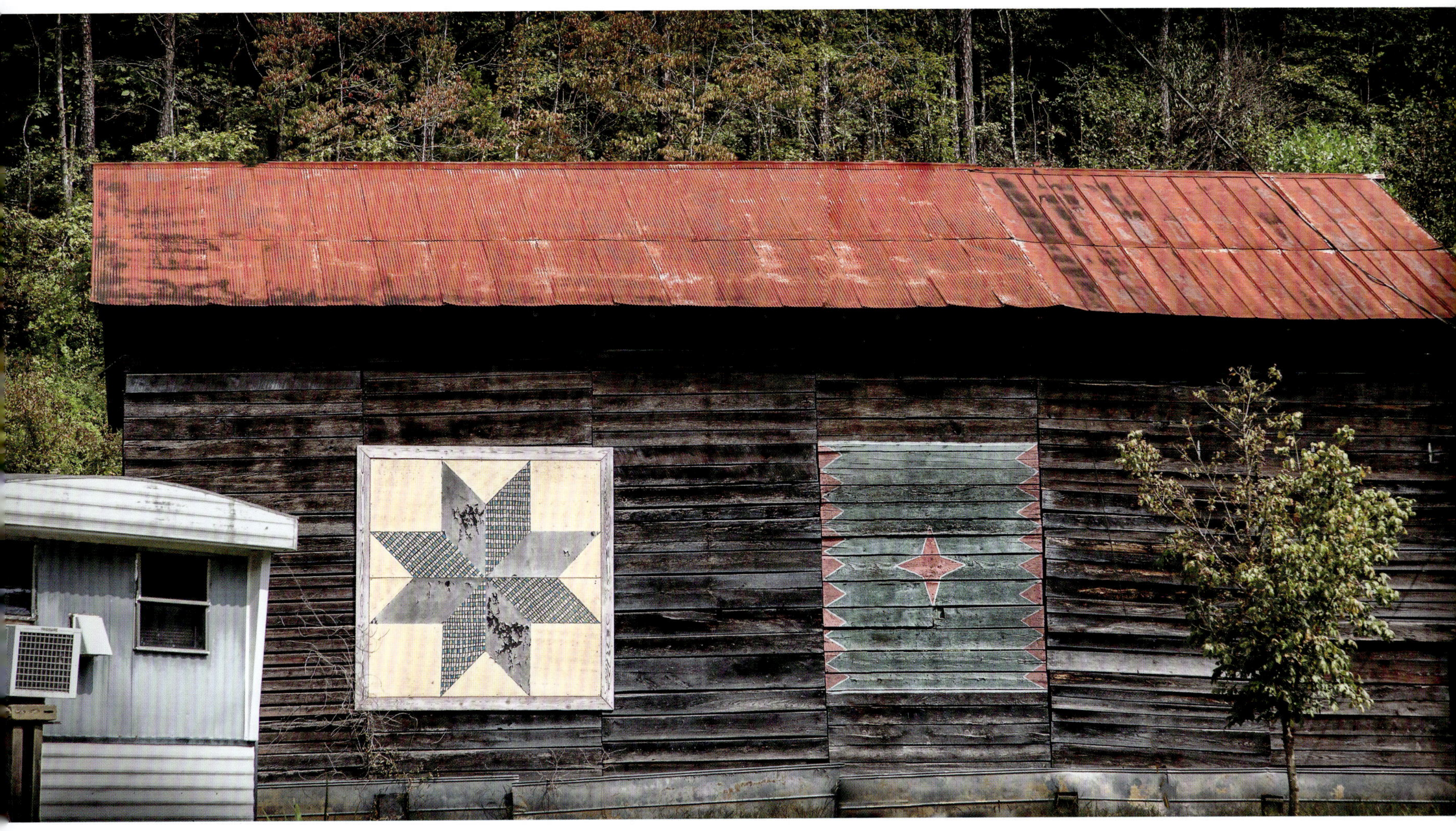

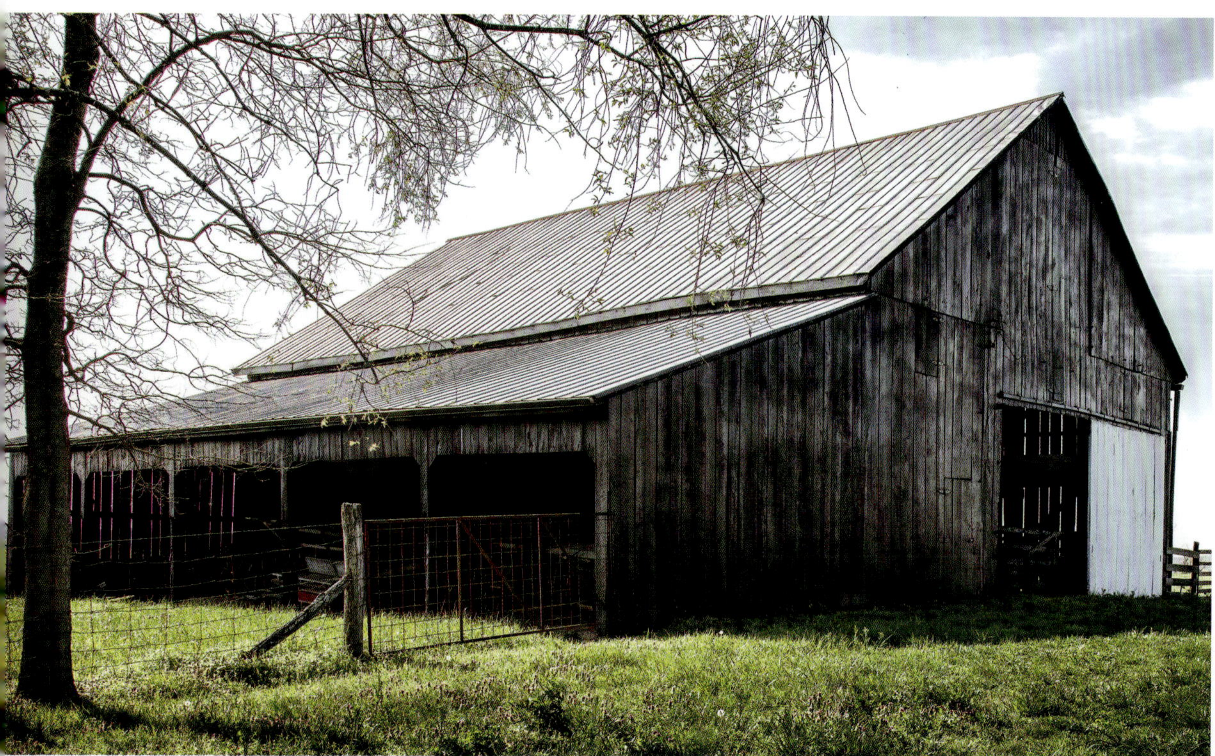

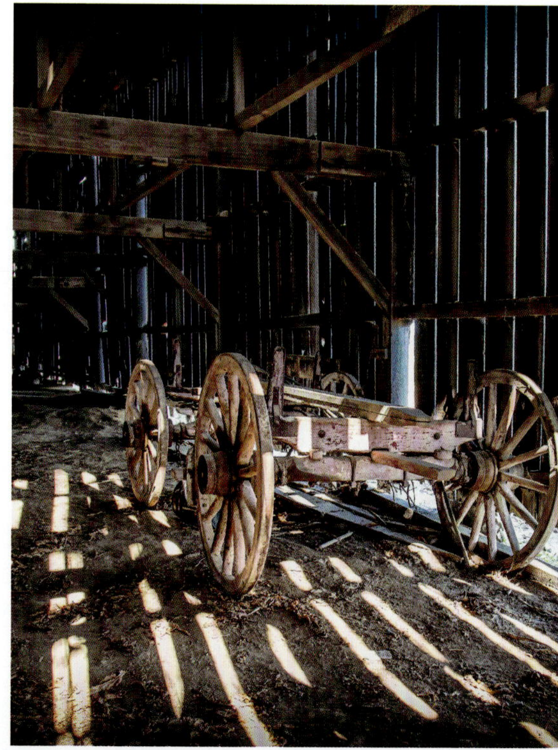

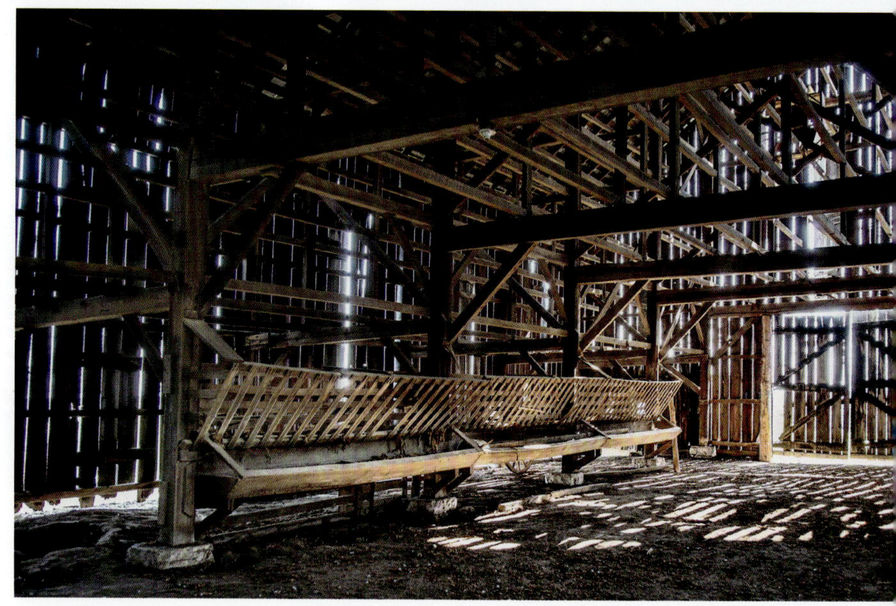

Clark County.

The Venable homestead features a wide selection of barns, a house, and multiple outbuildings. These images show a stock barn built around 1929; the interior of a tobacco barn built around 1948 along with the beam construction, hayracks, and a buggy stored in the aisle; a mule barn built around 1850; and a twentieth-century general use barn.

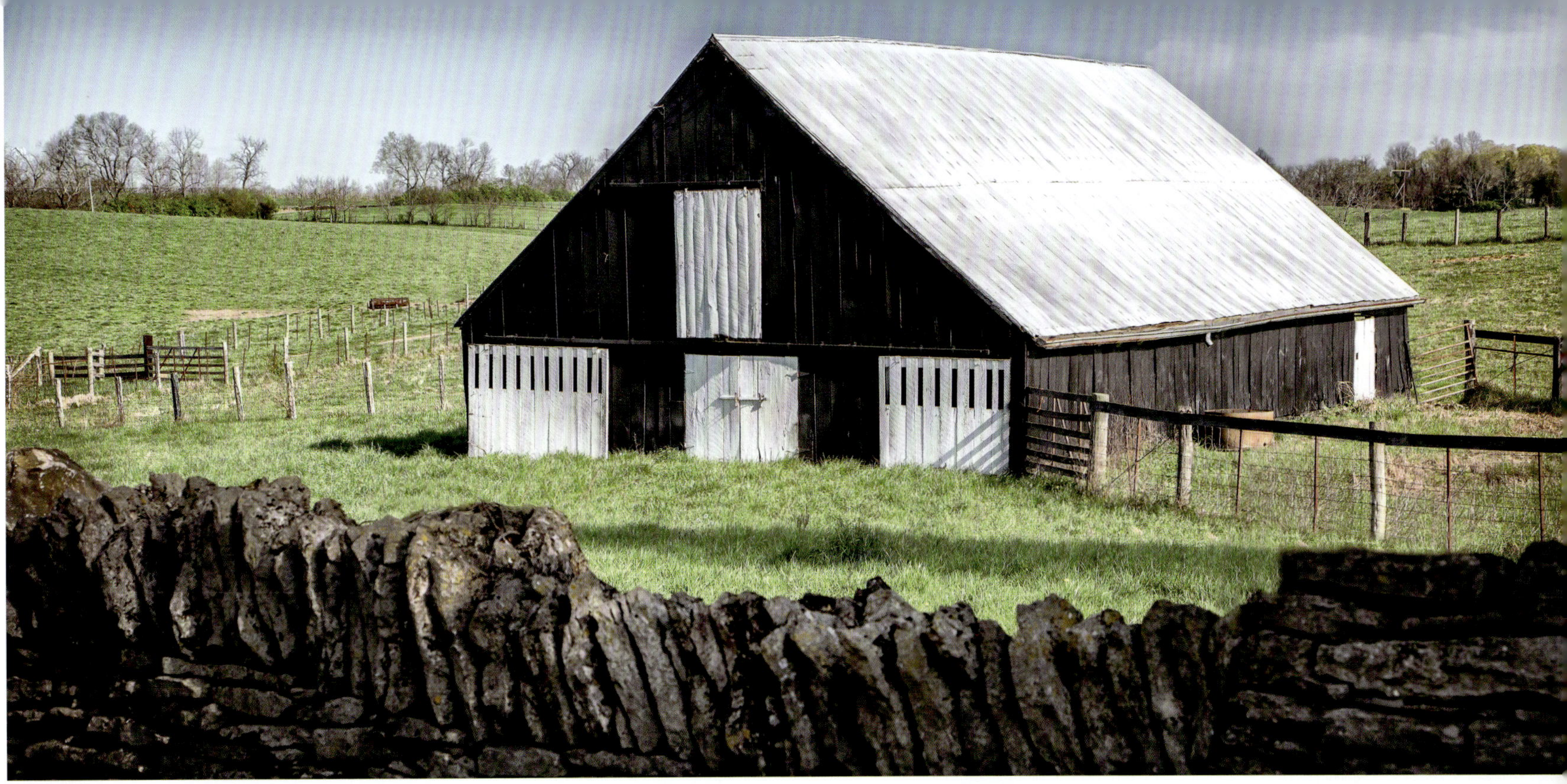

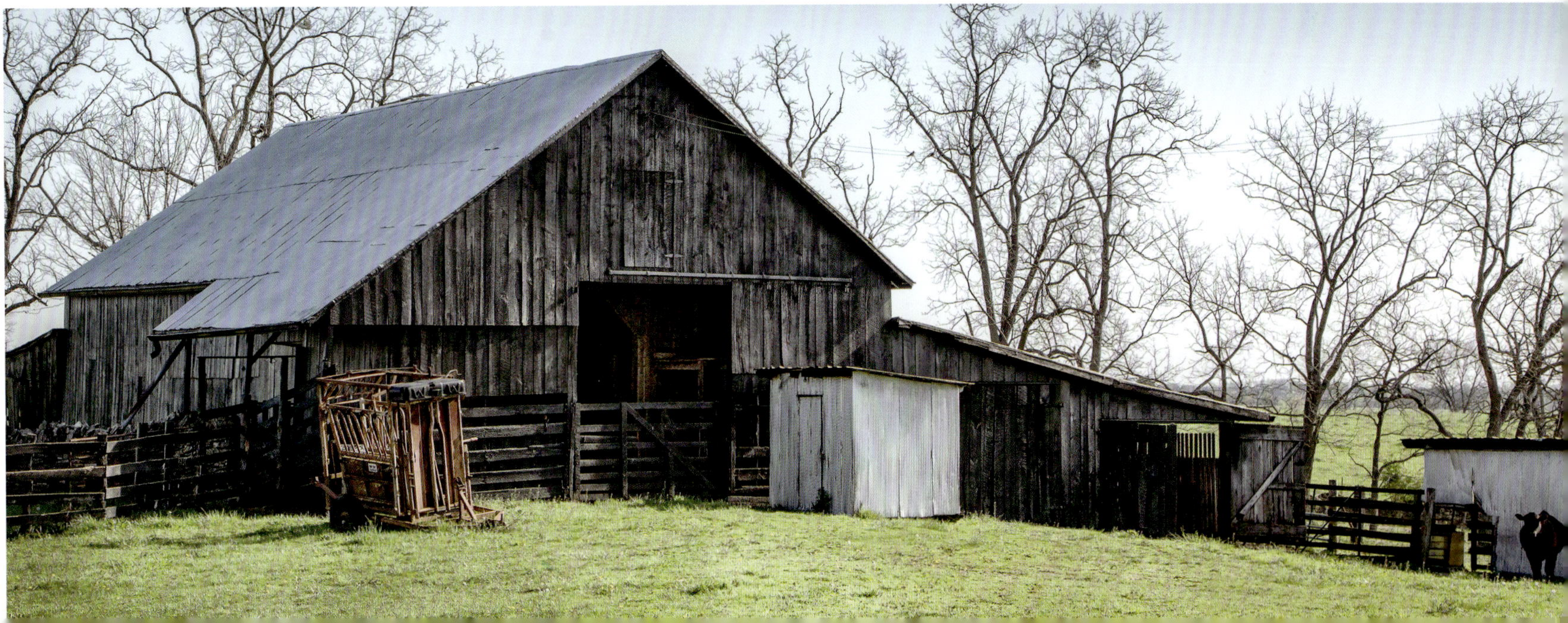

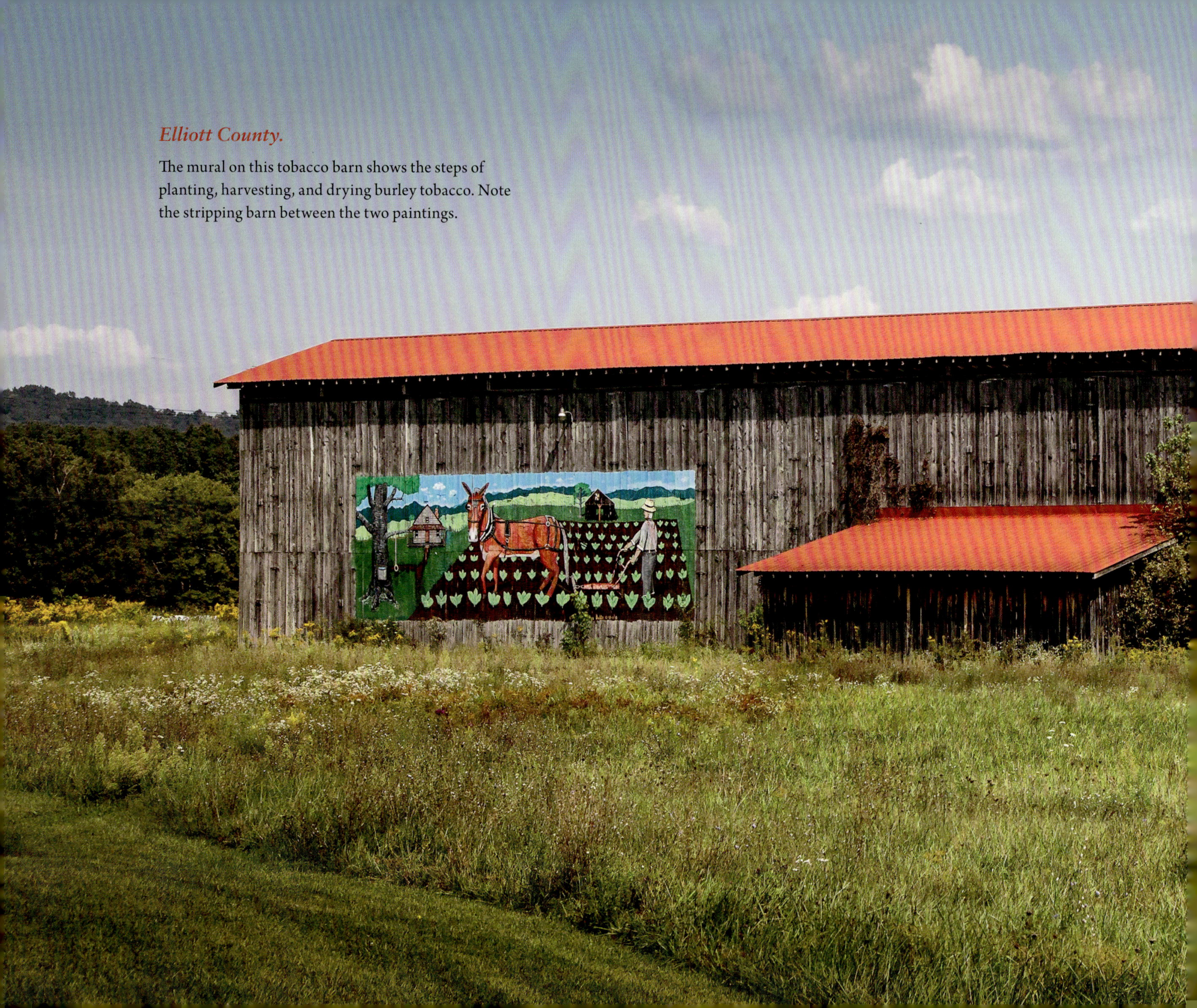

Elliott County.

The mural on this tobacco barn shows the steps of
planting, harvesting, and drying burley tobacco. Note
the stripping barn between the two paintings.

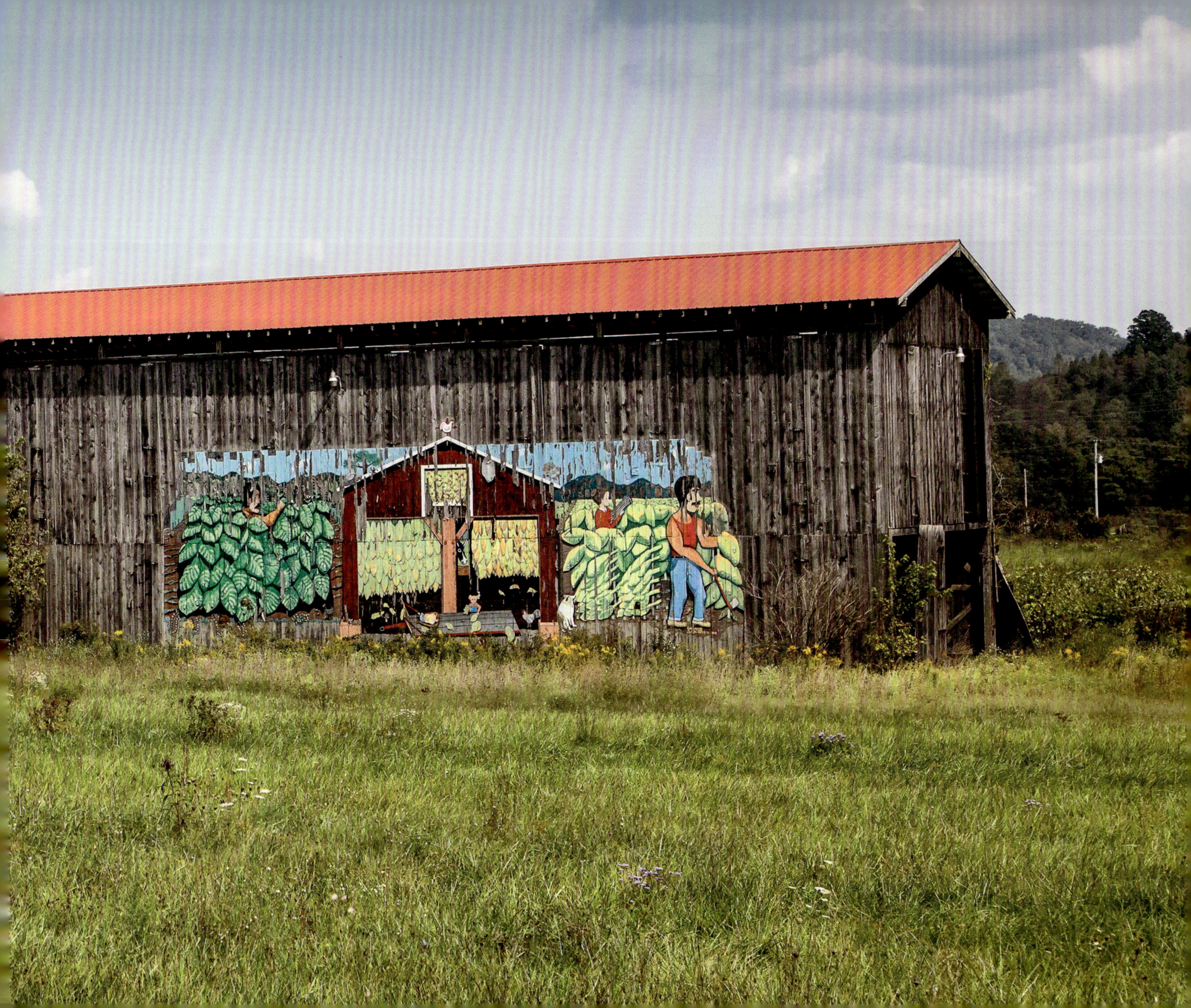

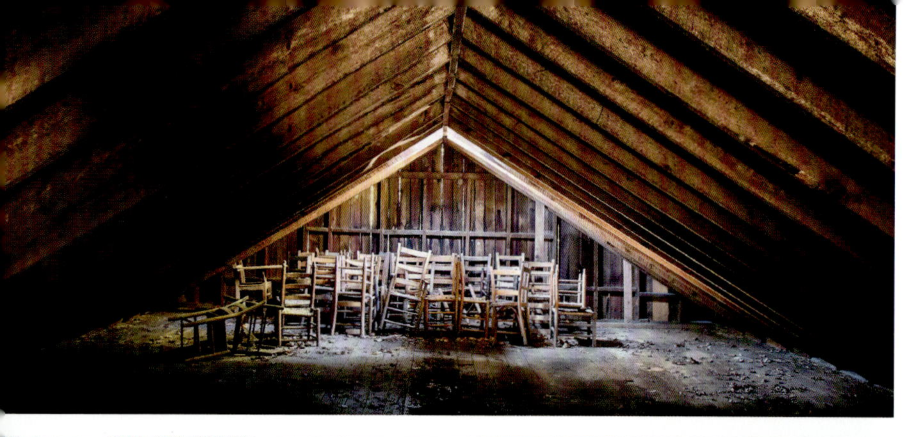

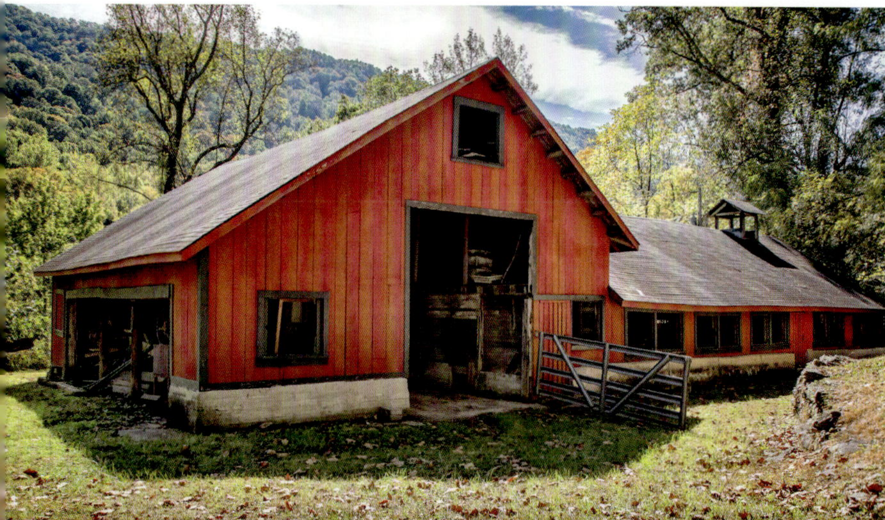

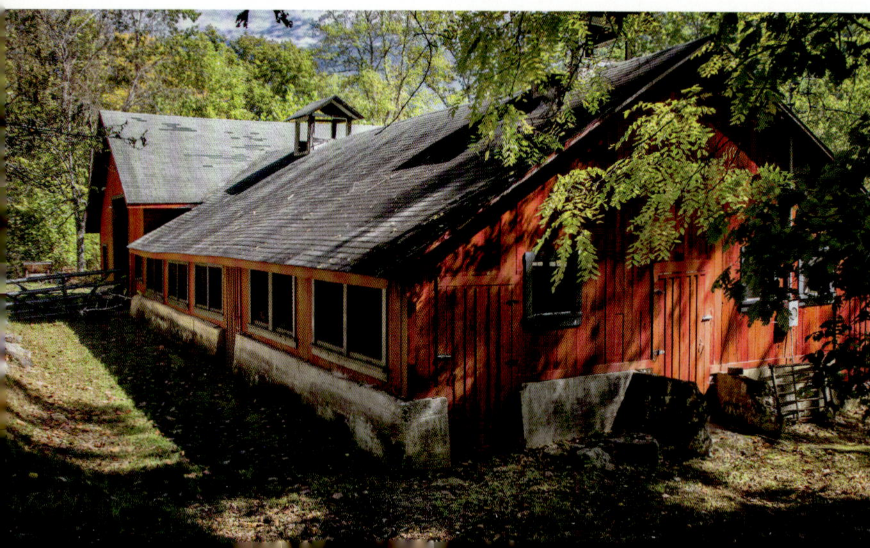

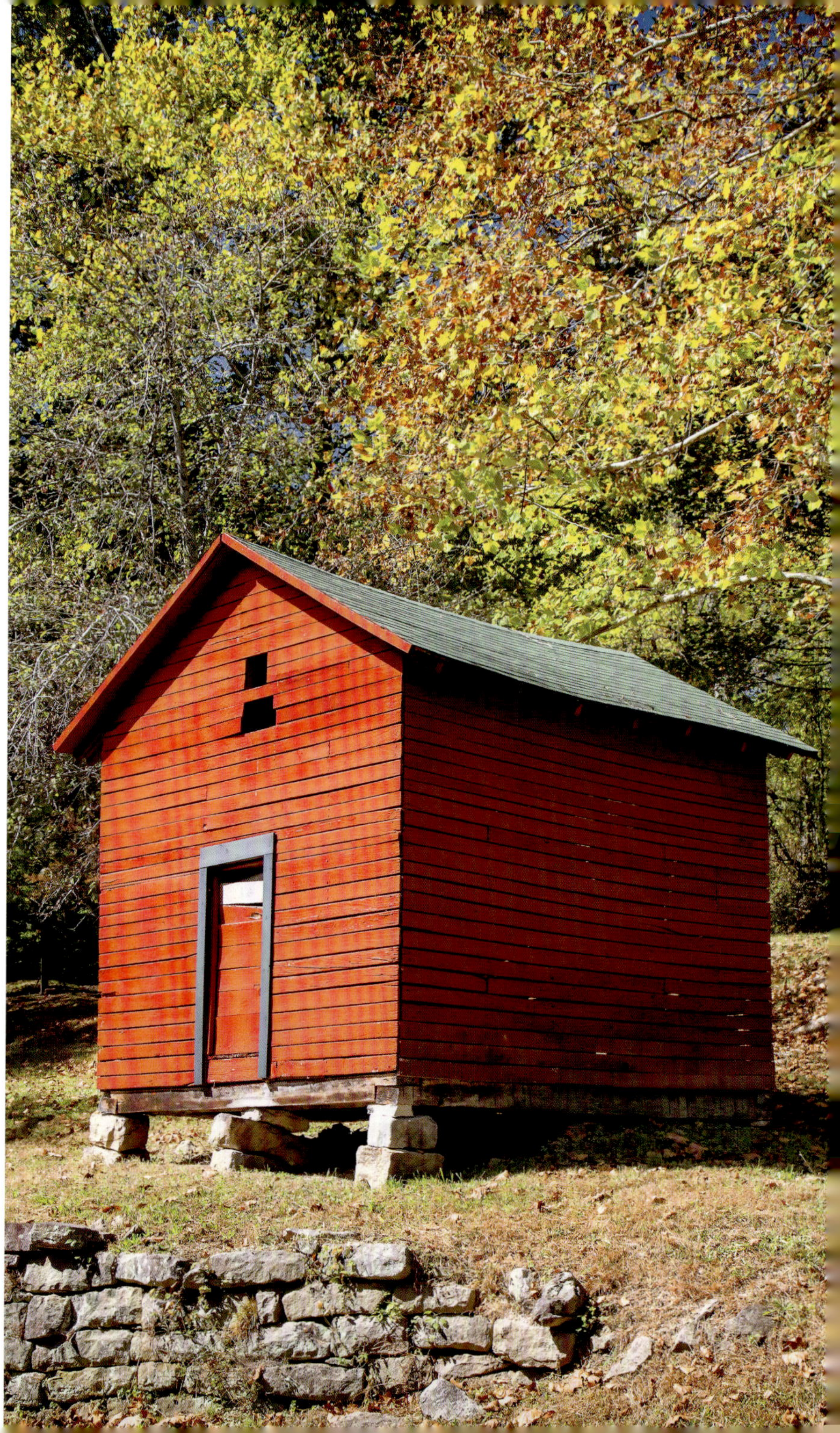

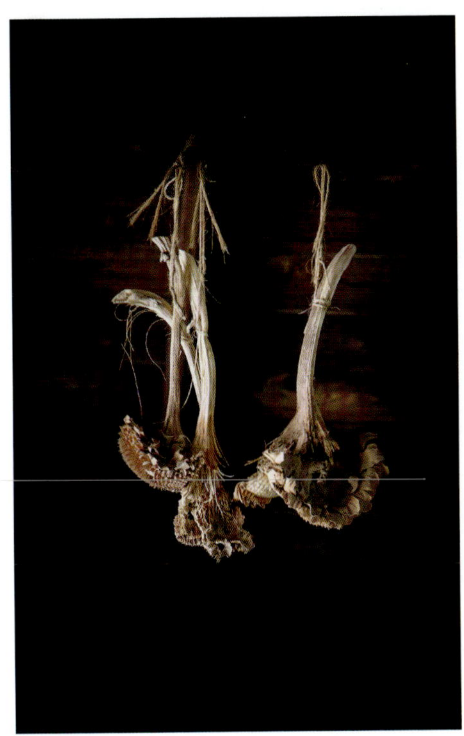 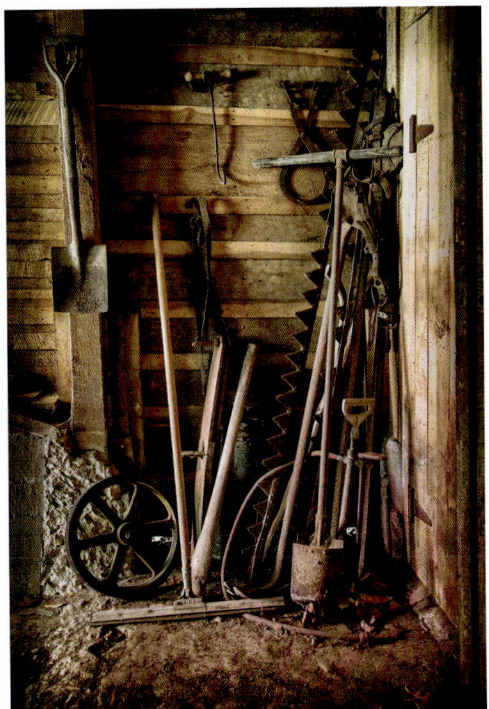 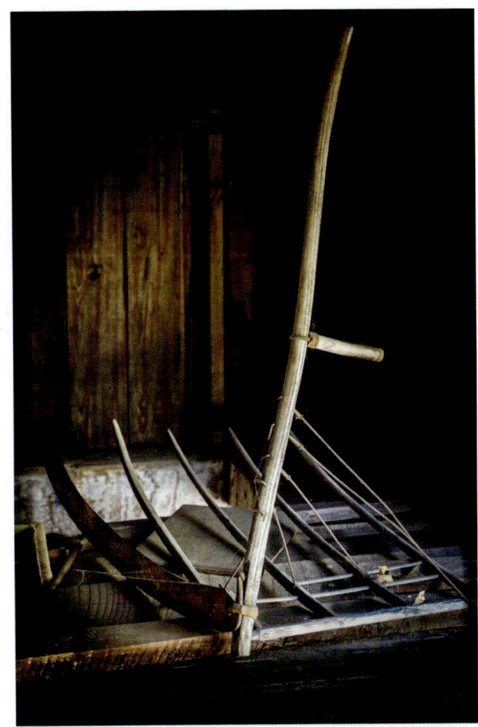

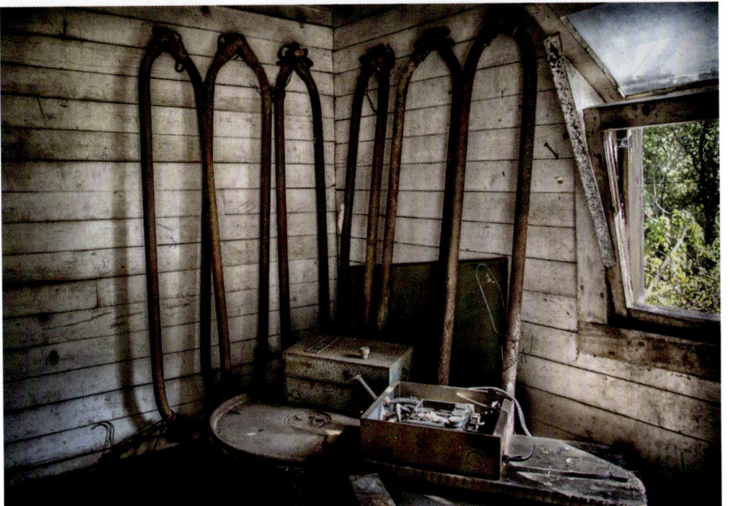

Harlan County.

The Pine Mountain Settlement School was founded in 1913 as a
school for children in Kentucky's remote southeastern mountain
communities. In 1915 Ethel DeLong started the construction of
the barn, which was designed with milk stalls for twelve cows,
four metal pens, three stalls for horses, a clothes-changing room,
a harness room, and two feed rooms. The settlement is still used
for educational purposes today. Old farm implements remain in
the barn, and a crib barn, now used for drying herbs, sits nearby.

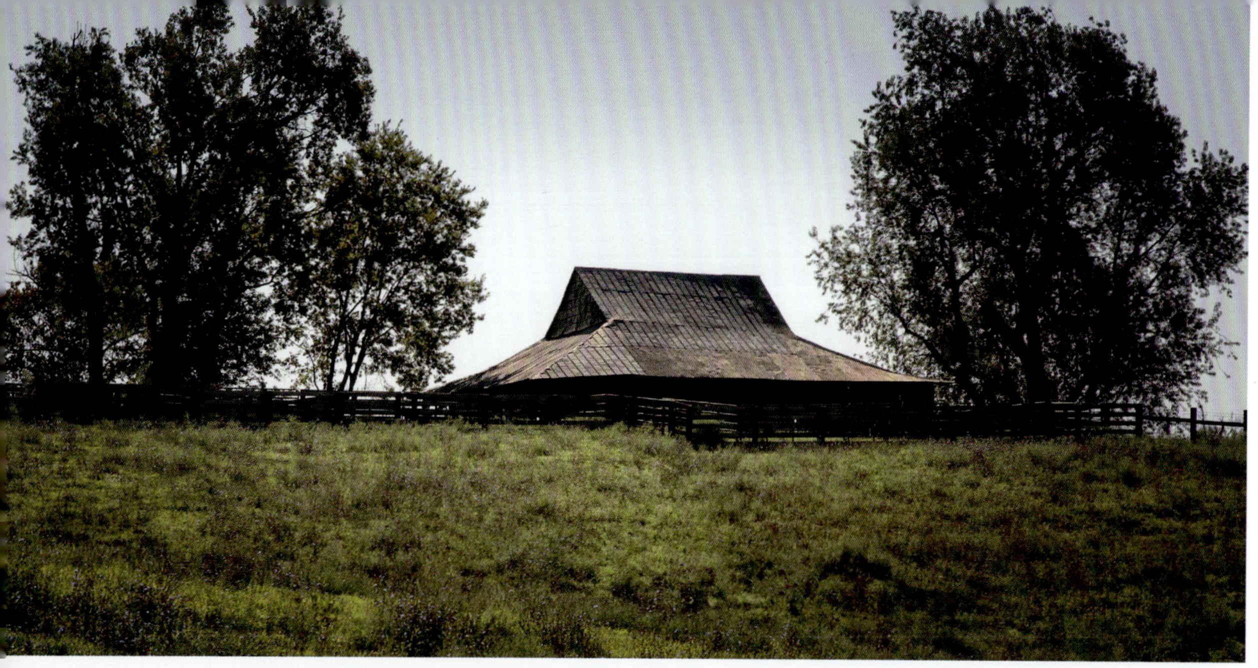

Madison County.

The design of the roof on this barn, known as a "gable on hip" barn, was popular as a tobacco barn roof in the nineteenth century.

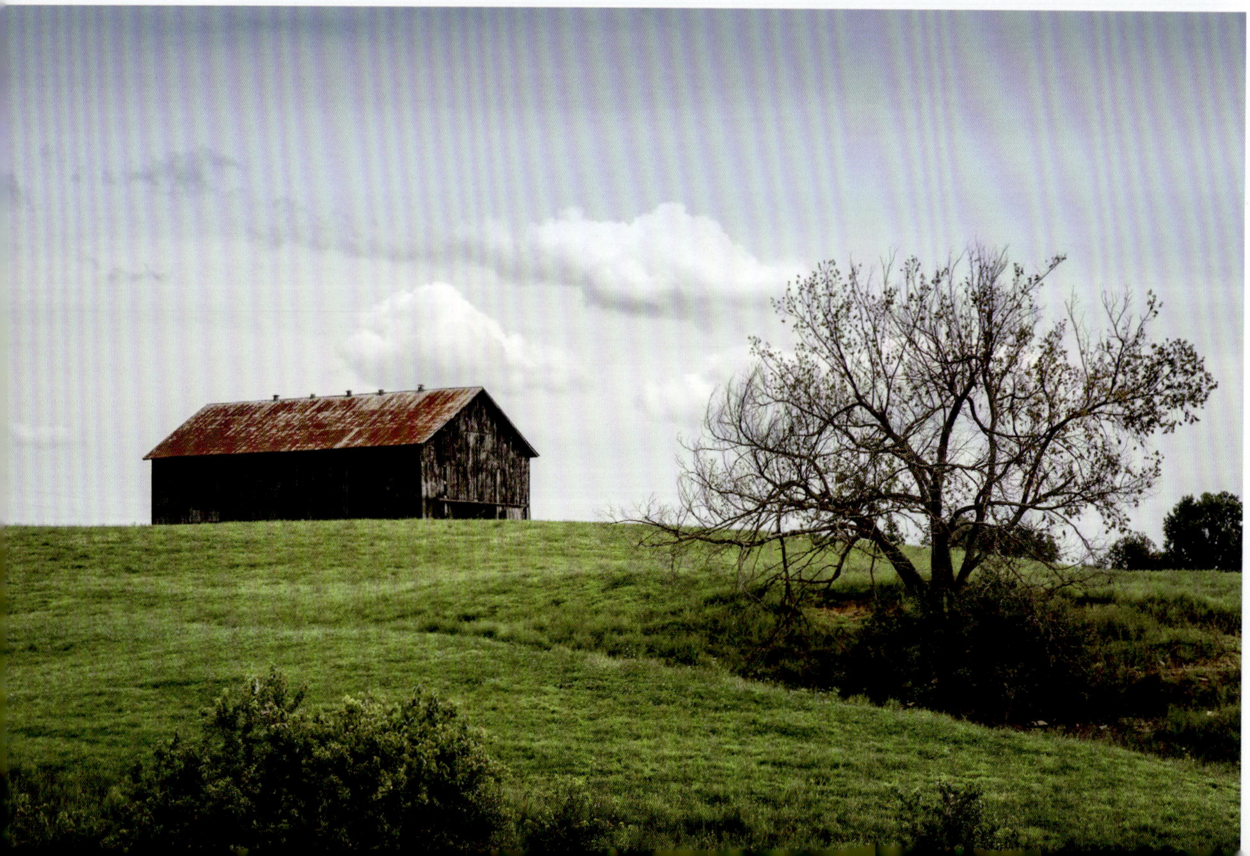

Madison County.

A barn overlooking rolling farmland was once a common sight. As farmland disappears, barns are more often found surrounded by urban development.

FACING PAGE

Rockcastle County.

This carriage barn now has the driveway curve past it.

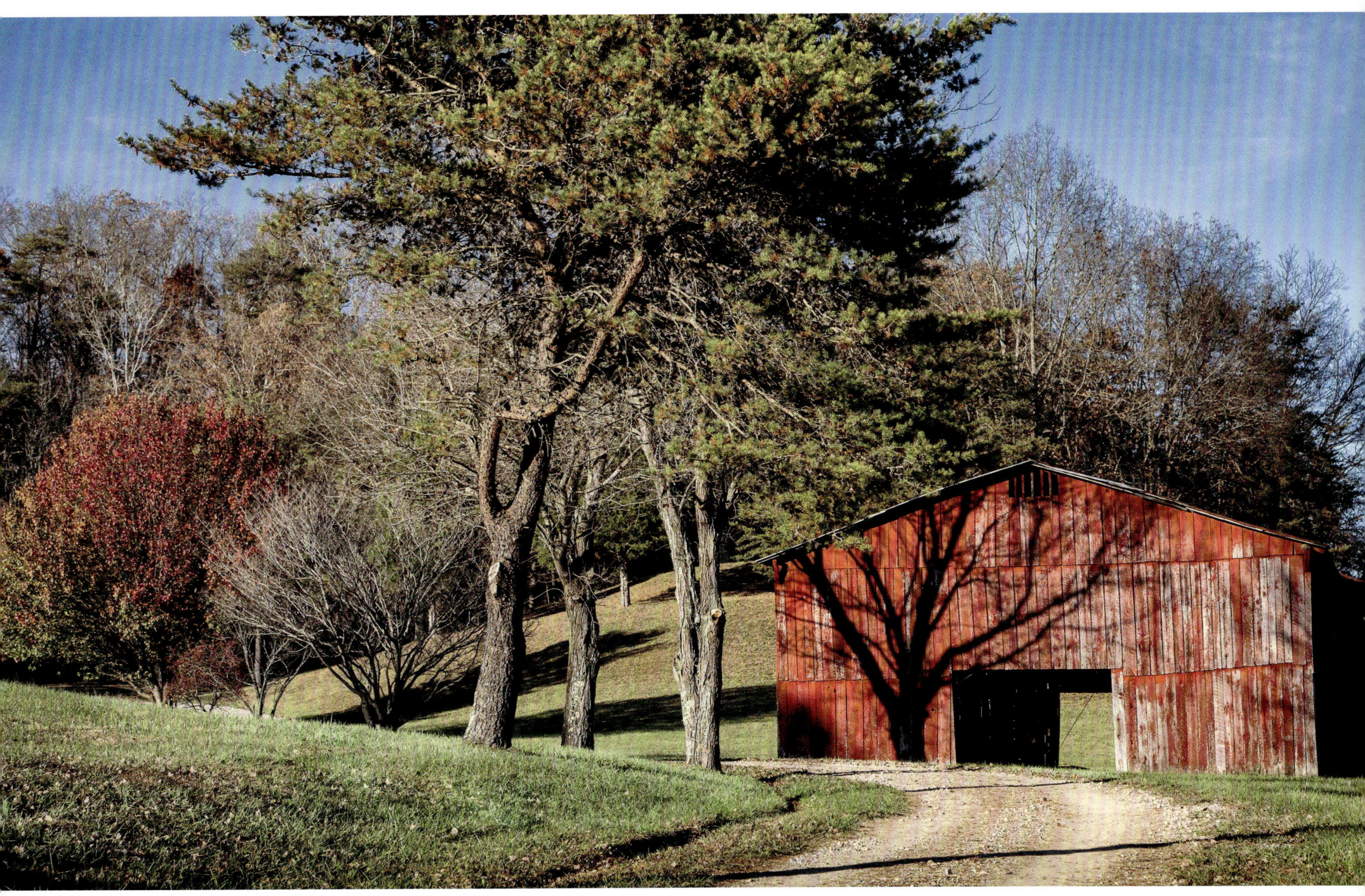

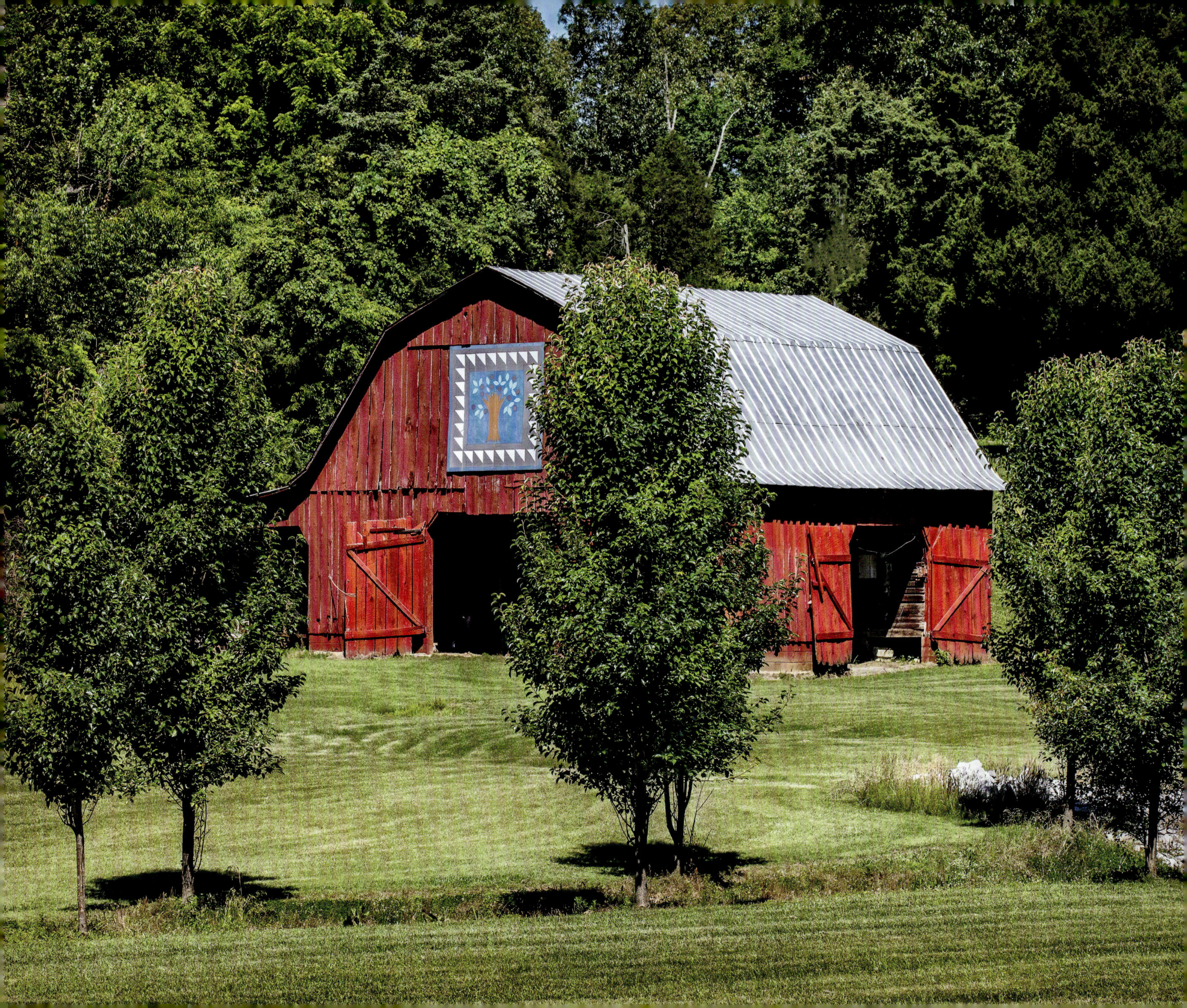

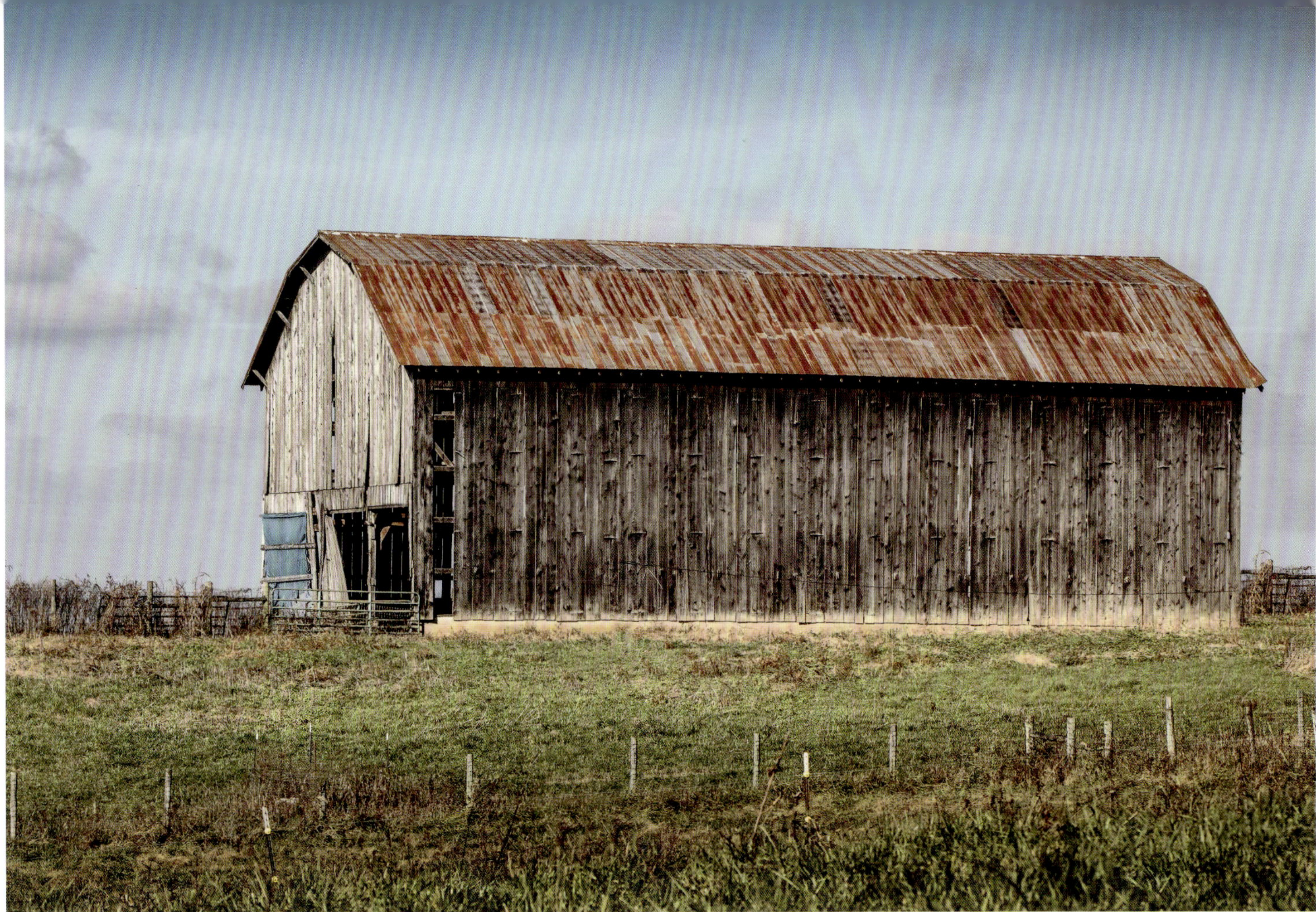

Livingston County.

The quilt peg barn pictured here has four pens with openings in the sides and ends. Although on property established over three hundred years ago, the Hill Springs Farm barn was built in the twentieth century.

Madison County.

This barn, currently on Berea College farm property, has a roof style common in the county, purportedly built by a Mr. Venable, who was an active local barn builder.

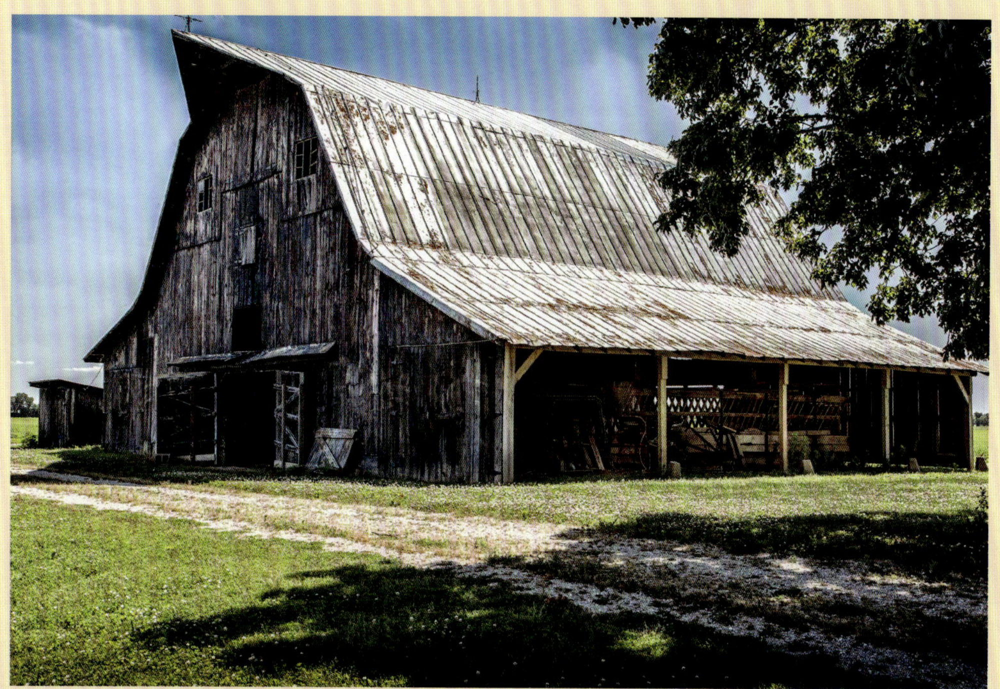

Daviess County.

The Sauer farm was established around 1894. Five generations have maintained the property over time. There are two barns close to the homestead house—this hay hood stock barn and a tobacco barn in the soybean field.

Christian County.

This burley tobacco barn sits in the middle of a cornfield, with logs keeping the side ventilation closed.

FACING PAGE

Madison County.

The cluster of outbuildings and barns on Silver Ridge Farm are typical of late nineteenth- and early twentieth-century farmsteads.

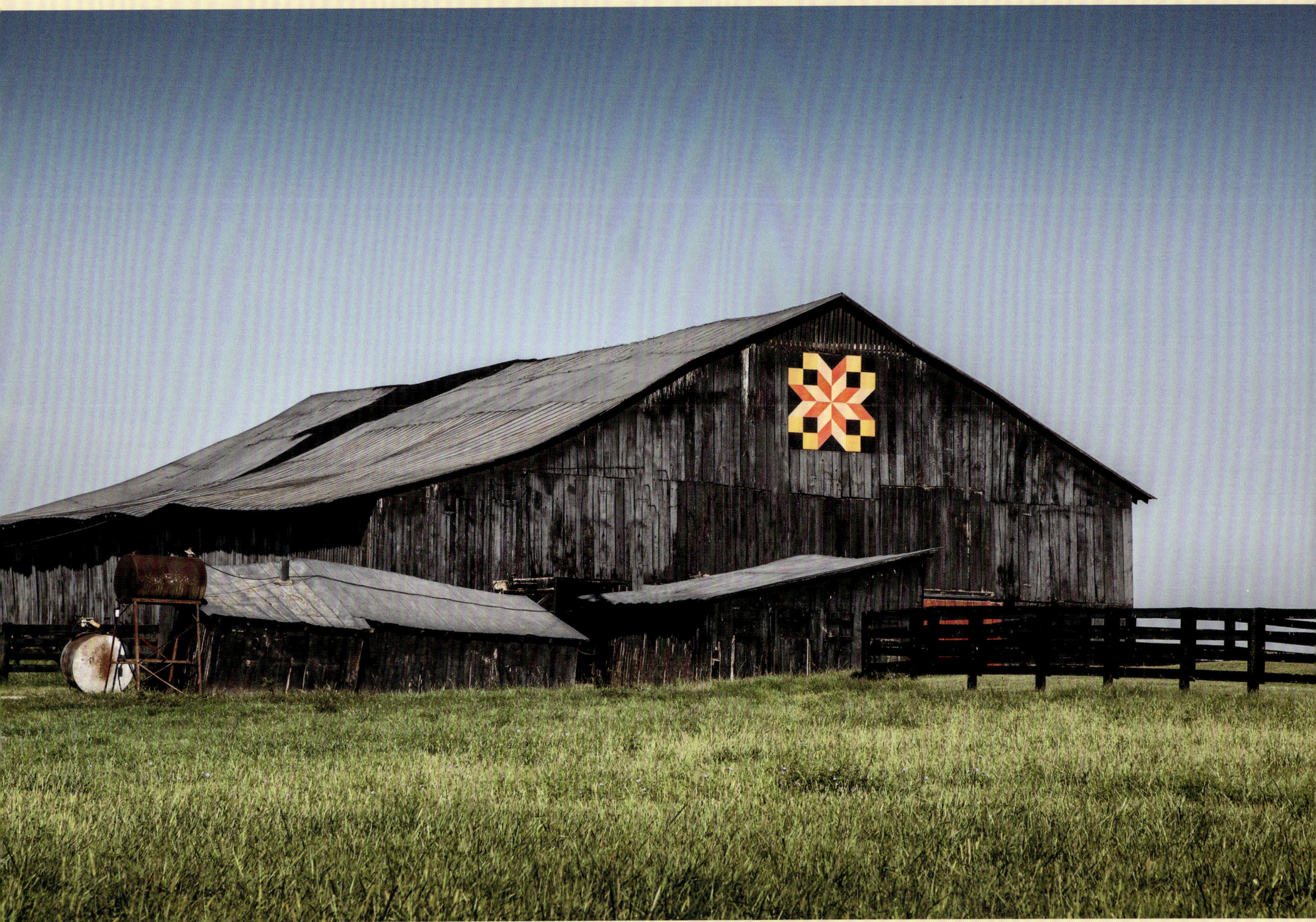

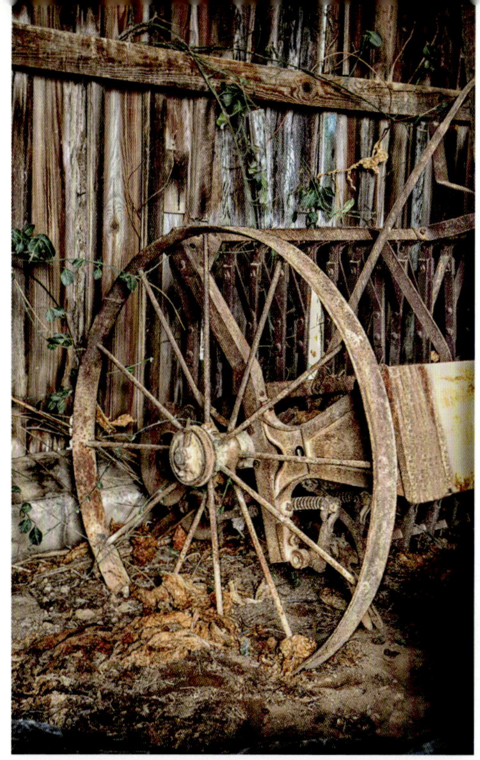
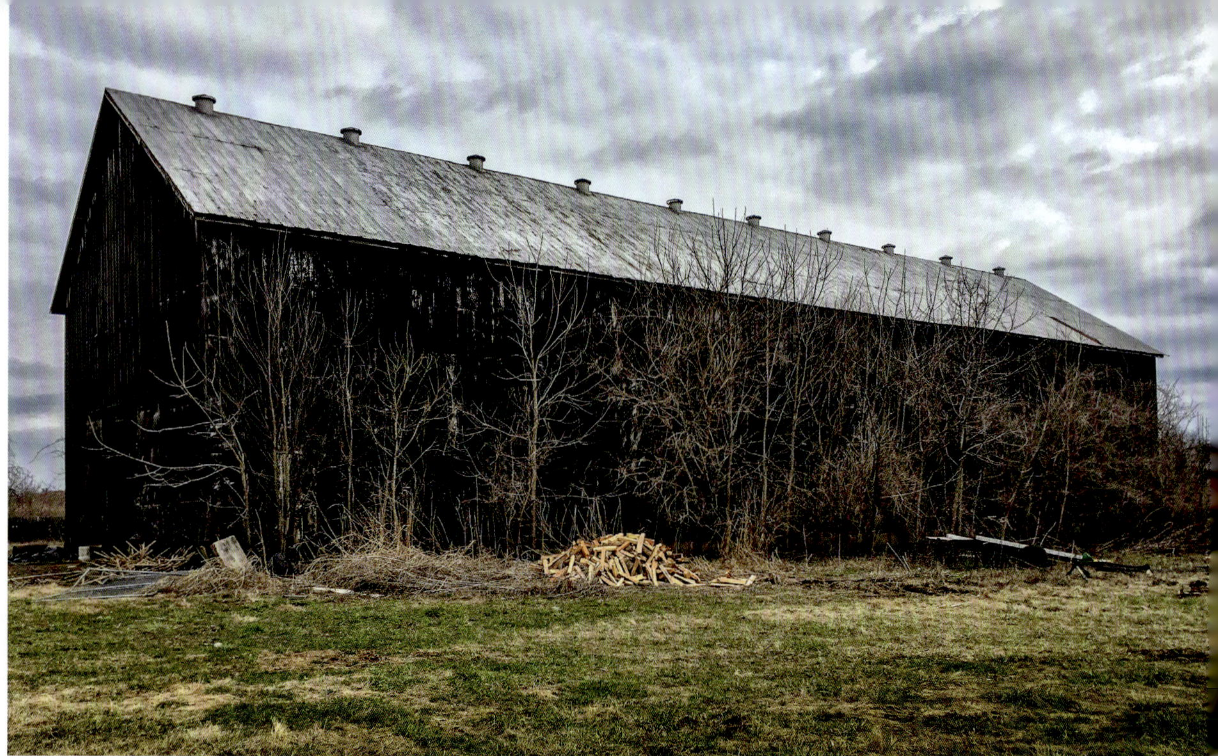
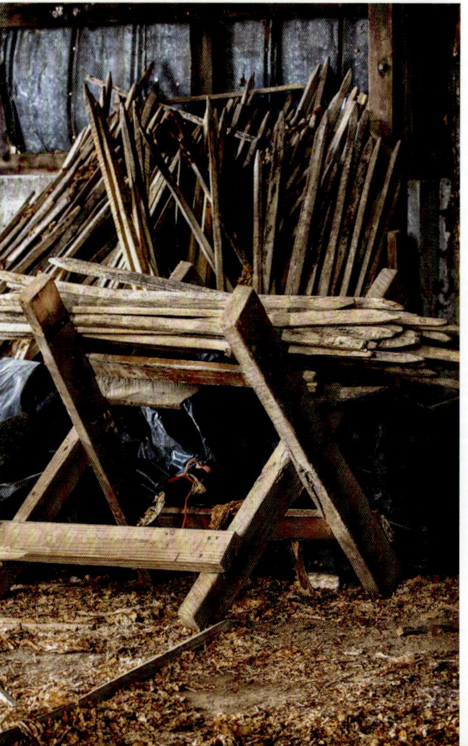
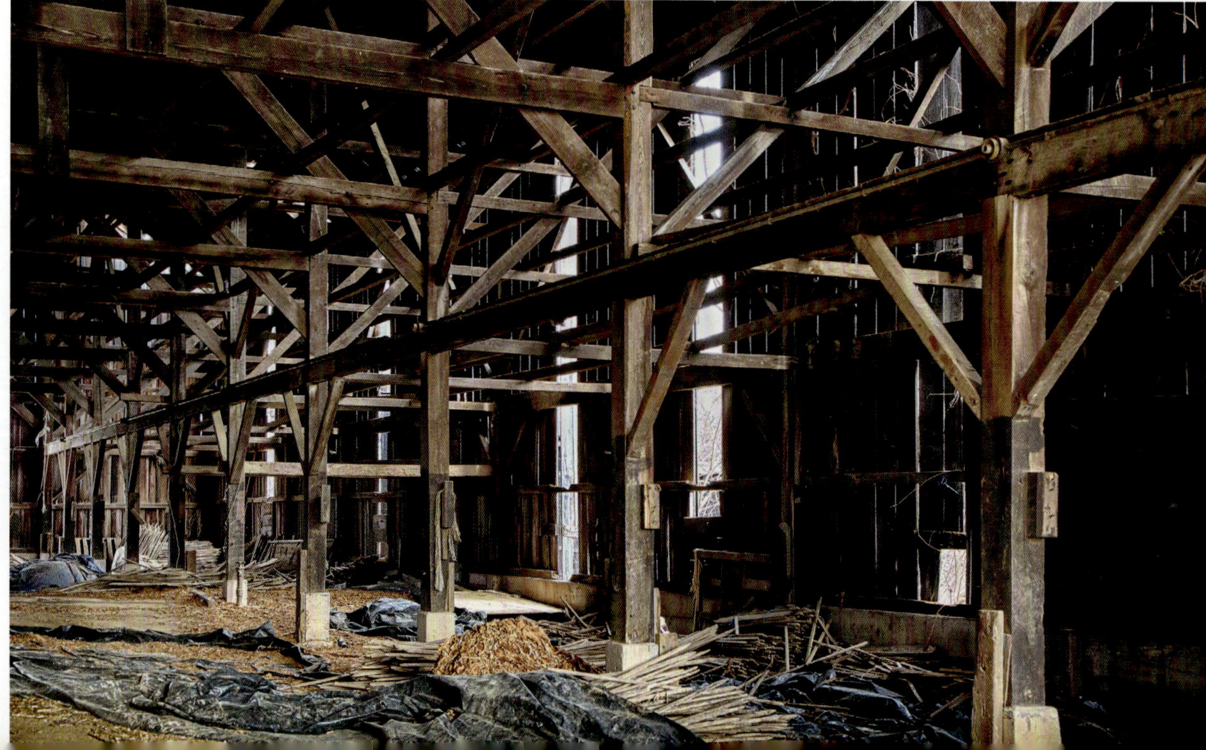

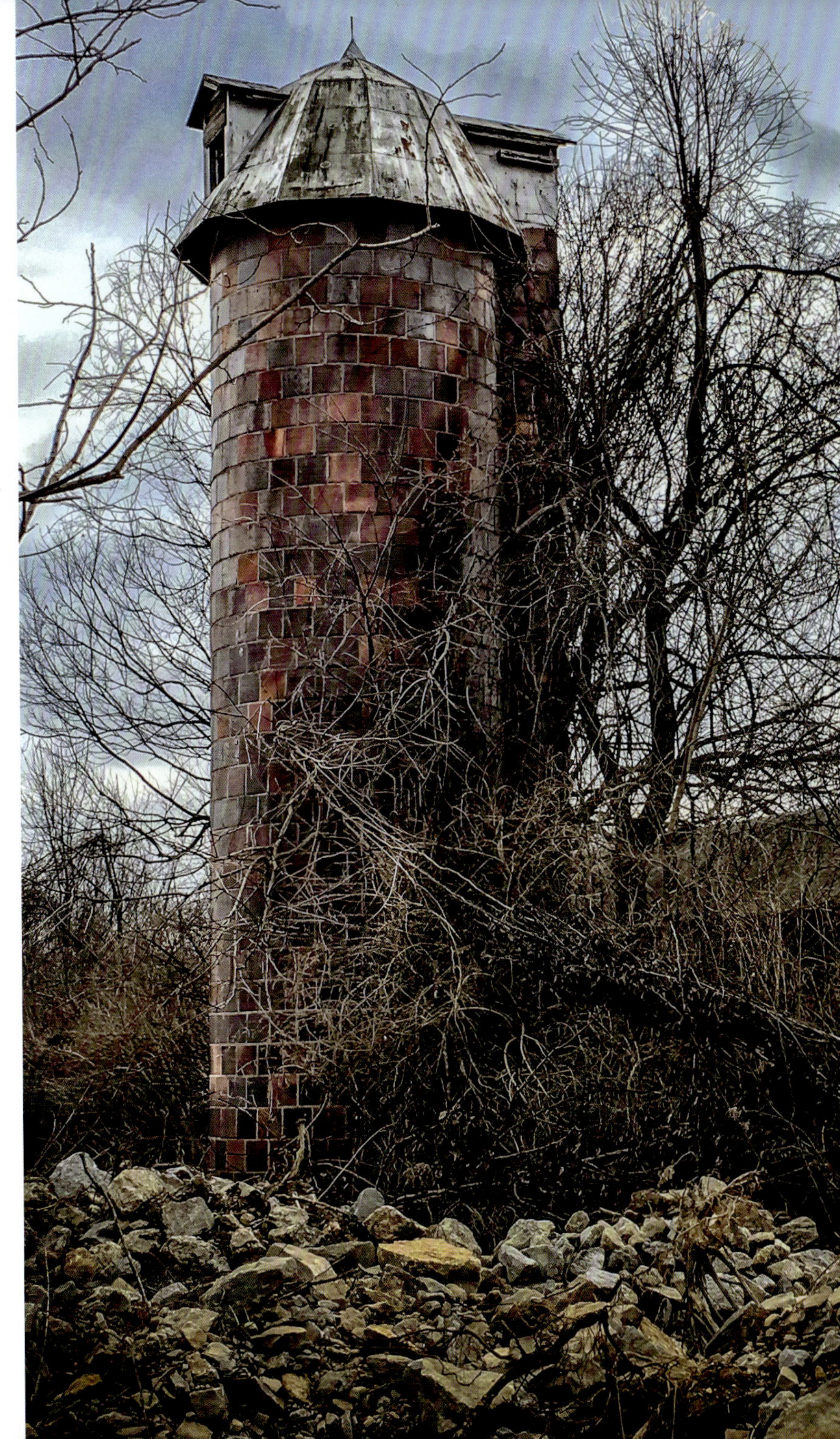

According to the cement footer inside this tobacco barn, the structure was built in 1930. No longer in use and scheduled for demolition, the tobacco barn was one of several barns on the property, with a tiled silo attached to a nearby storage building. The implements inside and the wide tobacco aisle are common for the era.

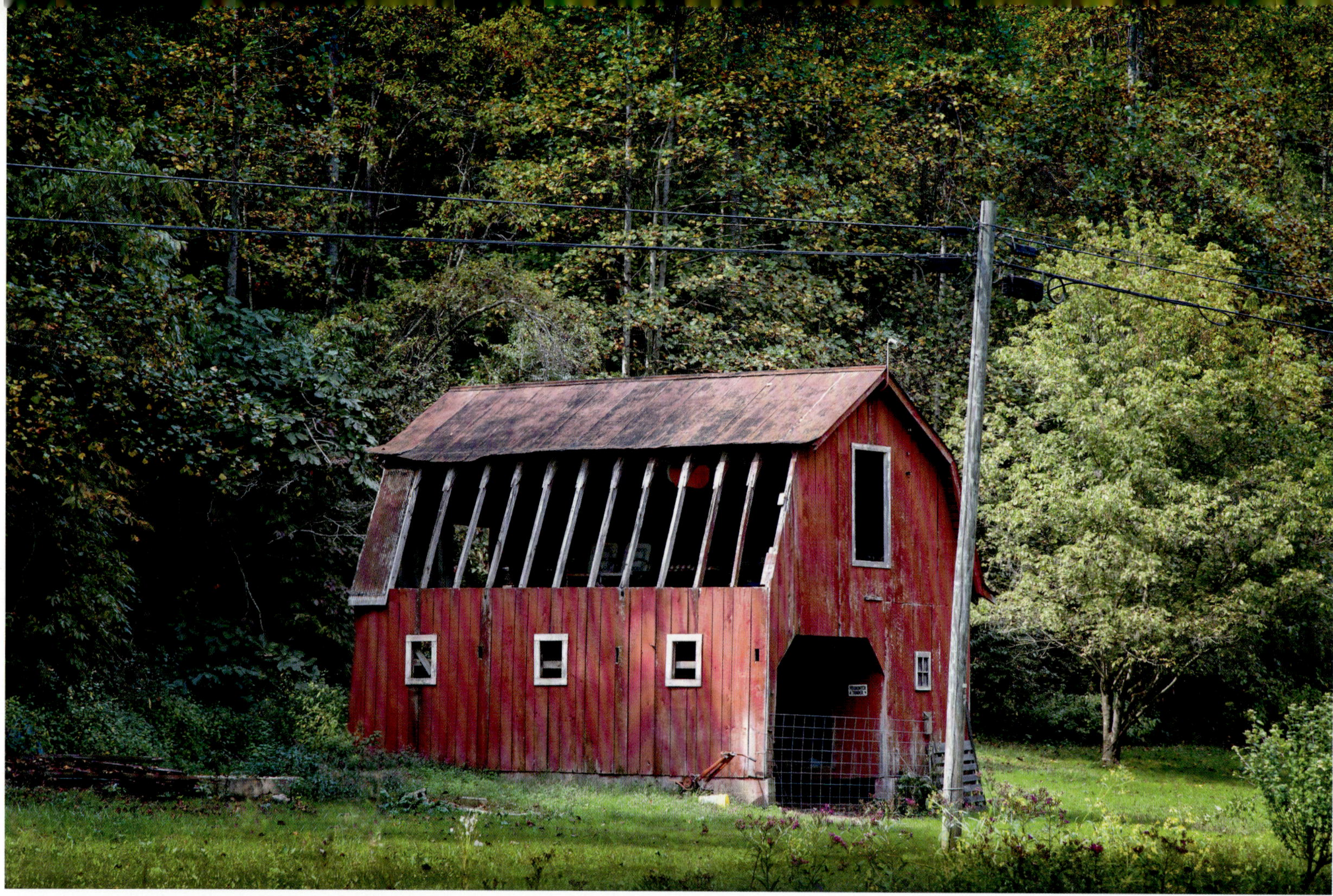

Lawrence County.

The roof on this barn has partially come off, giving the owners of
the barn an opportunity for unusual repurposing of the barn loft.
The addition of lawn chairs in the loft creates an
amusing porch for summer evenings.

FACING PAGE

Christian County.

Although originally intended for dark firing,
this barn is air-drying burley tobacco.

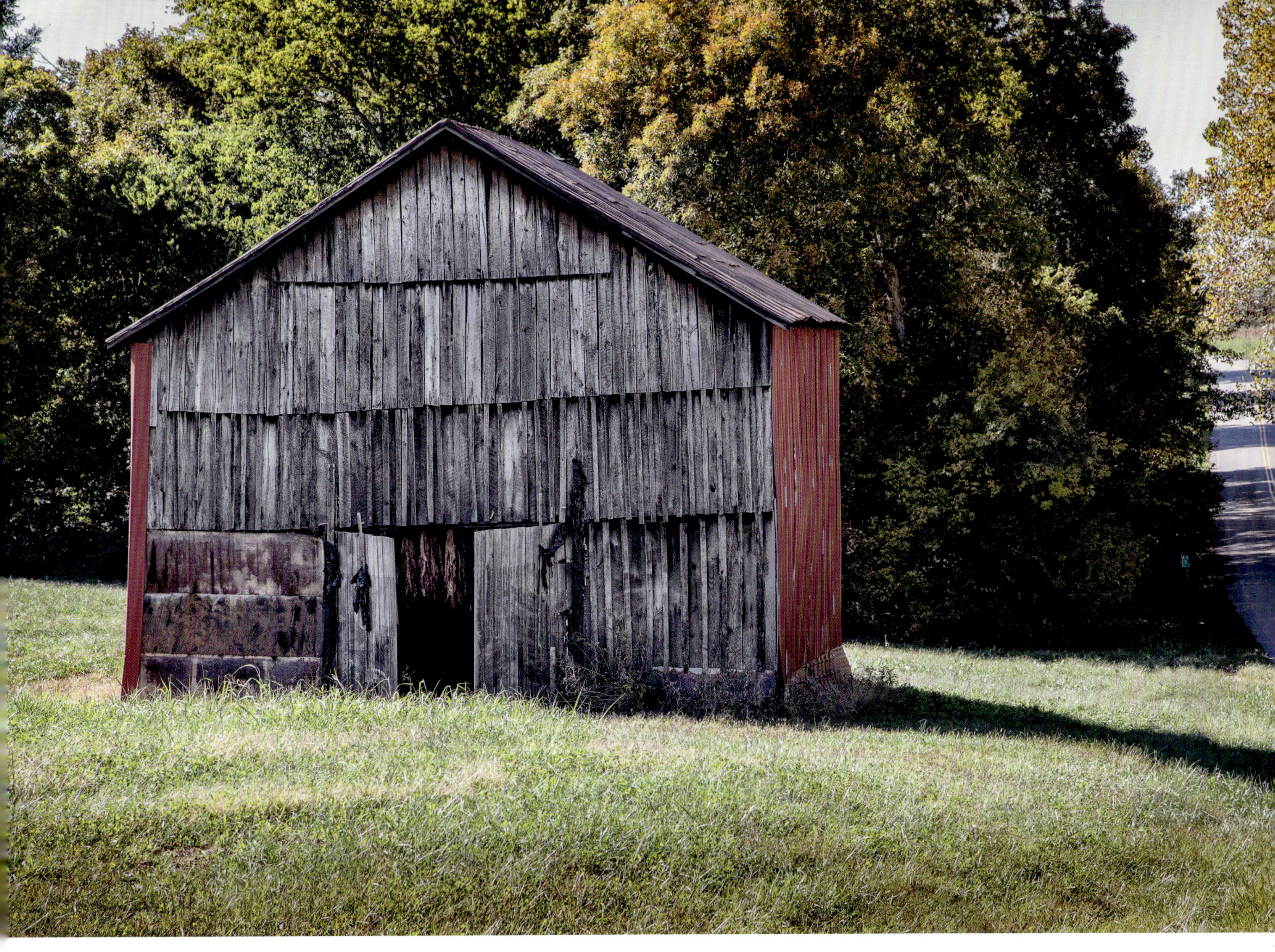

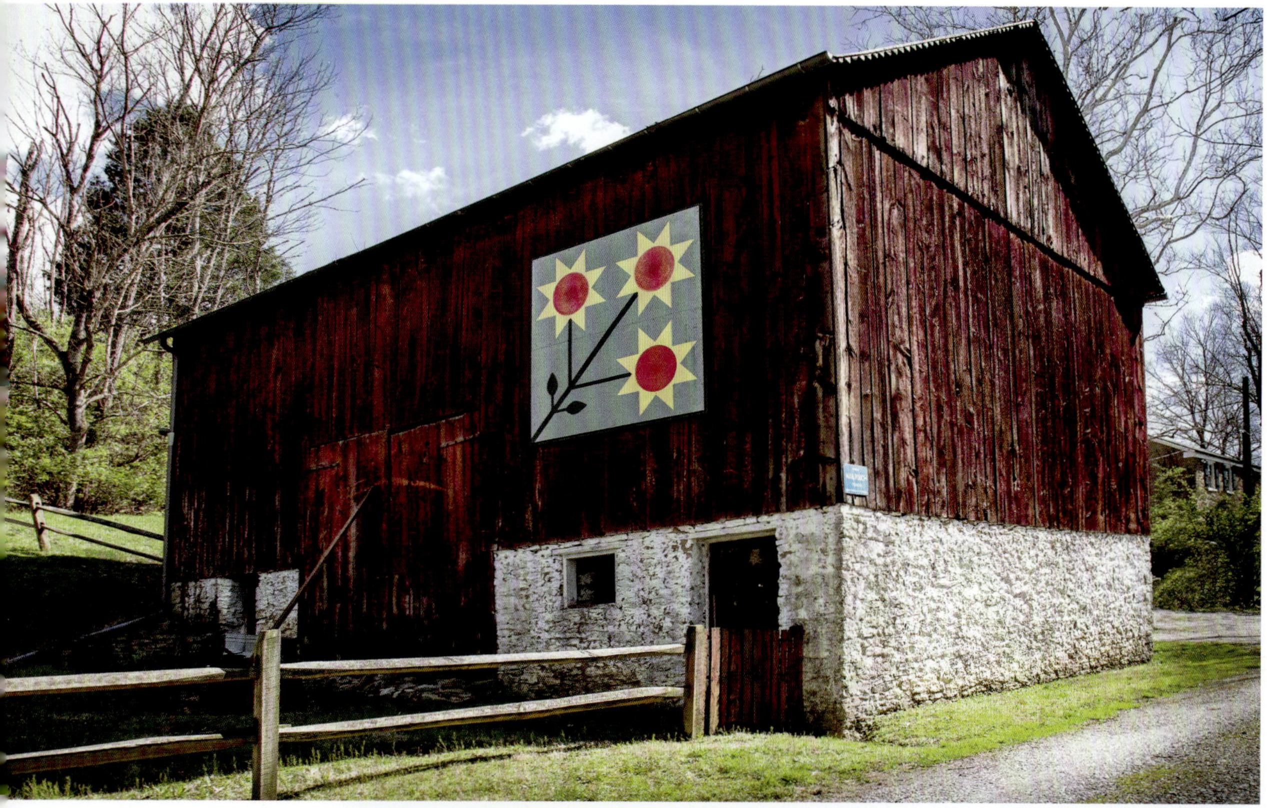

Campbell County.

The Faha barn, originally built by Catherine and Peter Faha in the mid-nineteenth century, is part of the German settlement area known as Camp Springs

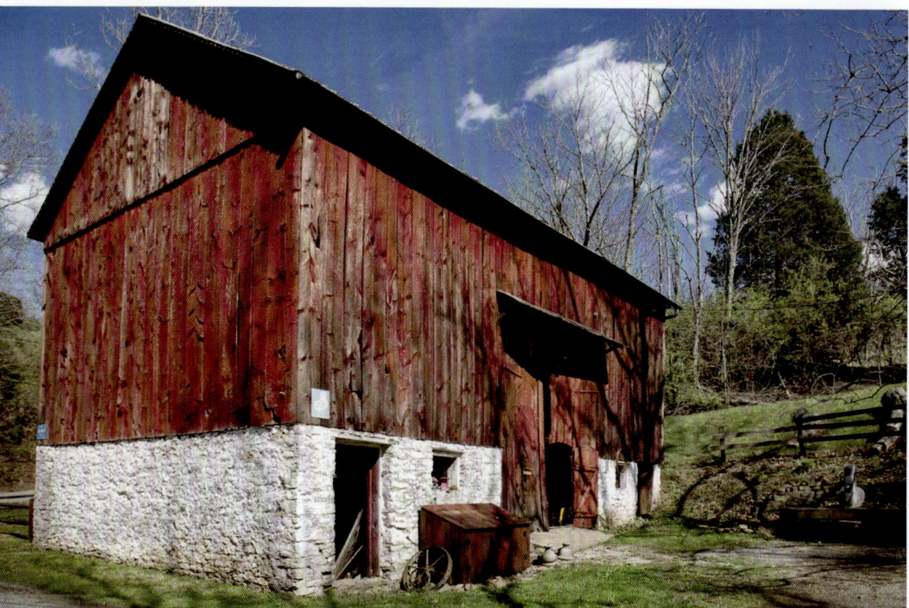

Campbell County.

The German settlers at Camp Springs brought their unique building design with them when they migrated. The sturdy stone ground floor housed livestock, and the upper levels of timber held hay that had been cut by horse-drawn reapers. Notice the unusual trussing visible in the attic.

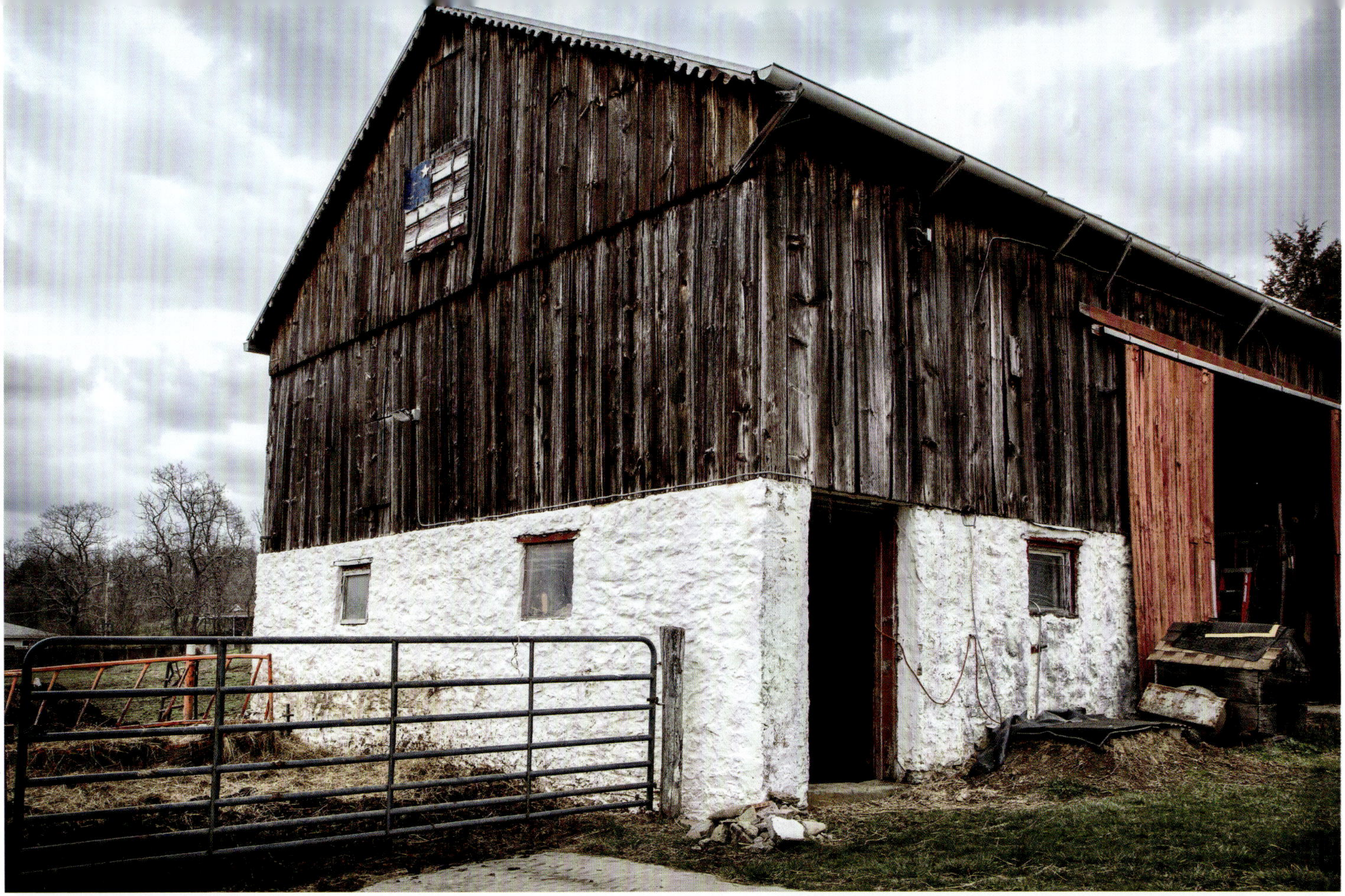

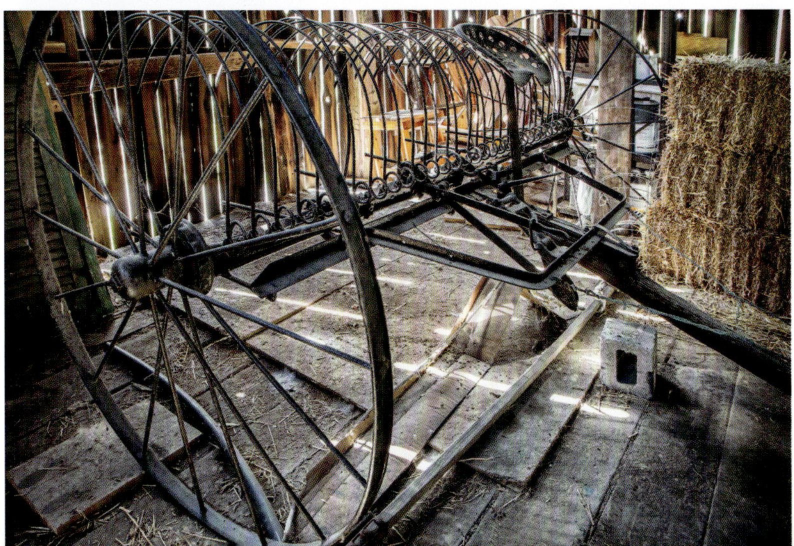

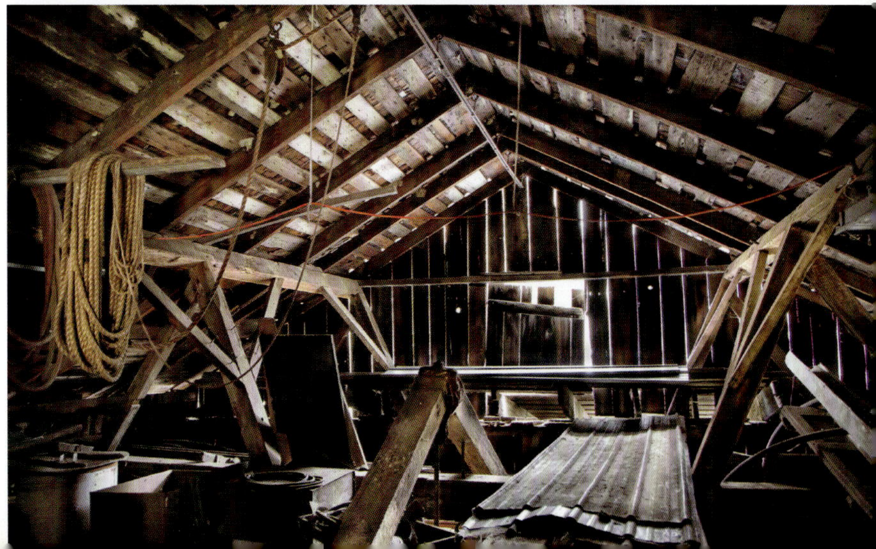

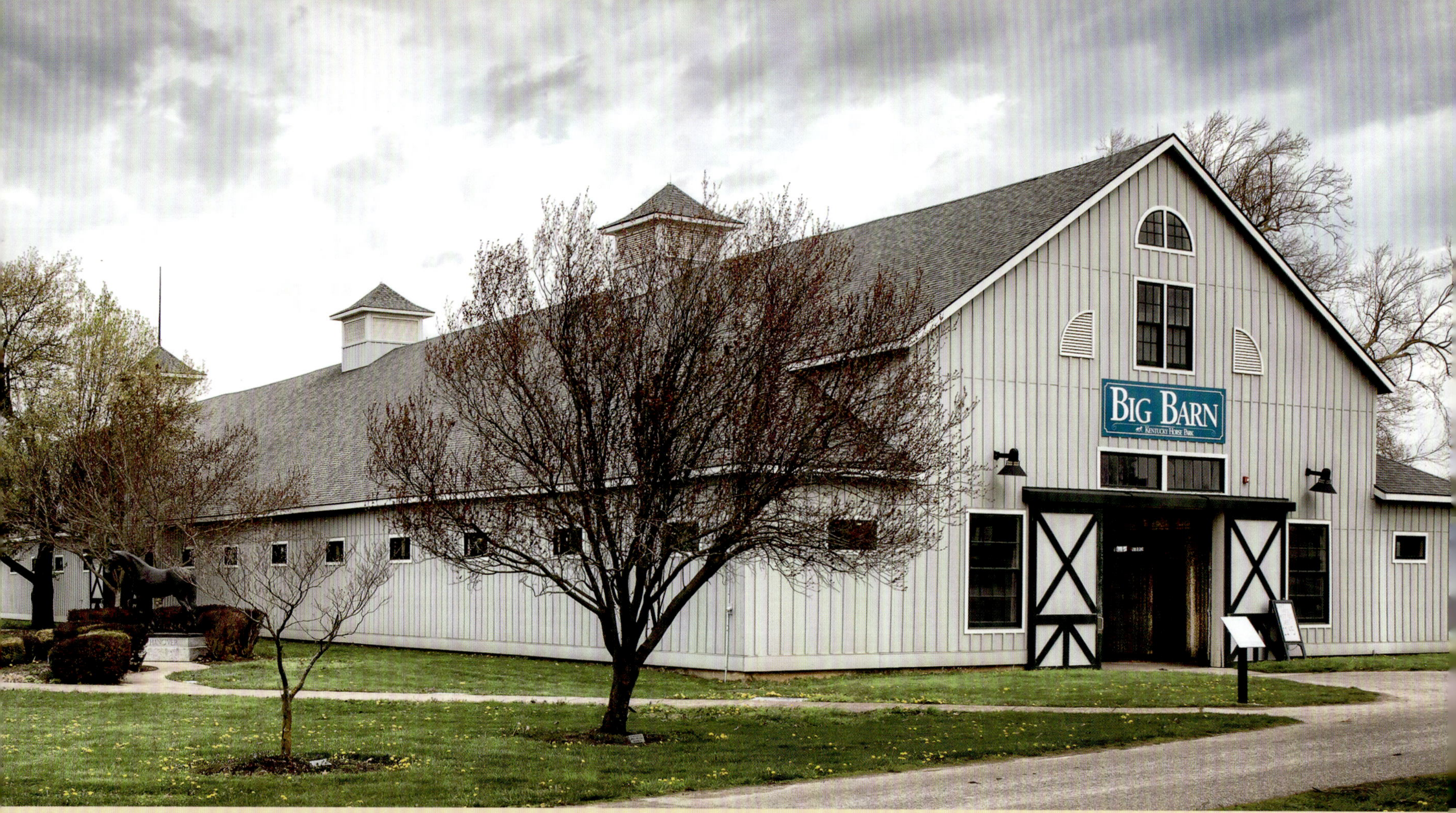

Fayette County.

The Kentucky Horse Park is a working horse farm, educational theme park, and equine competition facility. The Big Barn was built by John Creighton for his Standardbred horse operation. Later it was part of L. V. Harkness's Walnut Hall, a major Standardbred farm. Built circa 1897, the building is one of the largest horse barns ever built, at 476 feet long and holding fifty-two stalls. The width of the aisles is so great, three catwalks are built from side to side in the haylofts. The barn is now home to several draft horses.

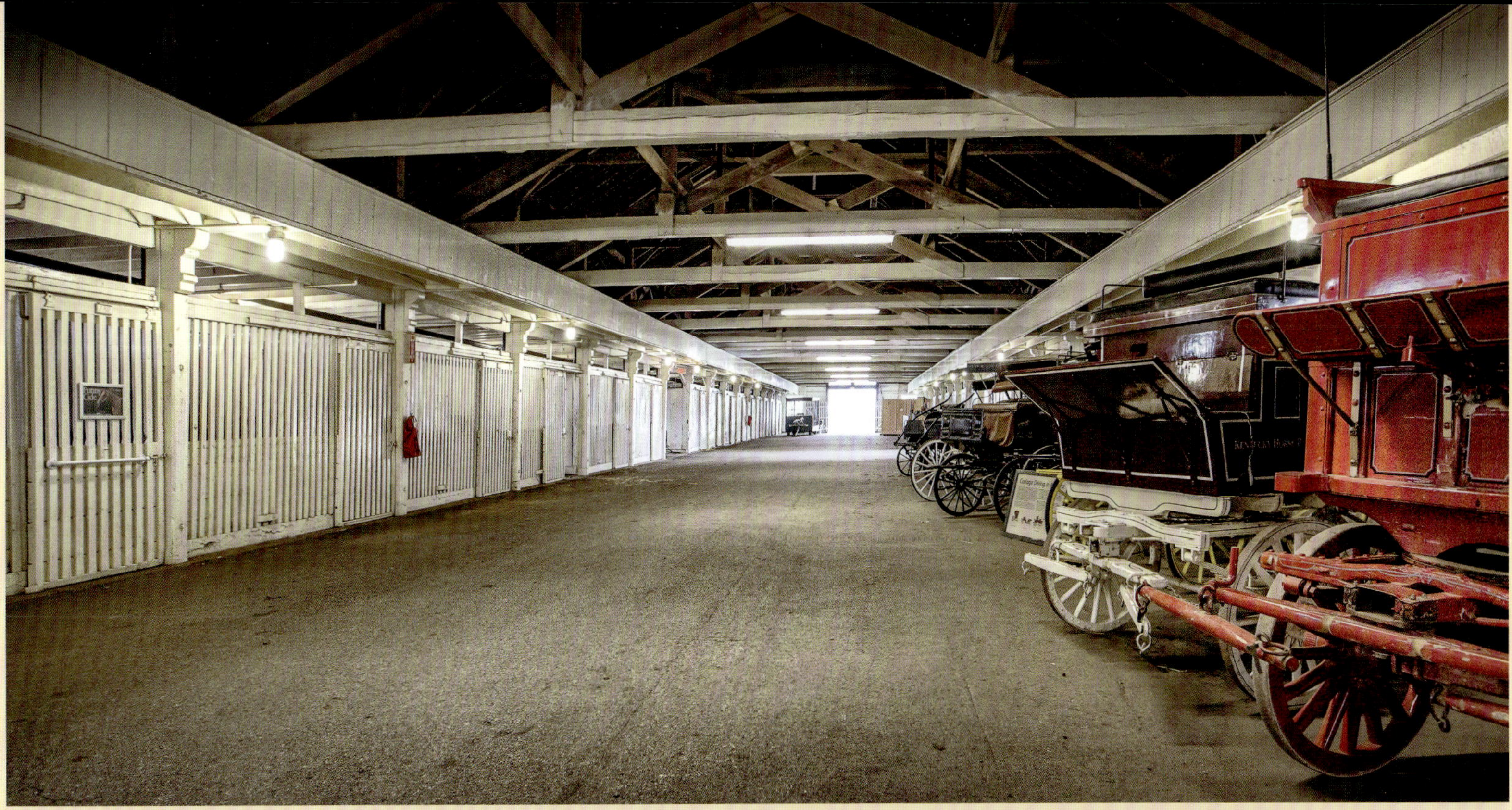

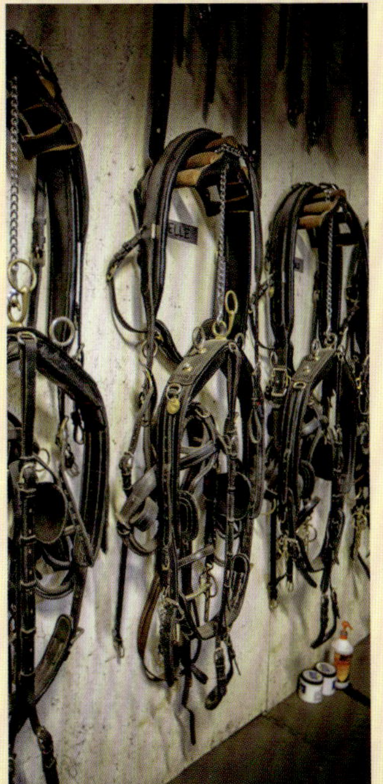

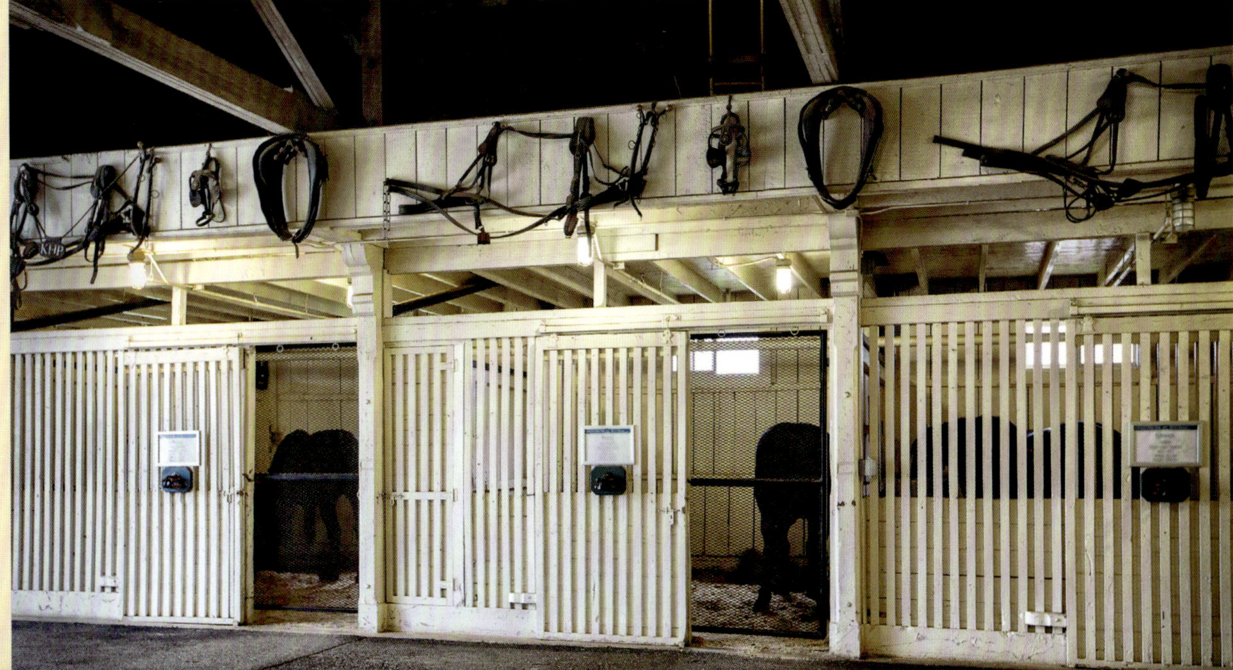

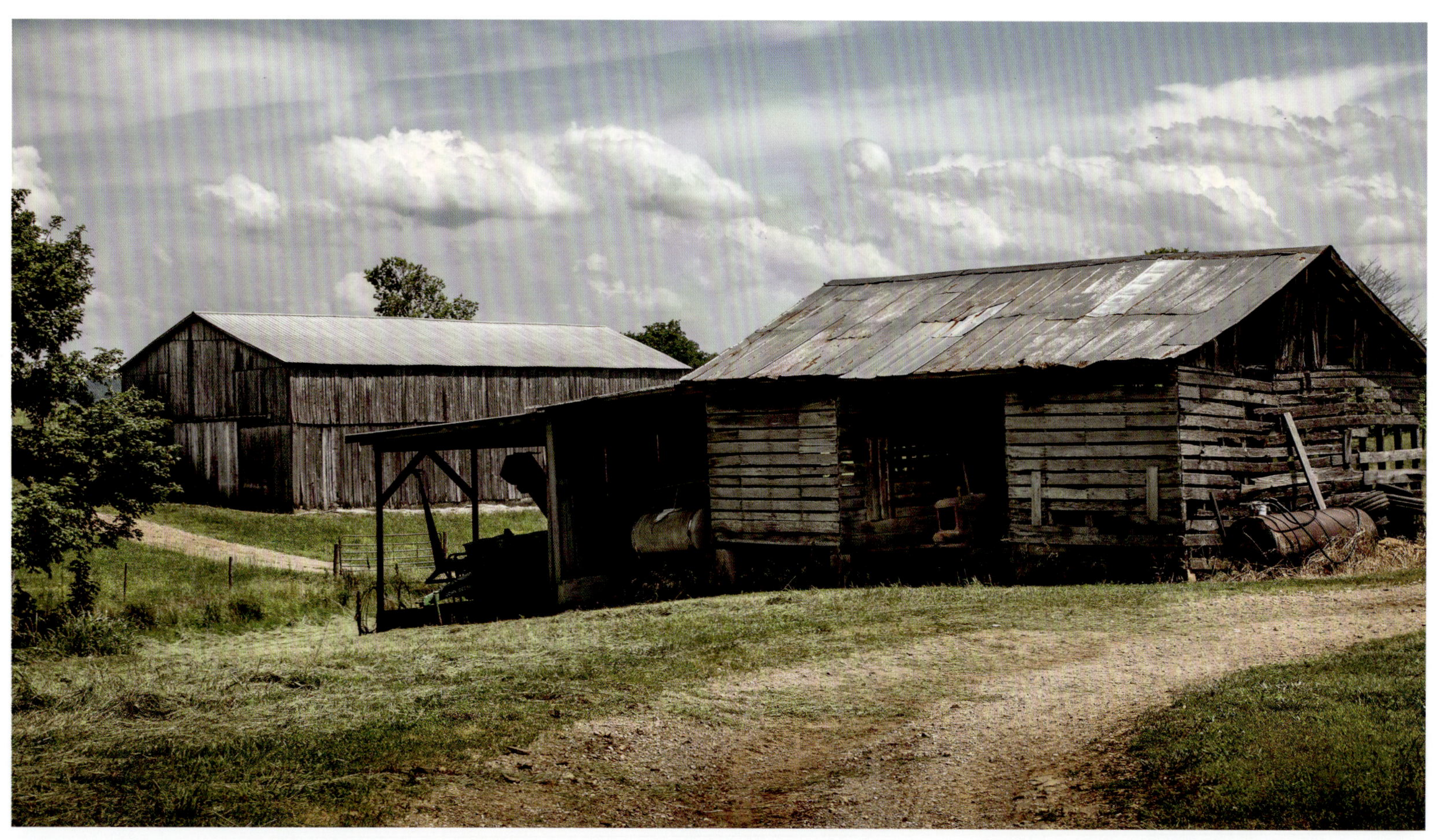

Alongside the road many barns can be seen without reference to their original farmland. This double-crib barn, with a tobacco barn in the background, still houses a tractor but does not hold corn or grains any longer.

Campbell County.

Part of the German settlement of Camp Springs, this bank barn is still used as a stock barn.

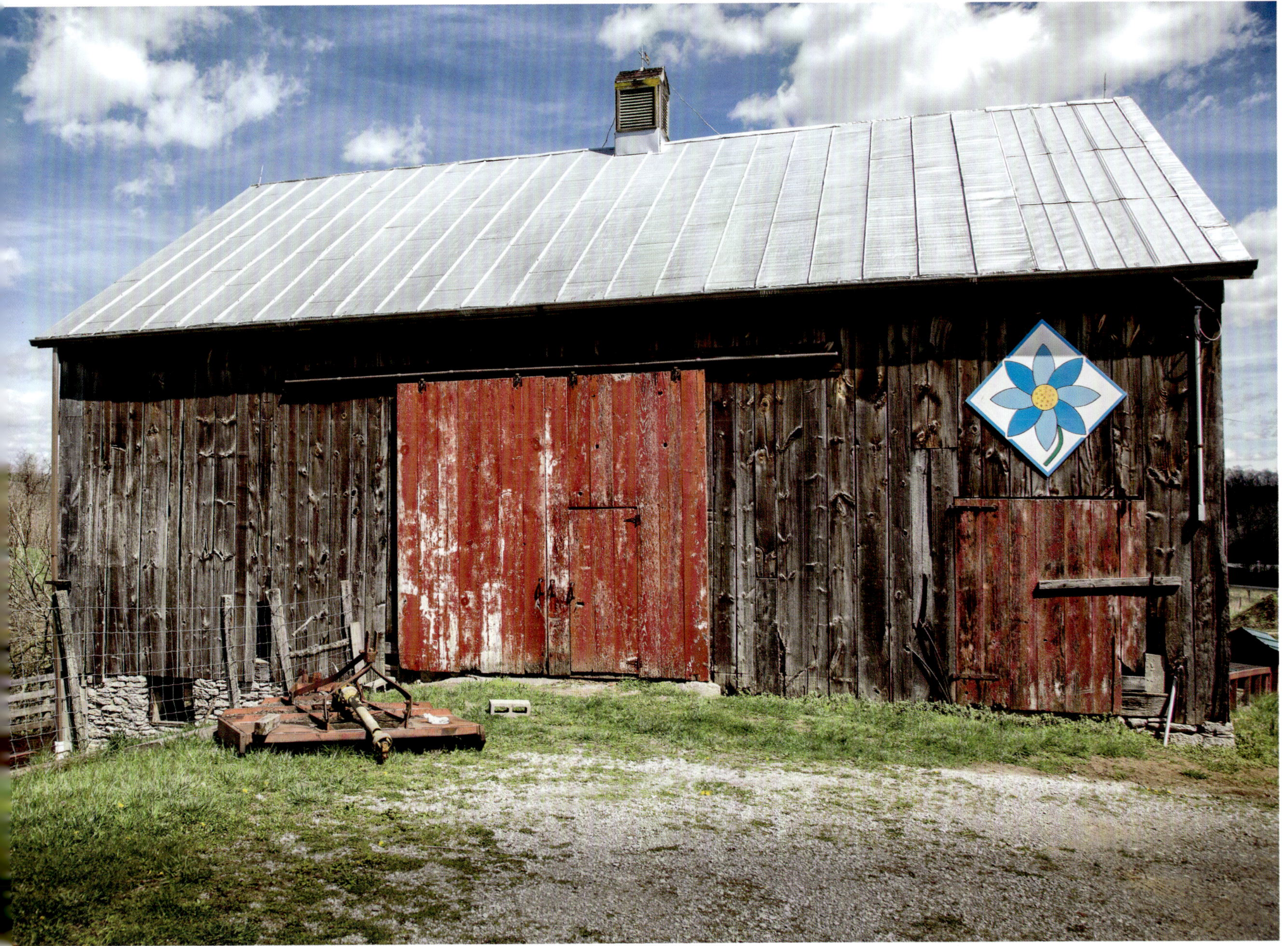

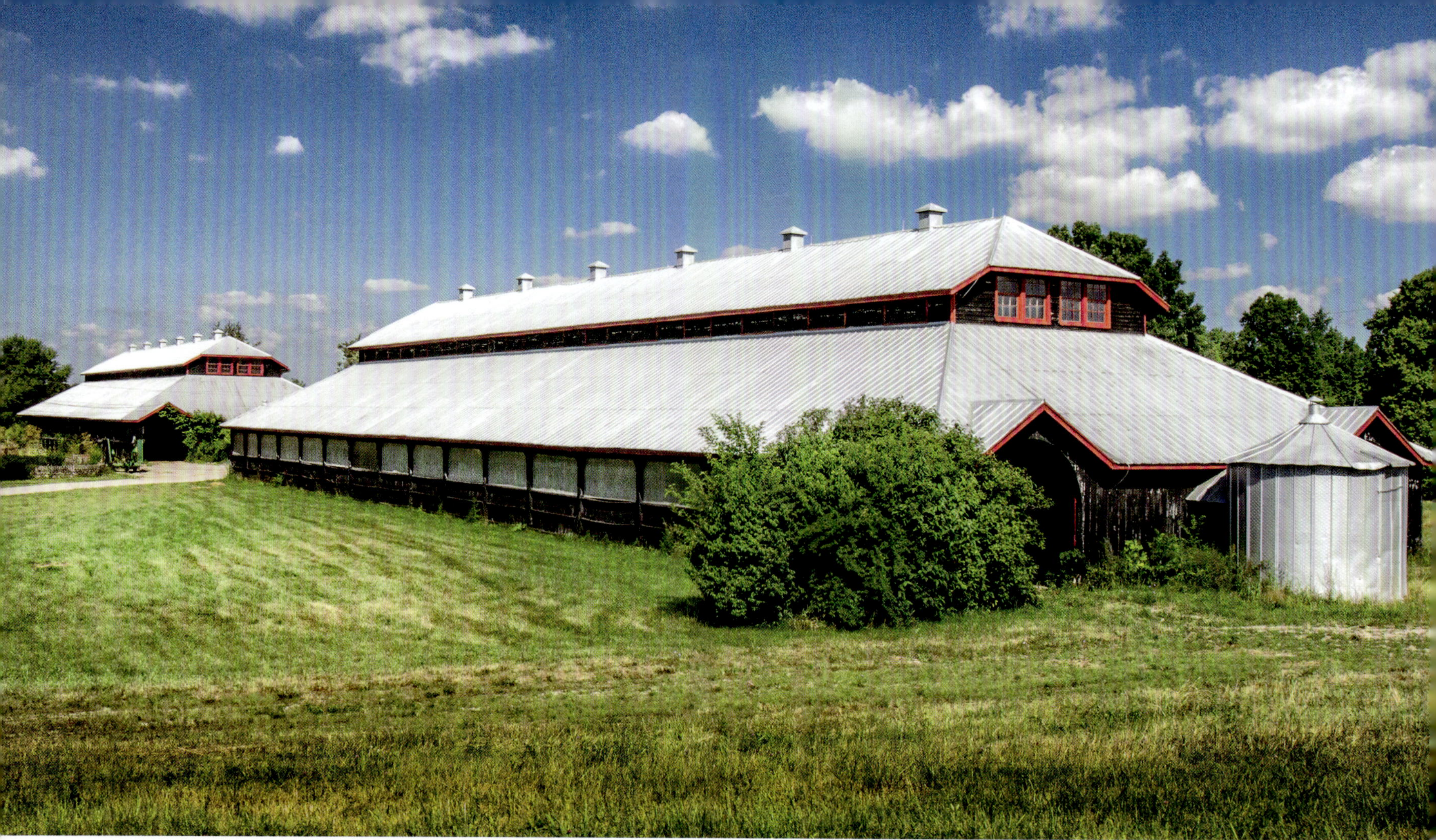

John E. Madden bred the first Triple Crown winner, Sir Barton, at Hamburg Place Farms in 1916. In the 1930s his son Edward built the two training barns in this image to replicate the barns his father built earlier in another location on the farm. Stored inside are items used for polo, harness racing, and pleasure driving. One of the barns, no longer used for horses, holds hay.

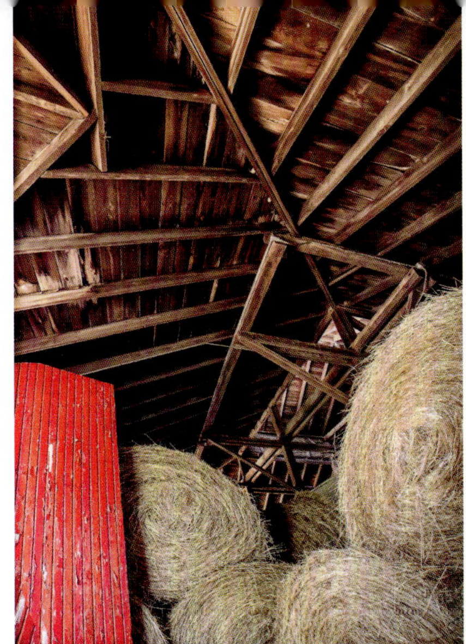
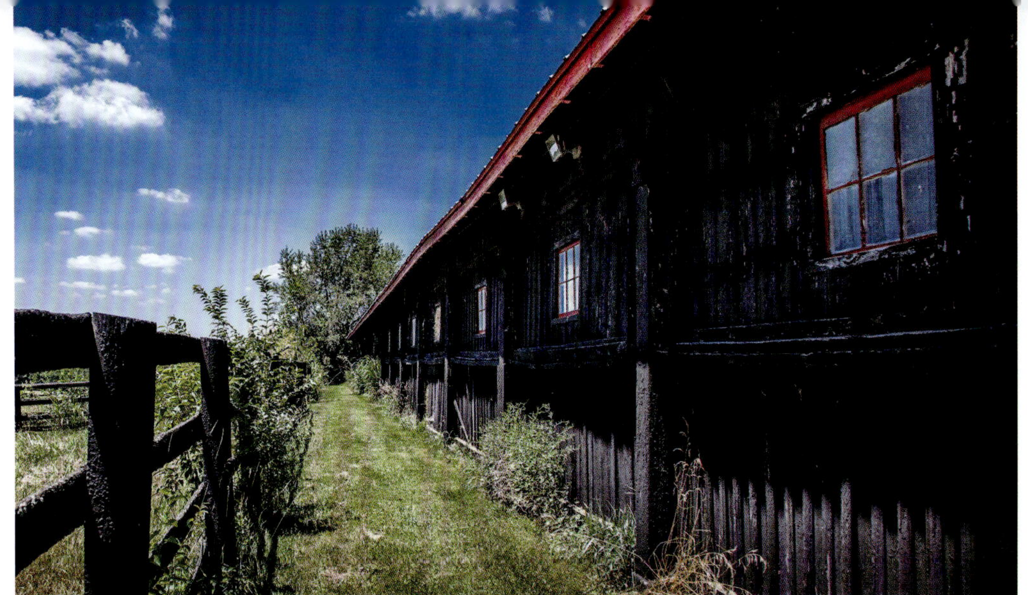
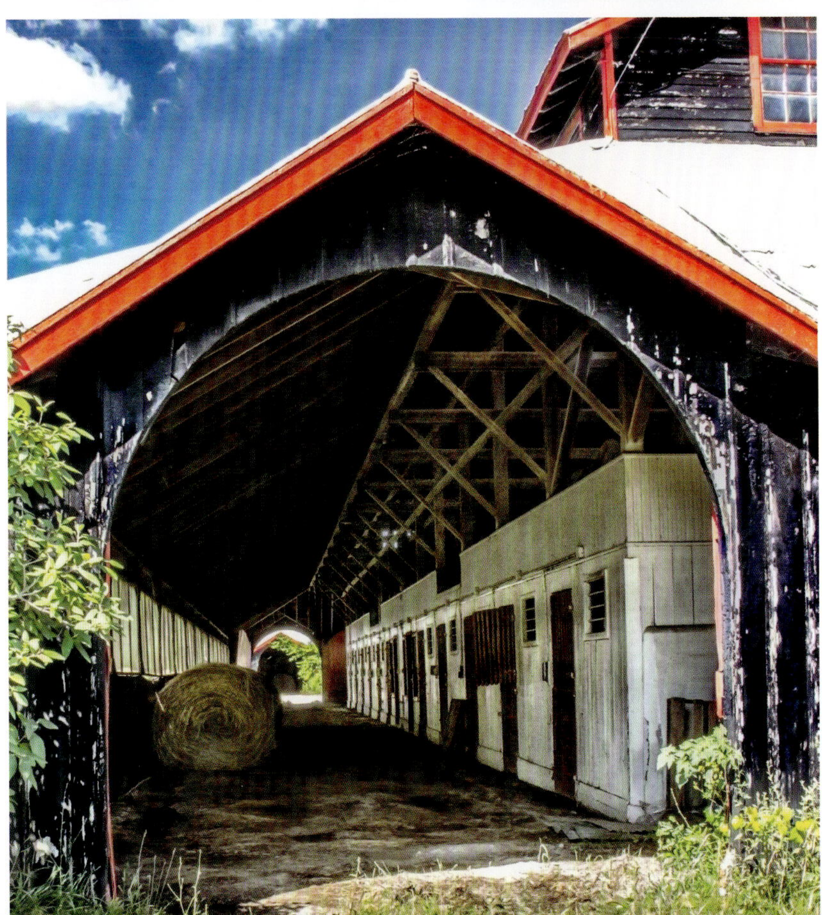
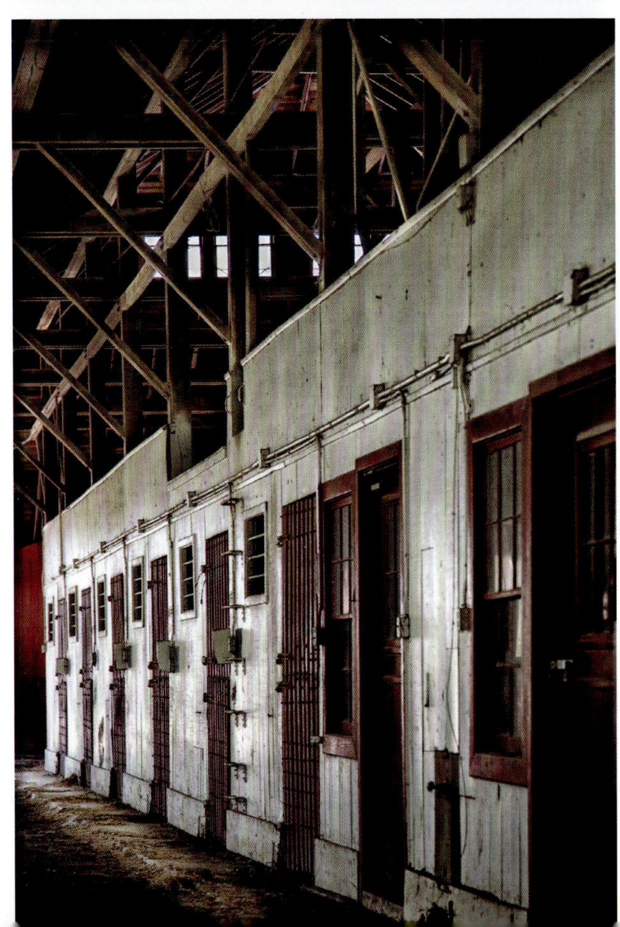

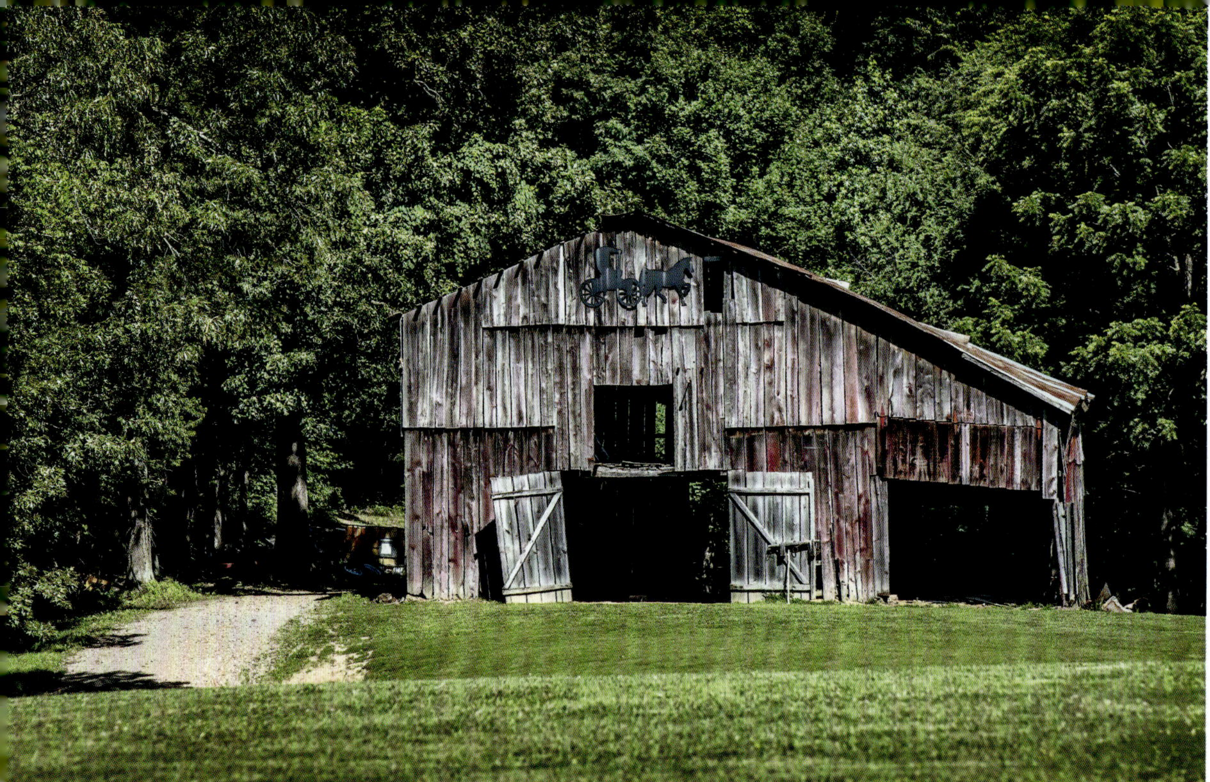

Livingston County.

Declaring itself for horse and carriage by its ornament, this roadside barn has a cover for the carriage extending off the stall area.

BELOW

Muhlenberg County.

The property this barn sits on was originally a land grant to the Lovell family and has remained in the family. Built over one hundred years ago, it was used for tobacco.

FACING PAGE

Livingston County.

When the rural farm landscape has vast fields, the barn sits as a solitary structure. When the barn serves livestock needs, it will more commonly be found as part of a cluster of buildings with close access to the house.

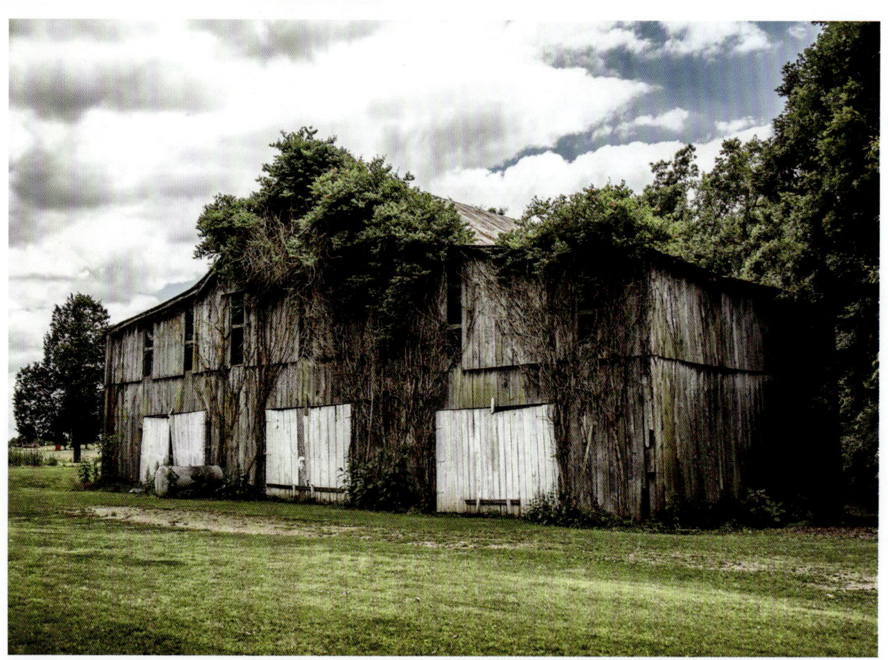

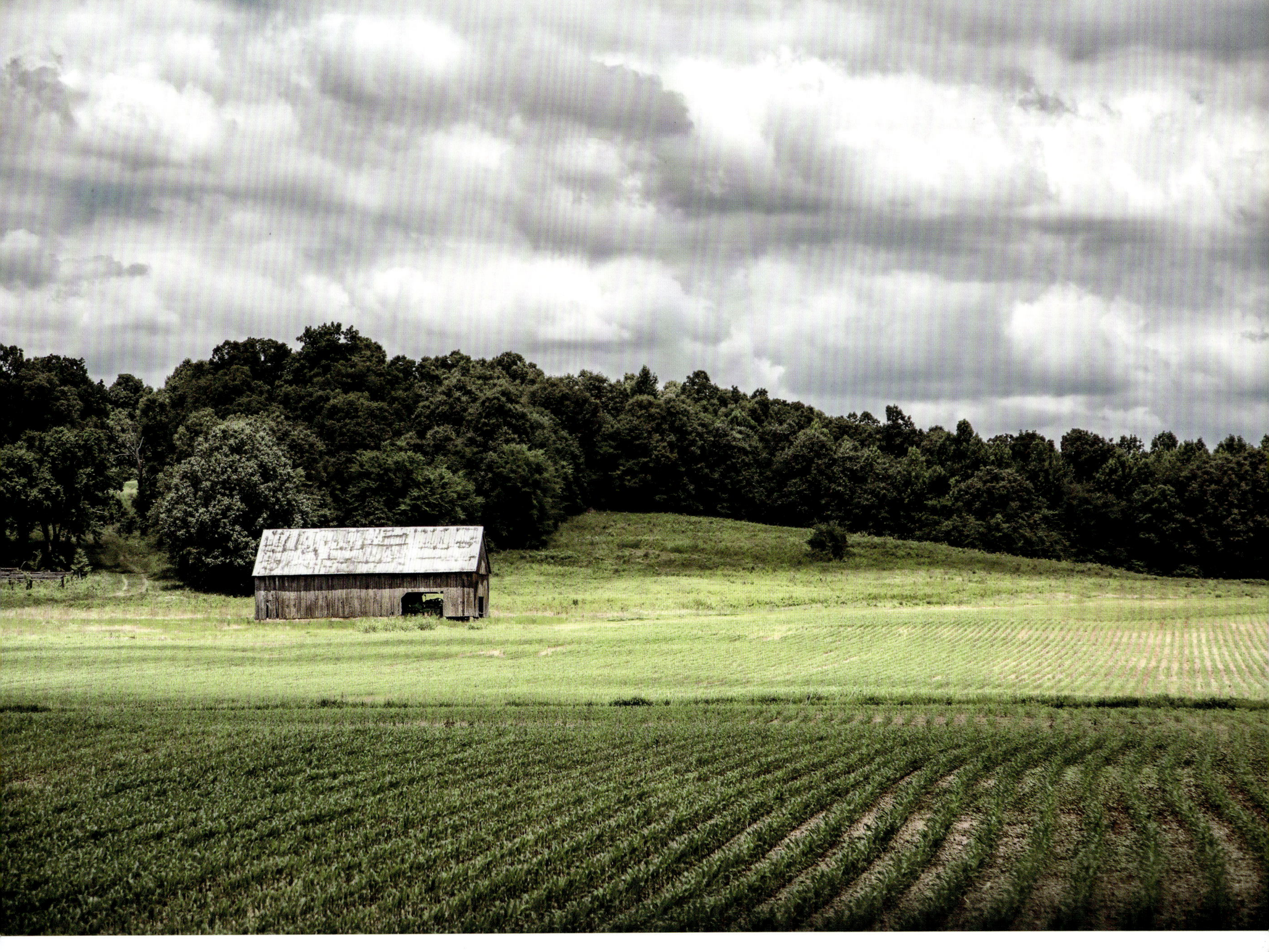

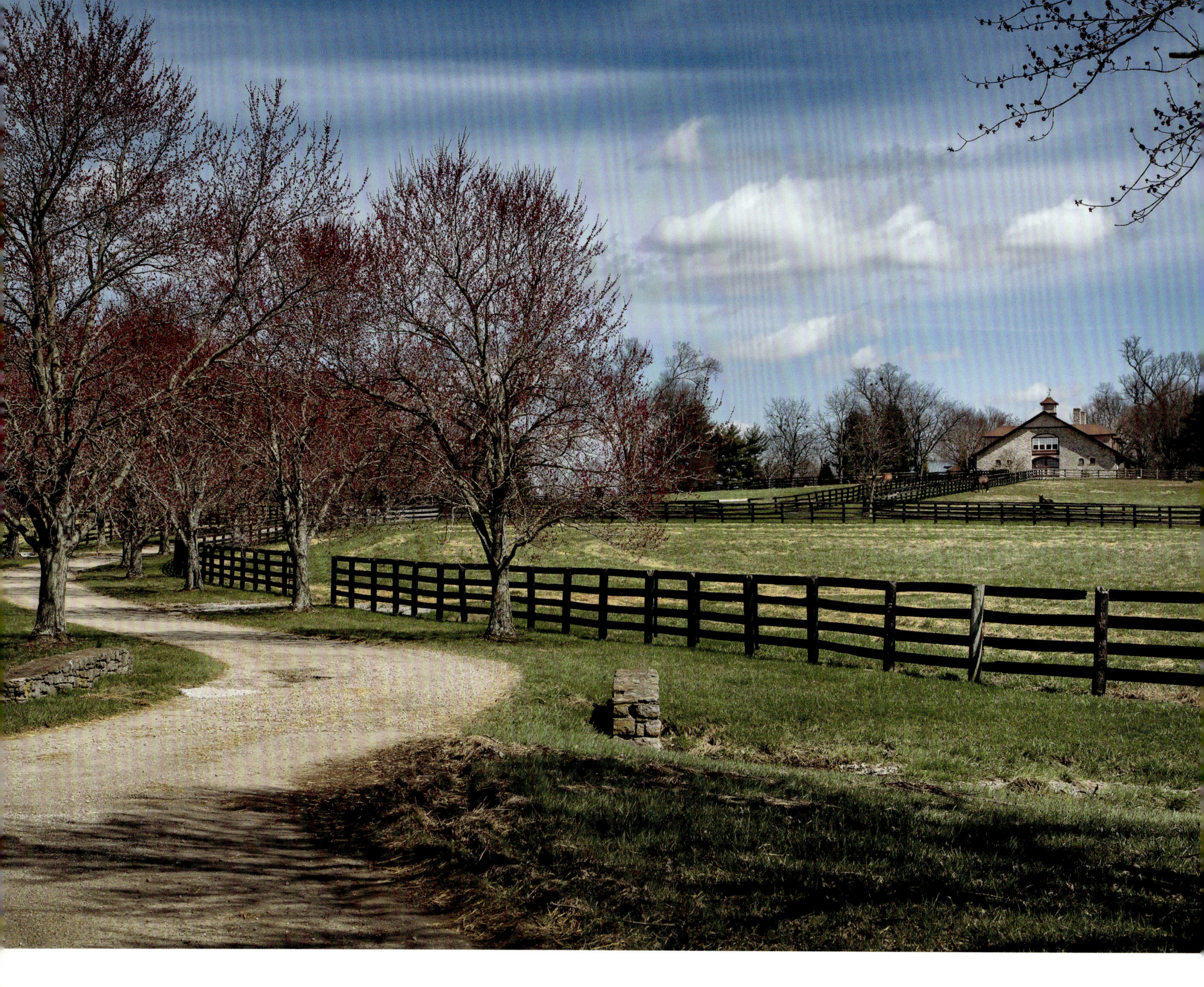

Evergreen Farm is located on the historic estate of Escondida. The quarried stone carriage barn, built circa 1900 and now used as a foaling barn, has fourteen stalls on the bottom, with a wide aisle to accommodate carriages. The outside of the barn has clipped-end gables, hipped-roof dormers, and bracketed eaves.

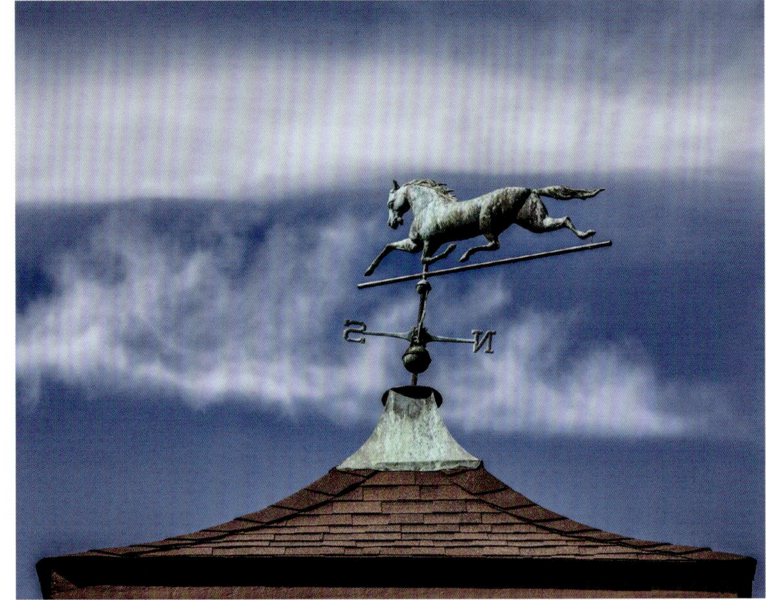

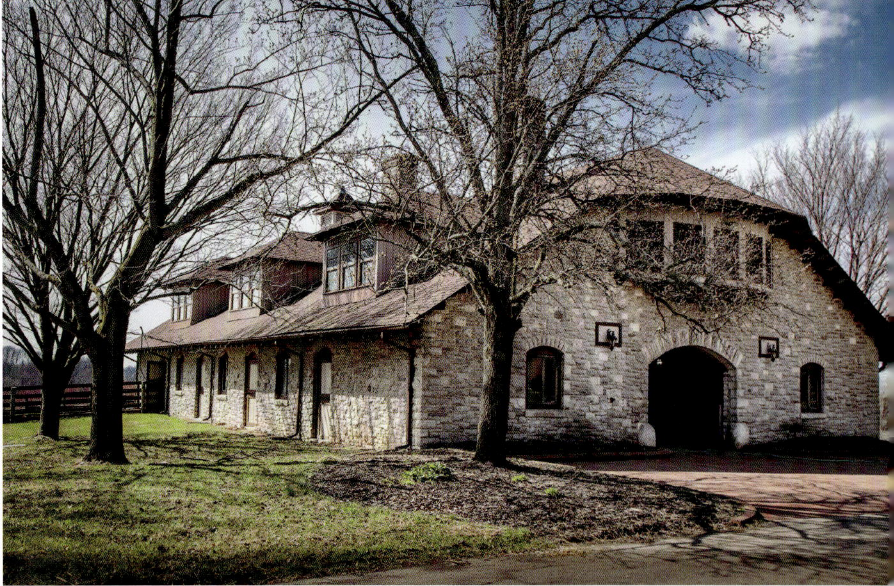

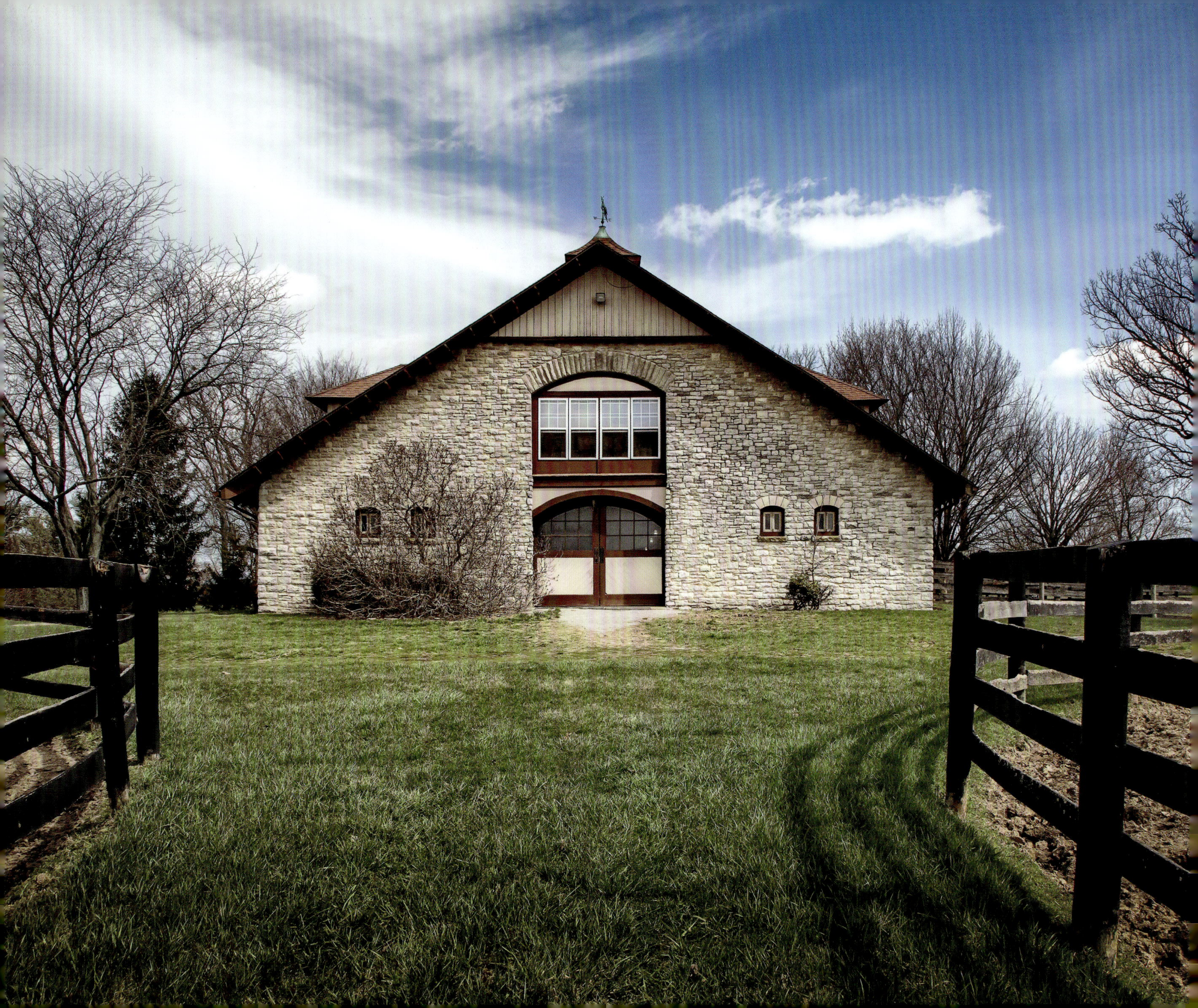

On the top floor of the Evergreen Farm's foaling barn are the works of the original carriage elevator. When the structure served as a carriage house in the early twentieth century, the platform would be lowered for the carriage and then raised to the upper level for storage. The farm also boasts several newer training and broodmare barns.

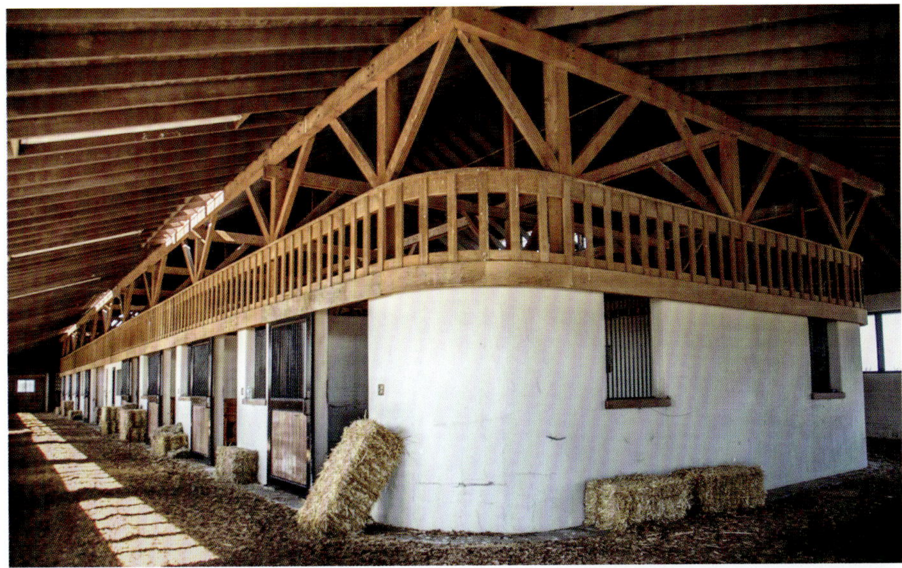

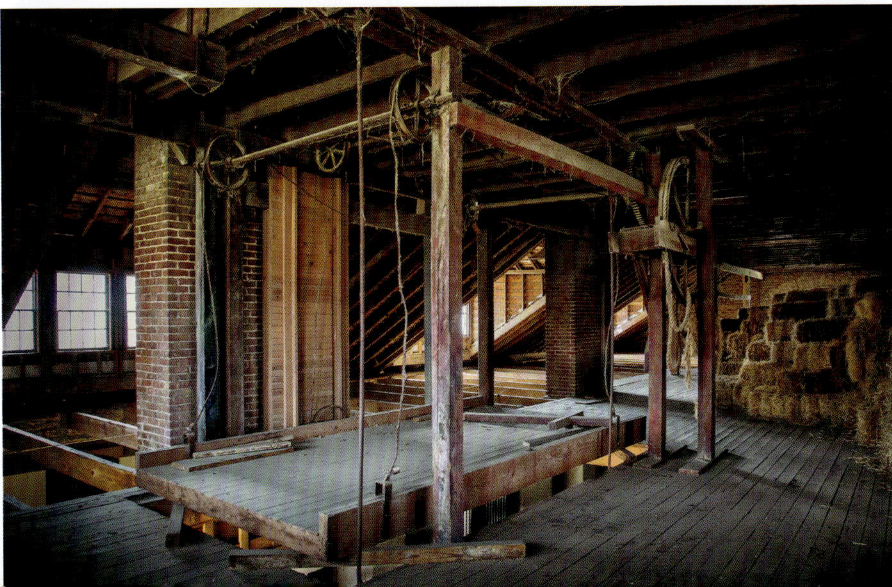

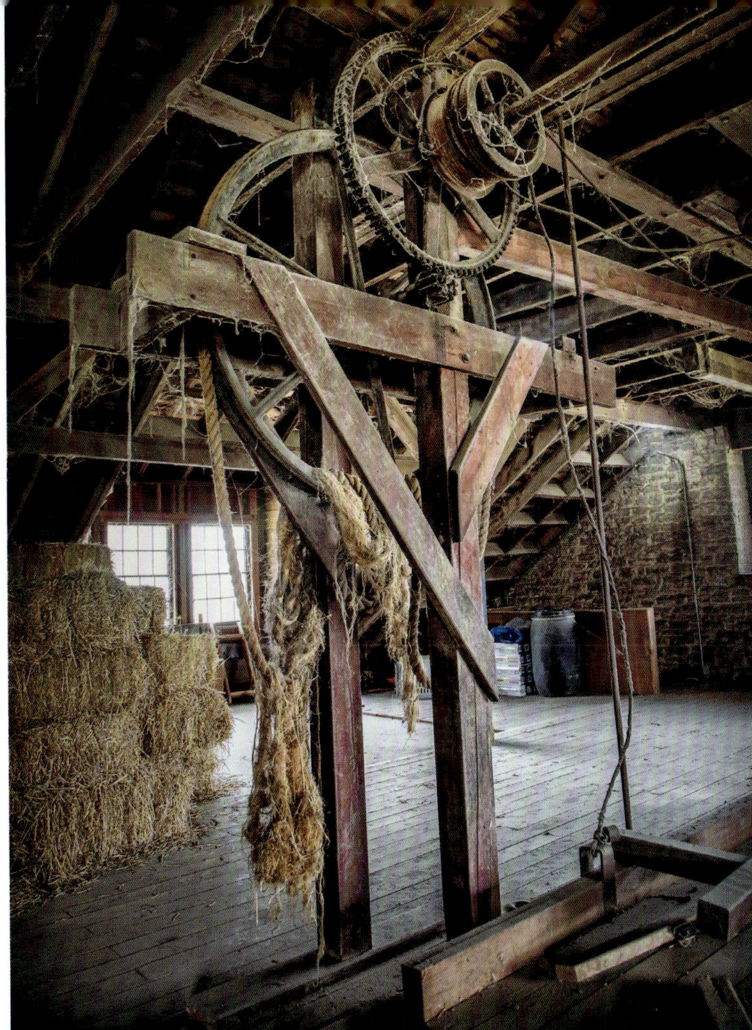

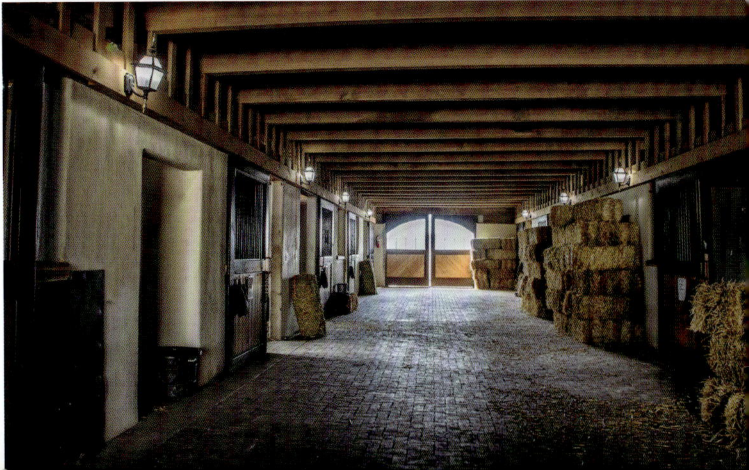

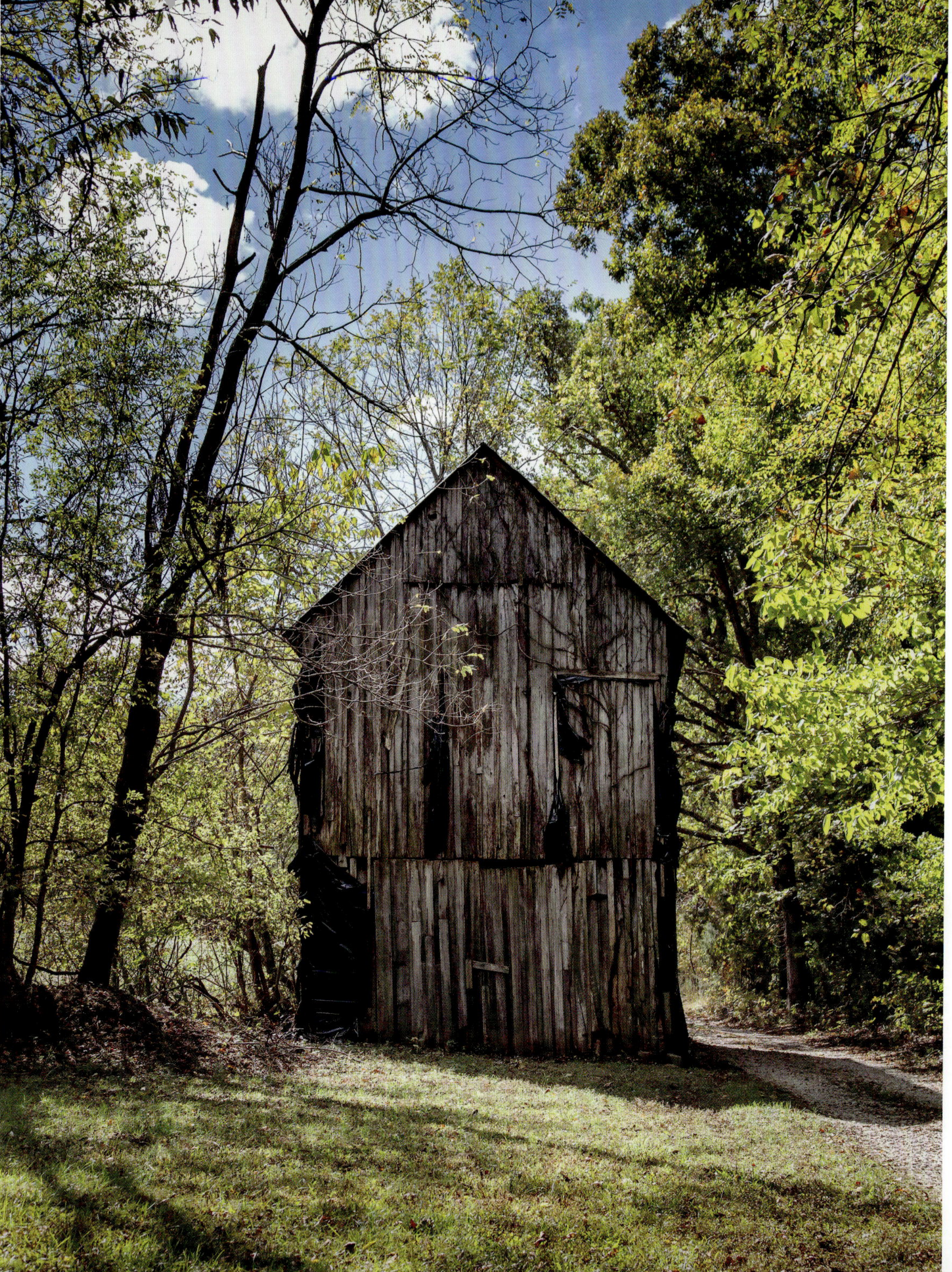

FACING PAGE

Rockcastle County.

This double-pen log barn, date unknown, has been altered many times to meet a variety of uses.

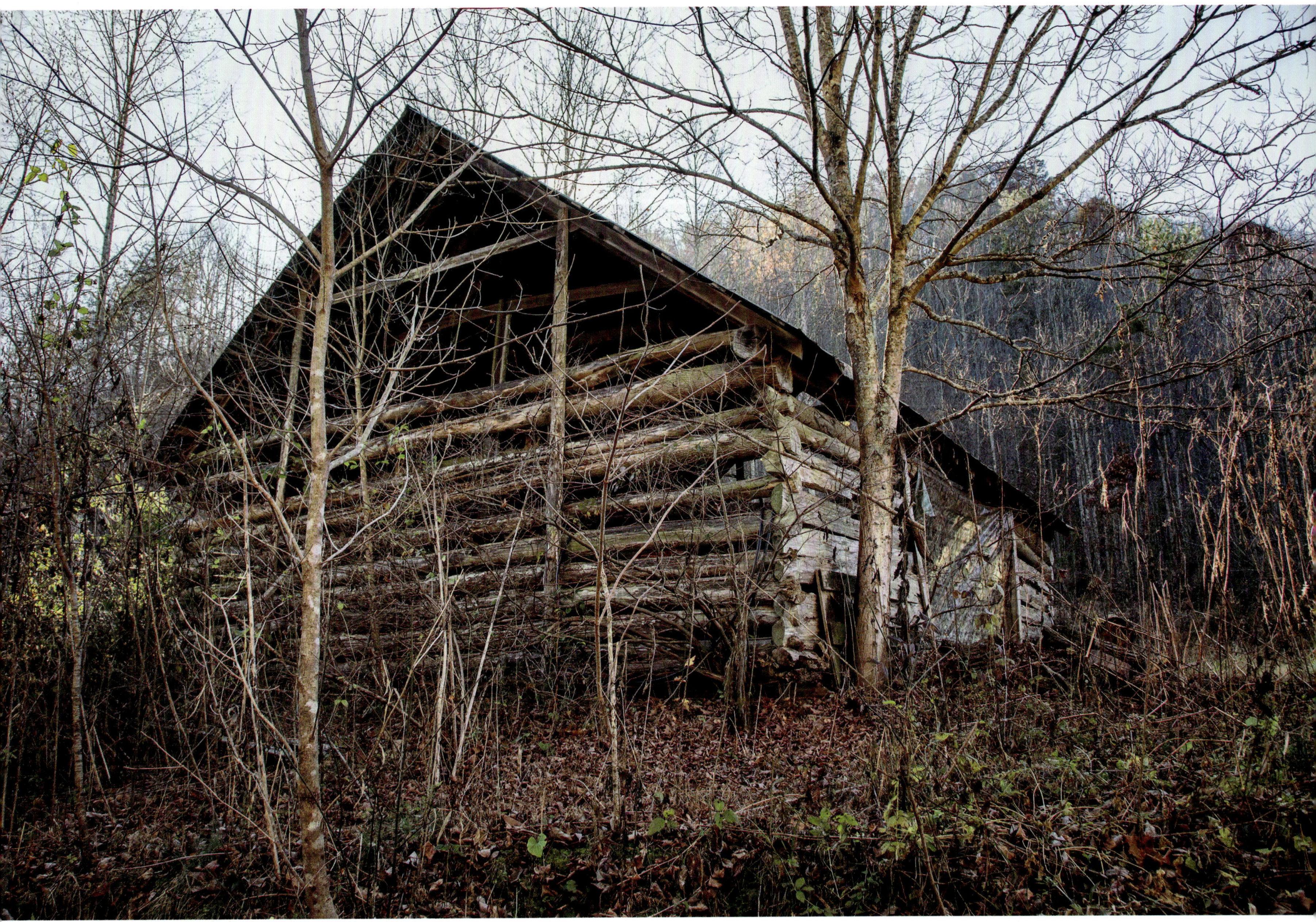

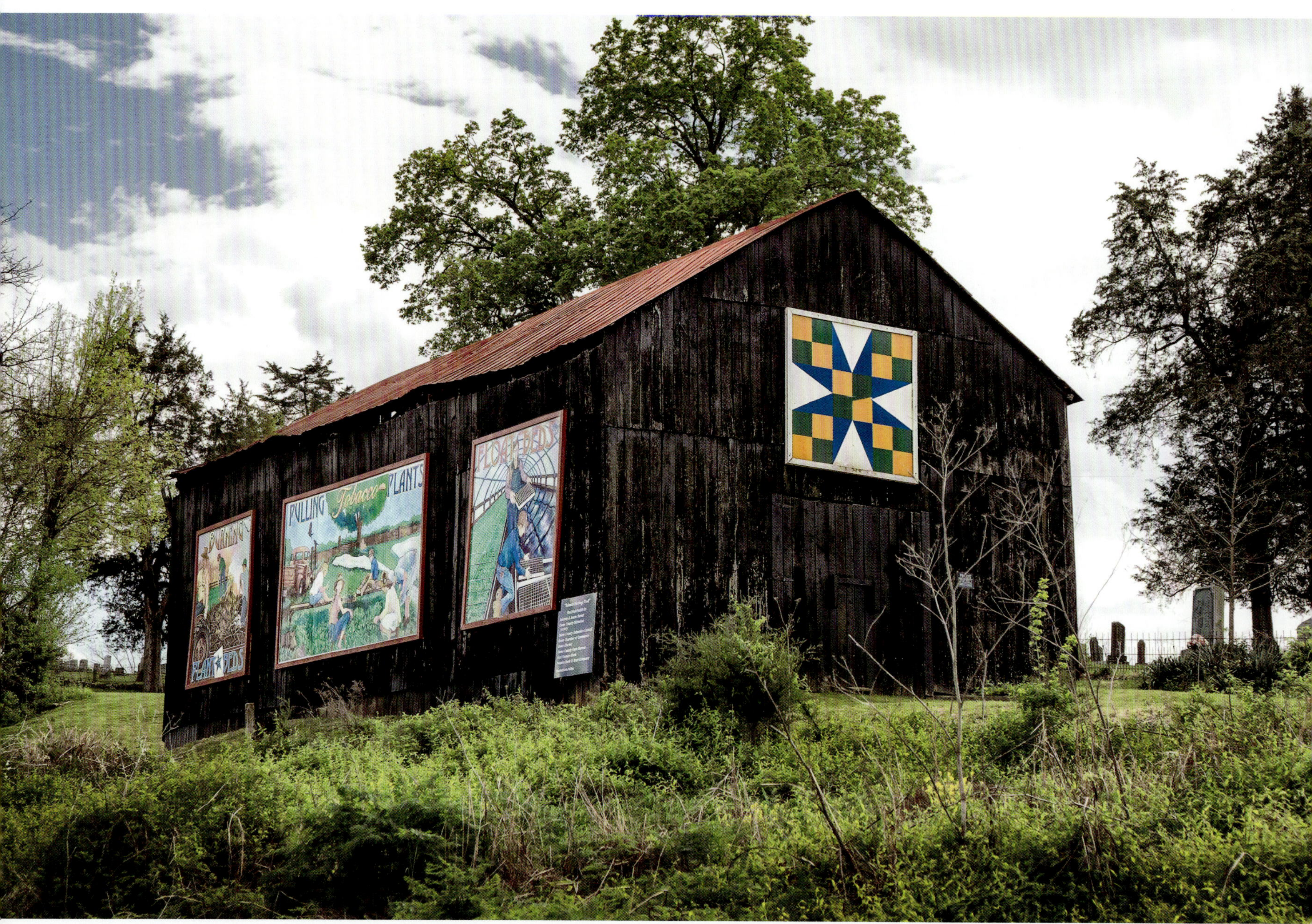

PULLING Tobacco PLANTS
FLOAT BEDS
BURNING
PLANT BEDS

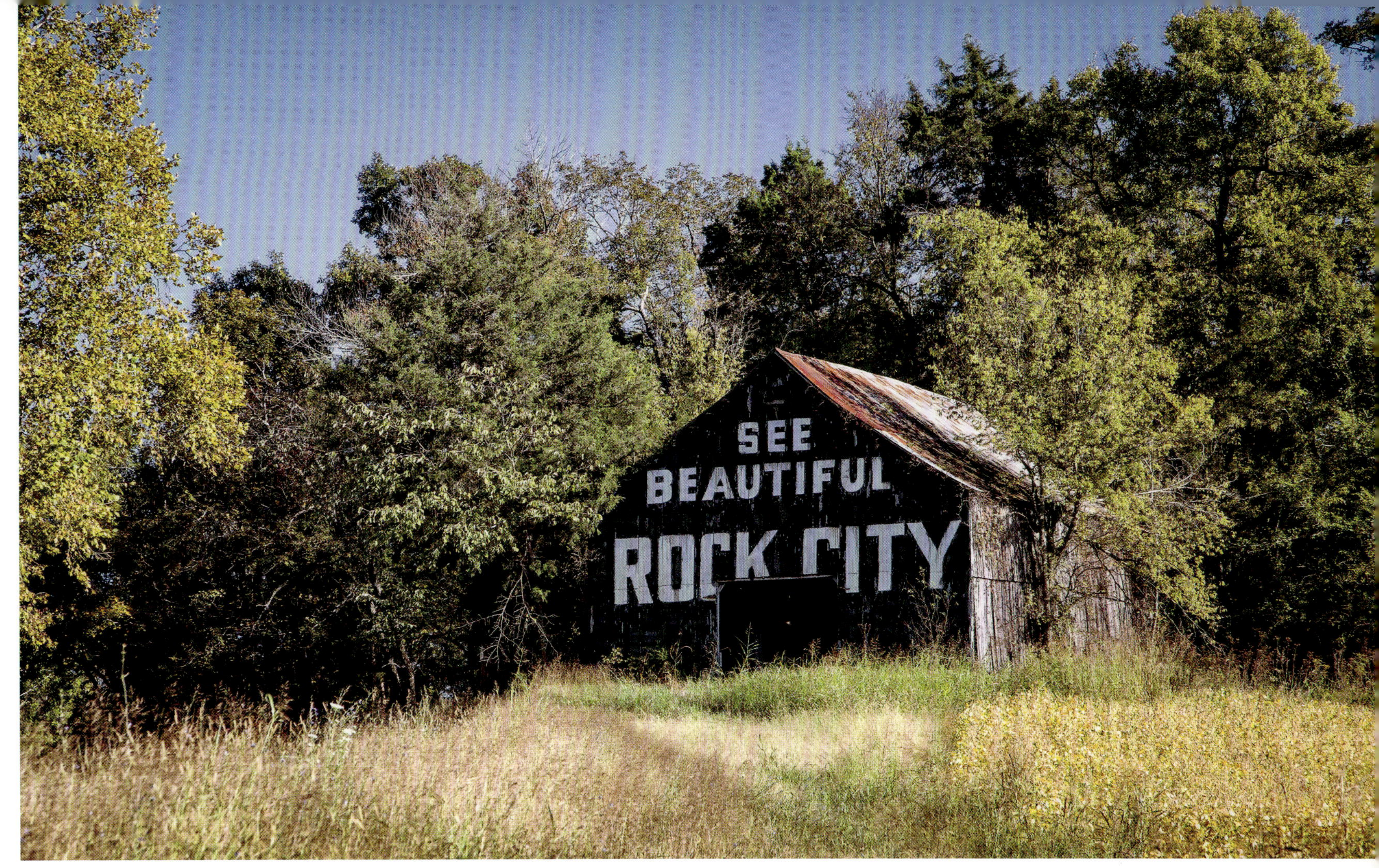

Owen County.

The signs near this barn indicate it is a Tobacco Heritage Trail Barn, one of several barns painted by the Agricultural Heritage Trail committee in the early twenty-first century.

Christian County.

Rock City barns, many painted by Clark Byers or his predecessors since 1935, are considered historic landmarks. This one is in Western Kentucky, close to Rock City in Chattanooga, Tennessee.

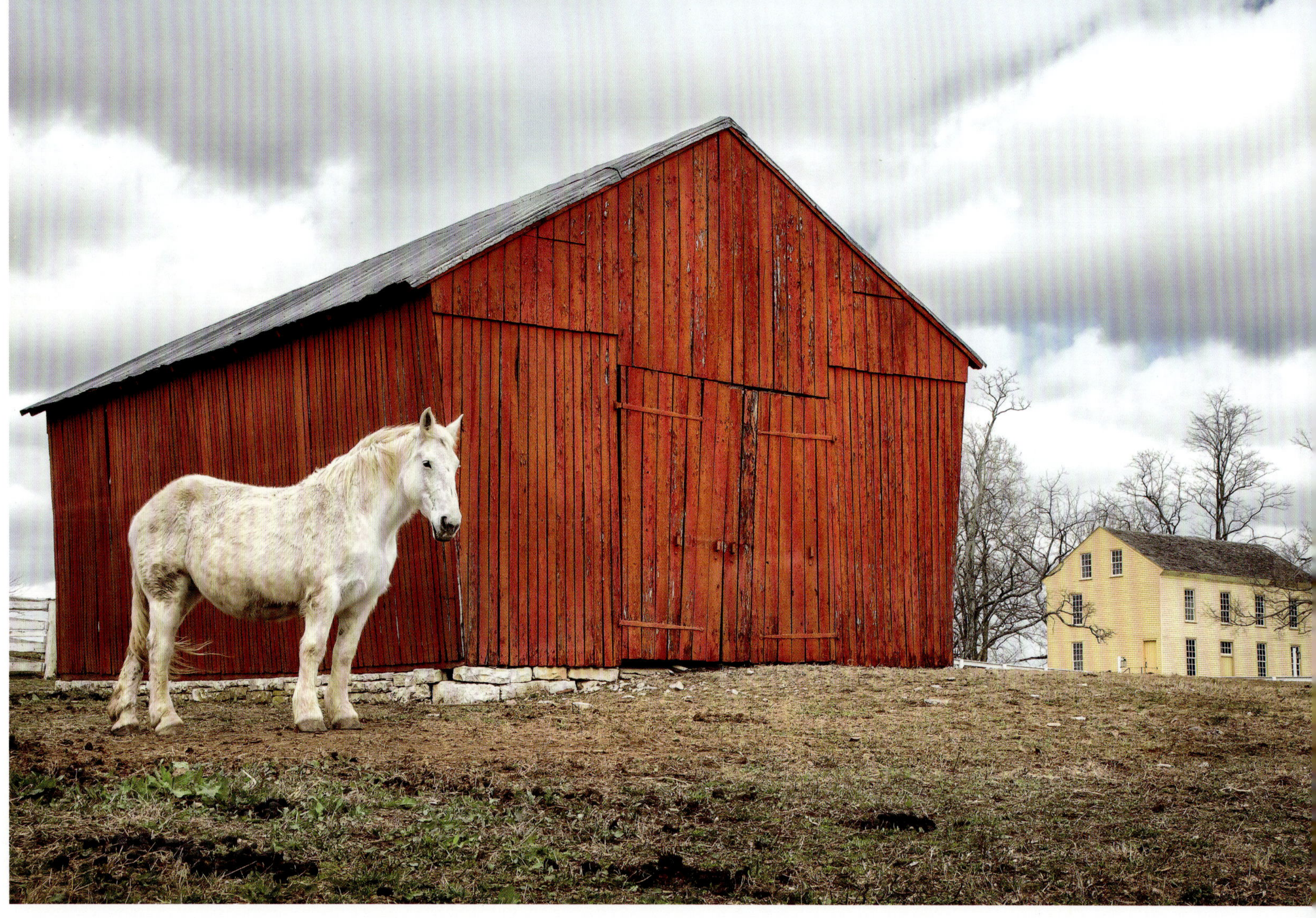

FACING PAGE

Mercer County.

The Shakers built this crib barn around 1854. It is still used today as part of the working agricultural program on site.

Mercer County.

Pleasant Hill was one of the sites that the Shakers established in Kentucky. The carriage and scale barn are located close to the dwellings. The stable barn, across the street, stores this amazing wooden trough, carved from a solid piece of hand-hewn beam. All are dated from the mid-nineteenth century.

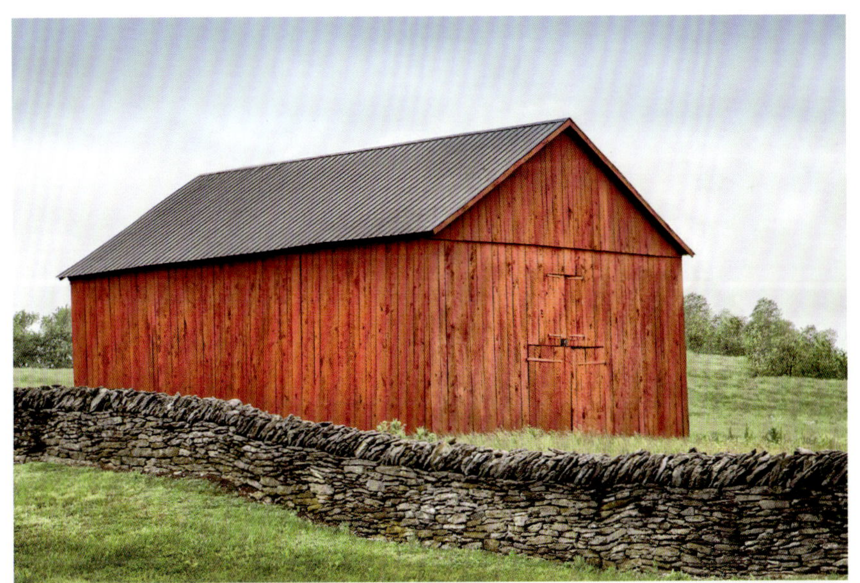
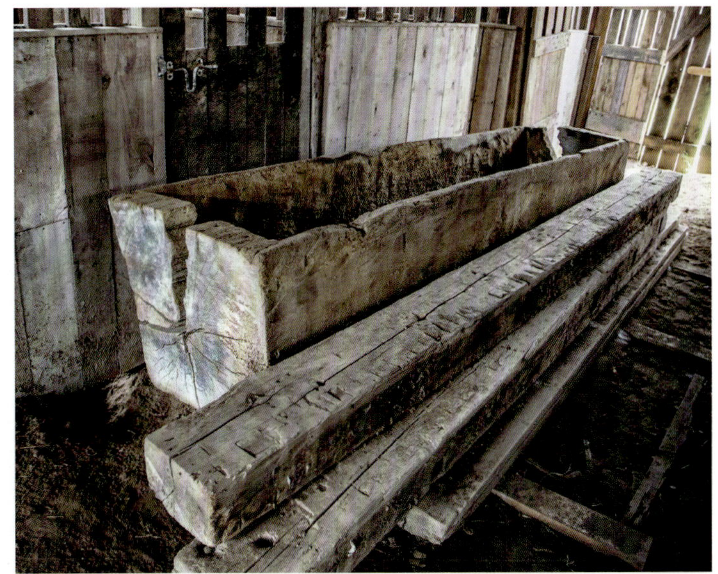

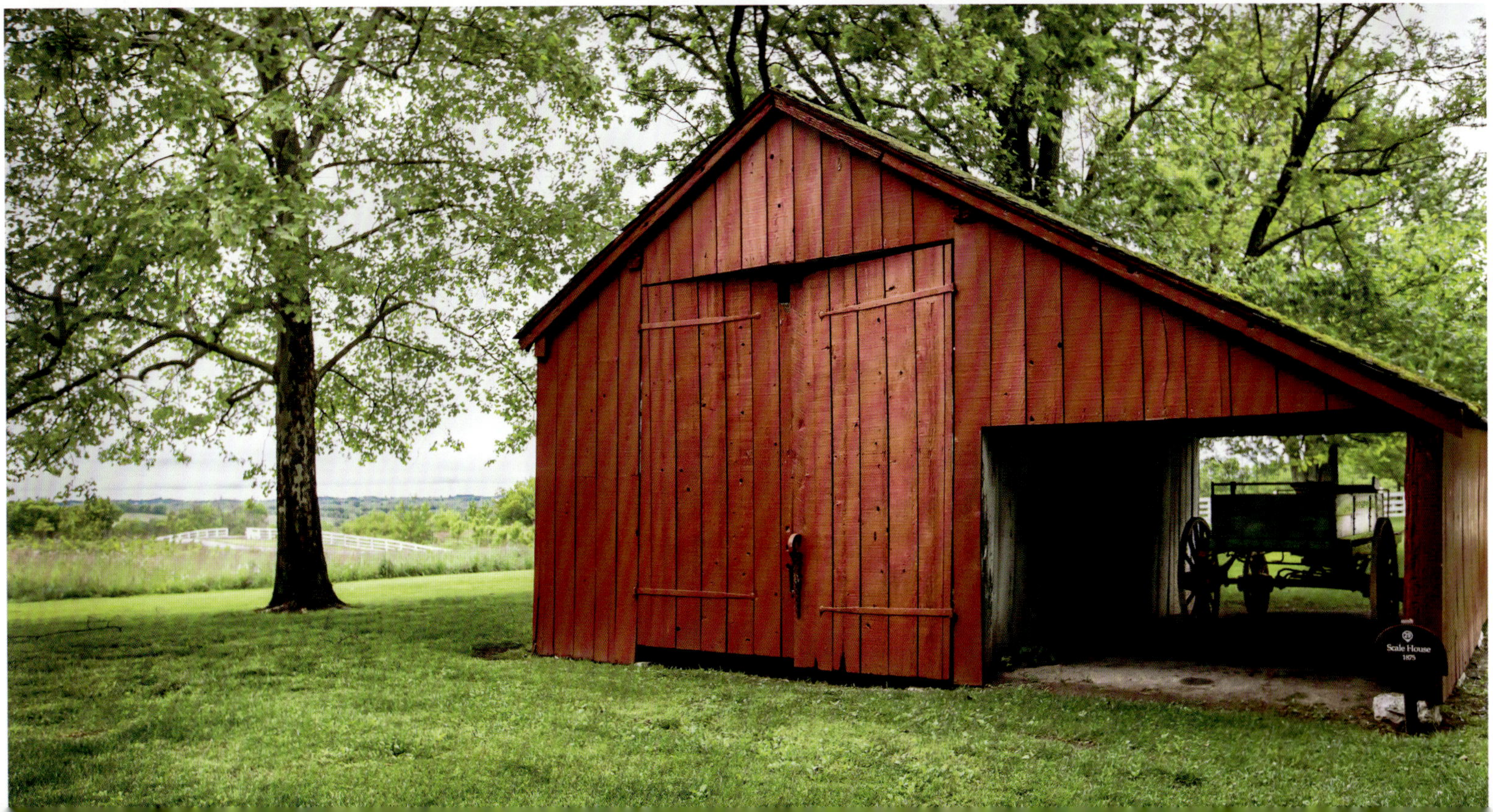

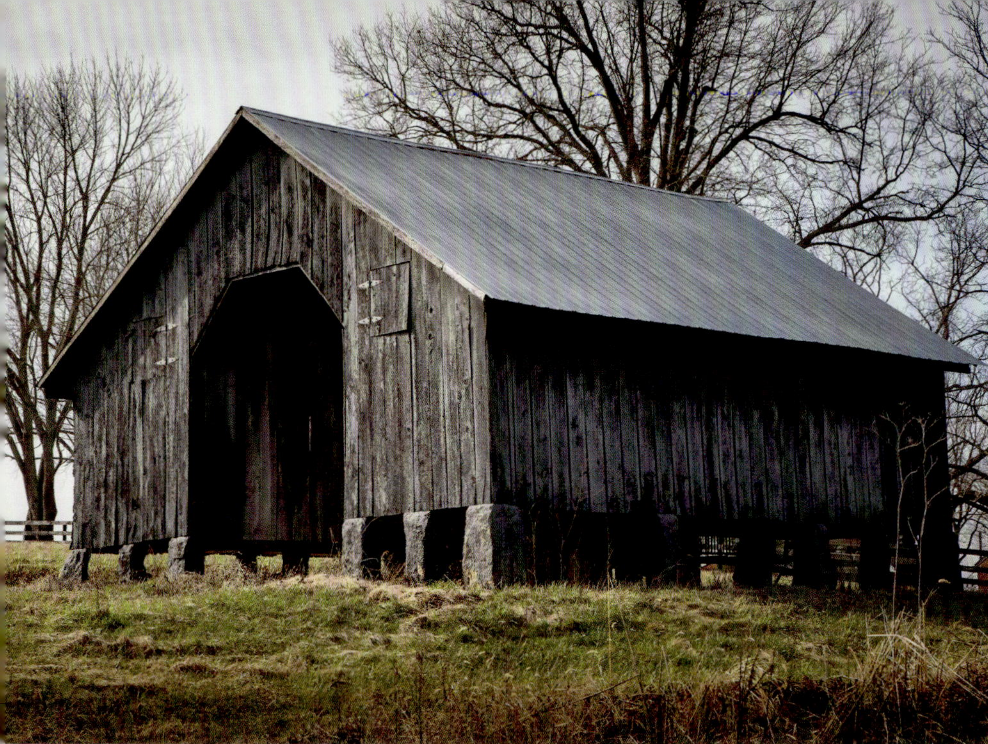

Bourbon County.

This drive-in double-crib barn on the Bucknore estate, built in the early twentieth century, has the classic opening for a cart or carriage.

Rockcastle County.

The barn in the foreground is an unusual design, with the sides on the bottom uncovered to allow easy access, like a pole barn. In the background is a corncrib barn, built in the late twentieth century.

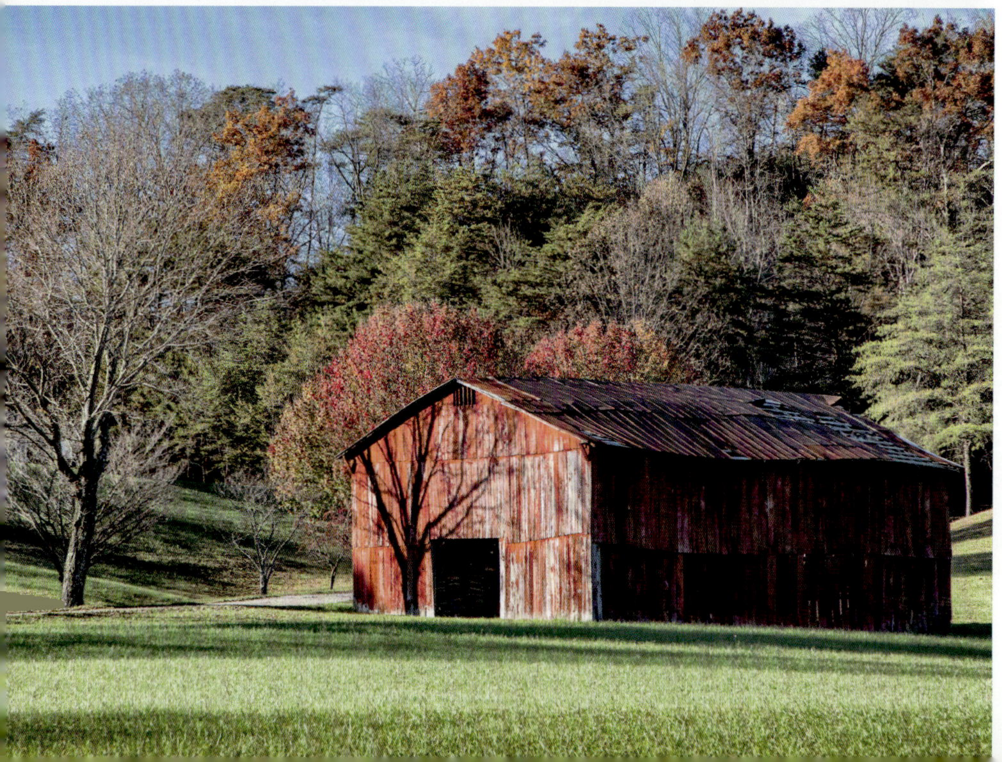

Bourbon County.

Walter Buckner established Bucknore Farm circa 1835–1841 and originally called the estate Locust Grove. Six generations later, the property remains in the family. Near to the Greek Revival mansion is this barn, clearly altered over time. Two early nineteenth-century brick buildings are the starting point. The springhouse with a carriage house on top is at one end, and a brick general-purpose barn with a stone foundation is in the center. Later, timber and stone additions were built at either end of the central barn and used for dairy and stock. Implements of various sorts, such as this cauldron and saw, are still inside.

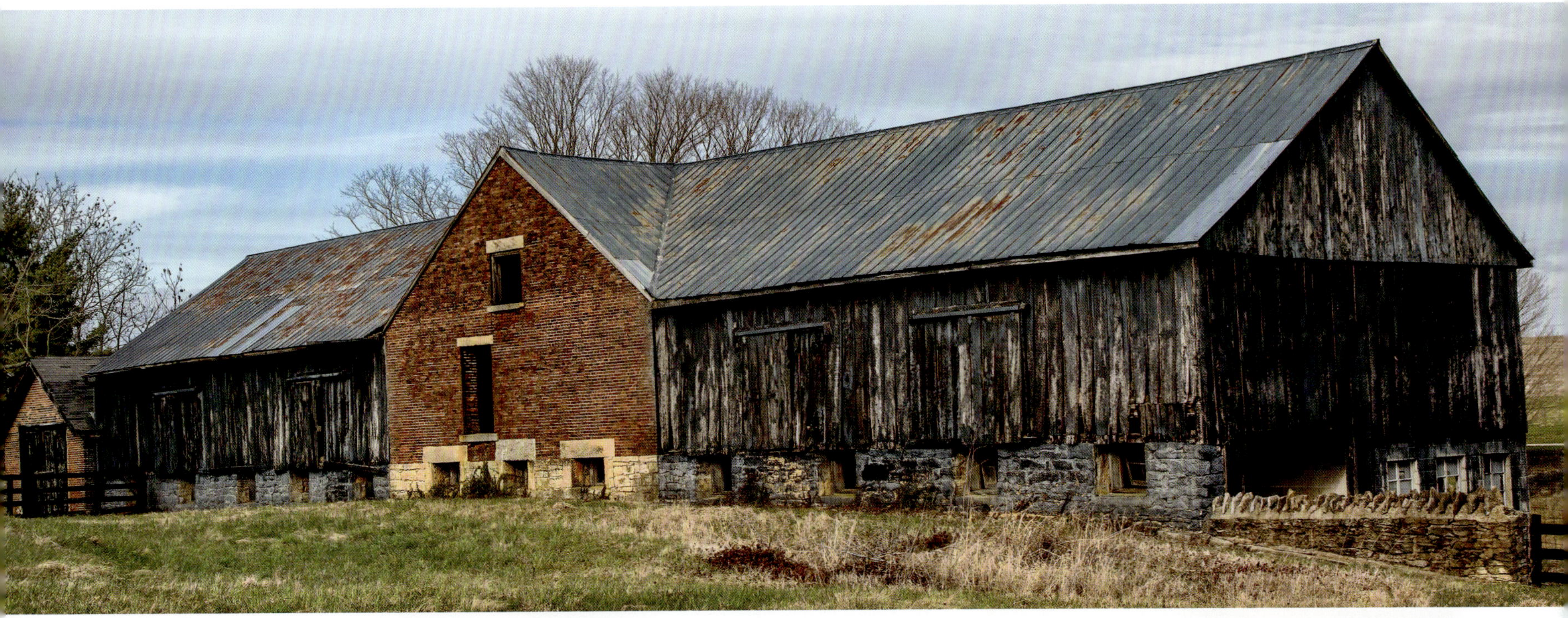

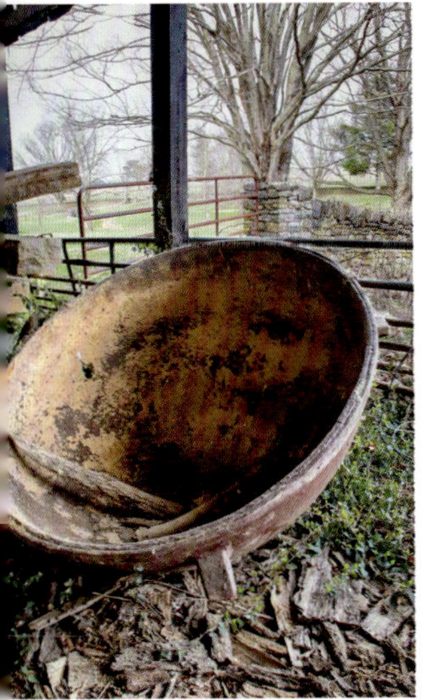

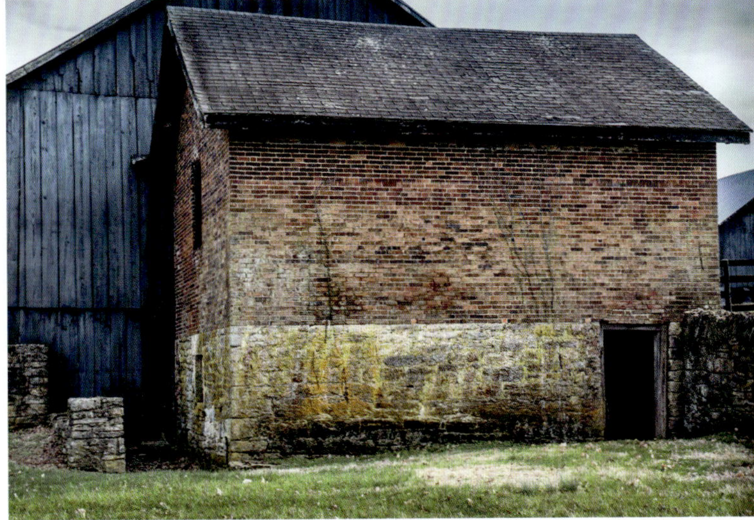

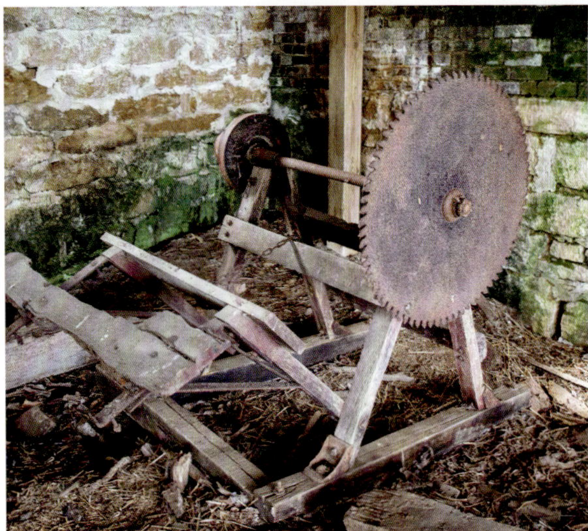

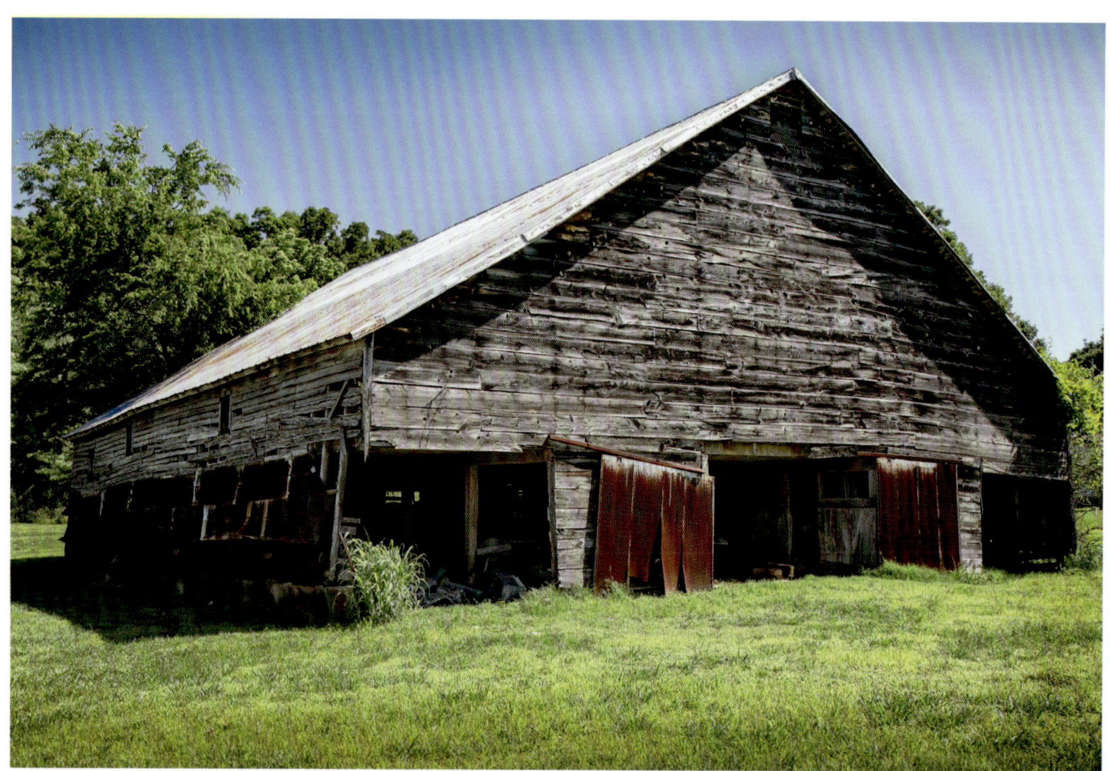

Livingston County.

Both of these barns belong to the Crouch family and were built circa 1930. Once used for stock, they now store equipment.

FACING PAGE

Franklin County.

As the stone chimney outside of this barn suggests, part of the barn was once a house. Inside the structure, timber surrounds a log house to create this barn. Later an extension was added to the back. This barn was built in the eighteenth century.

200

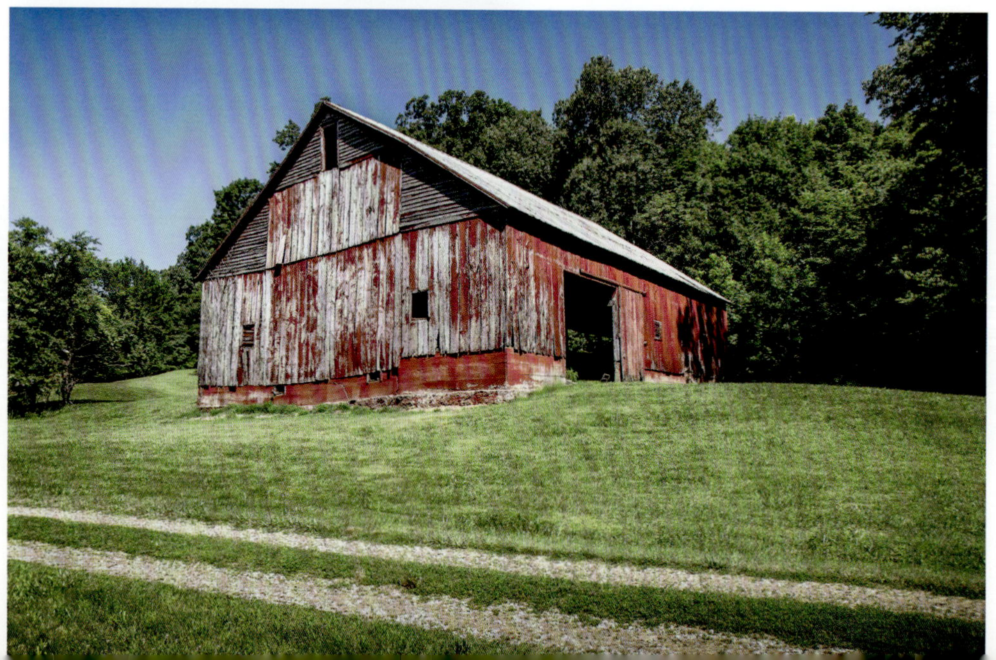

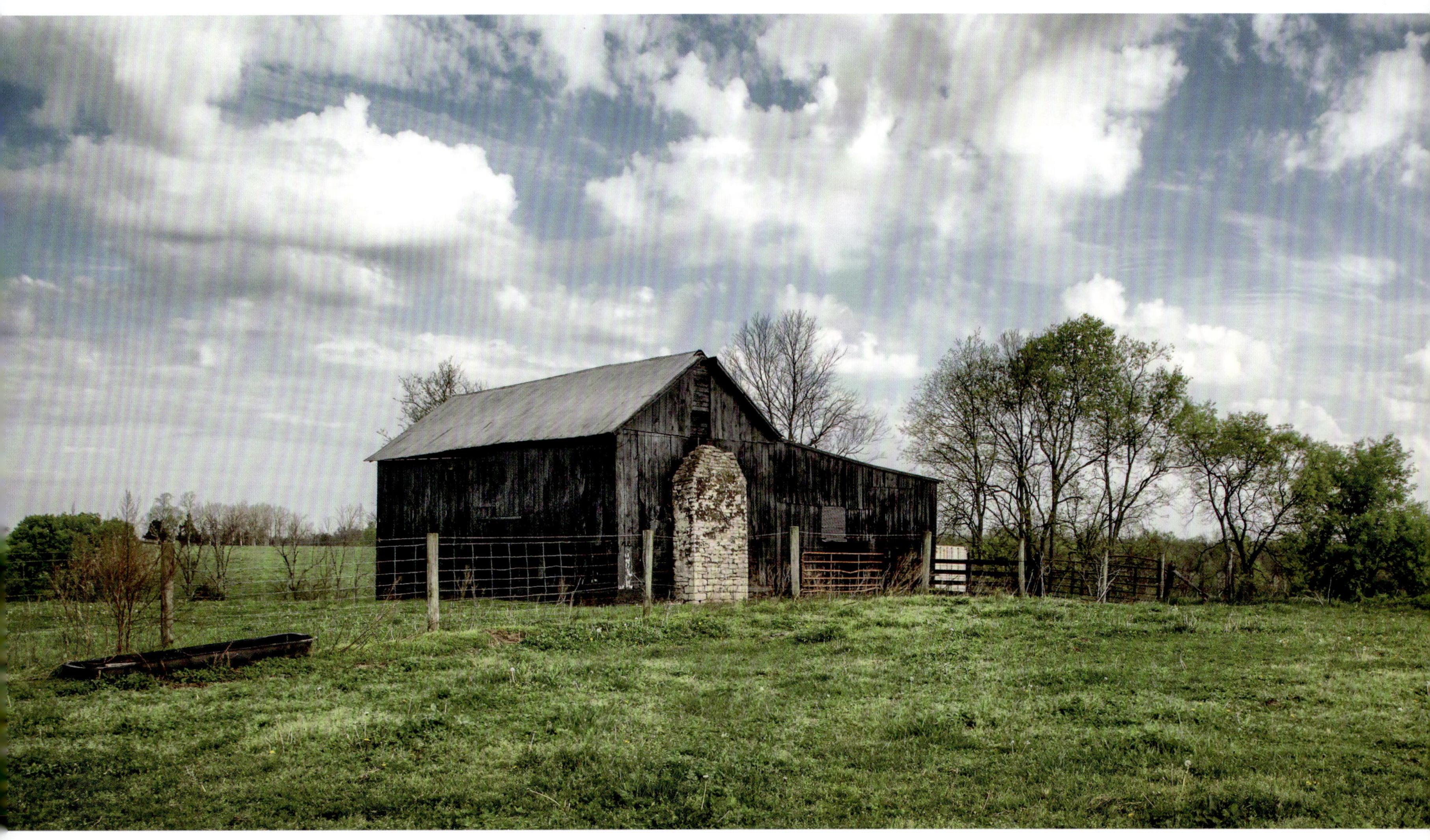

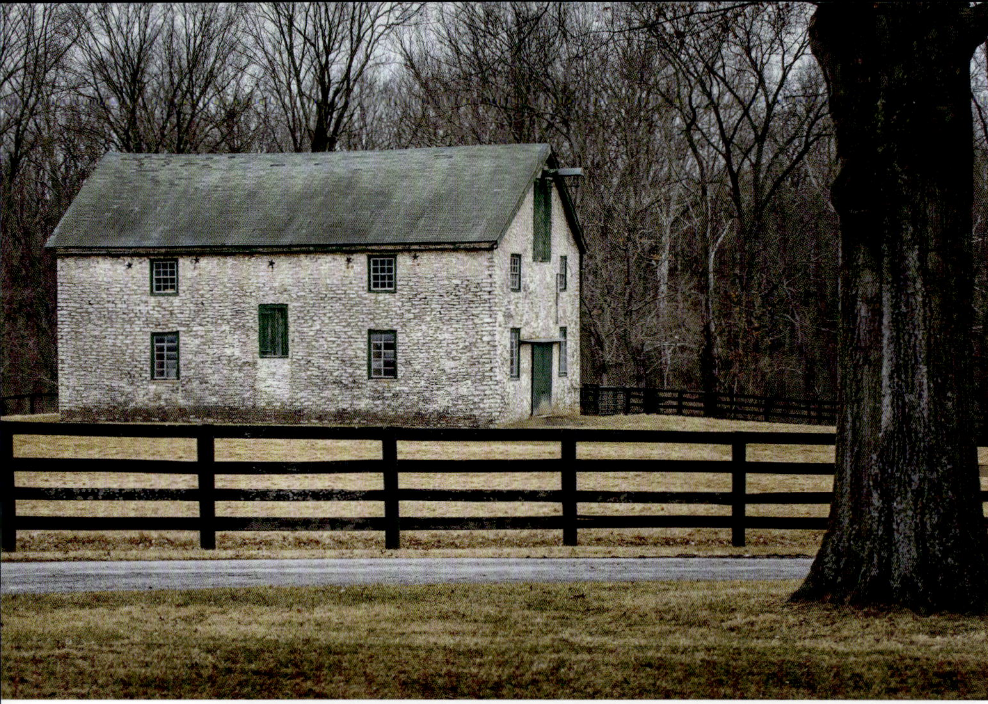

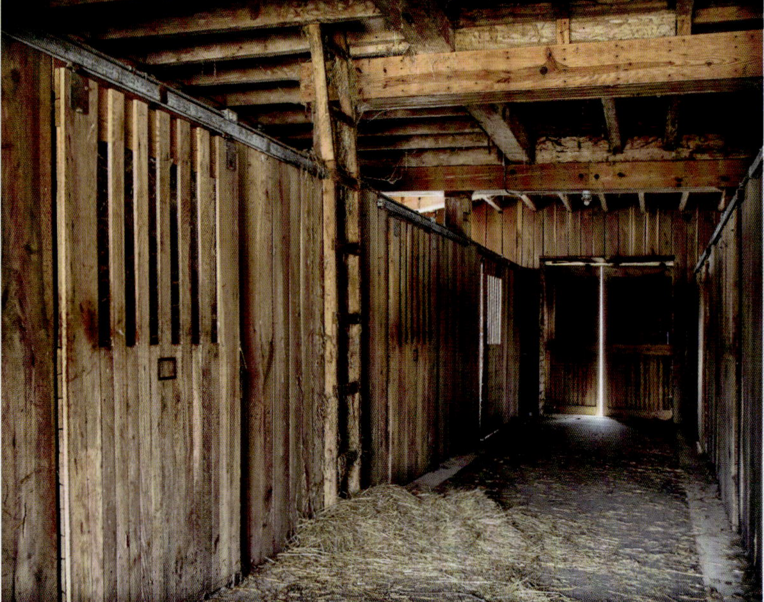

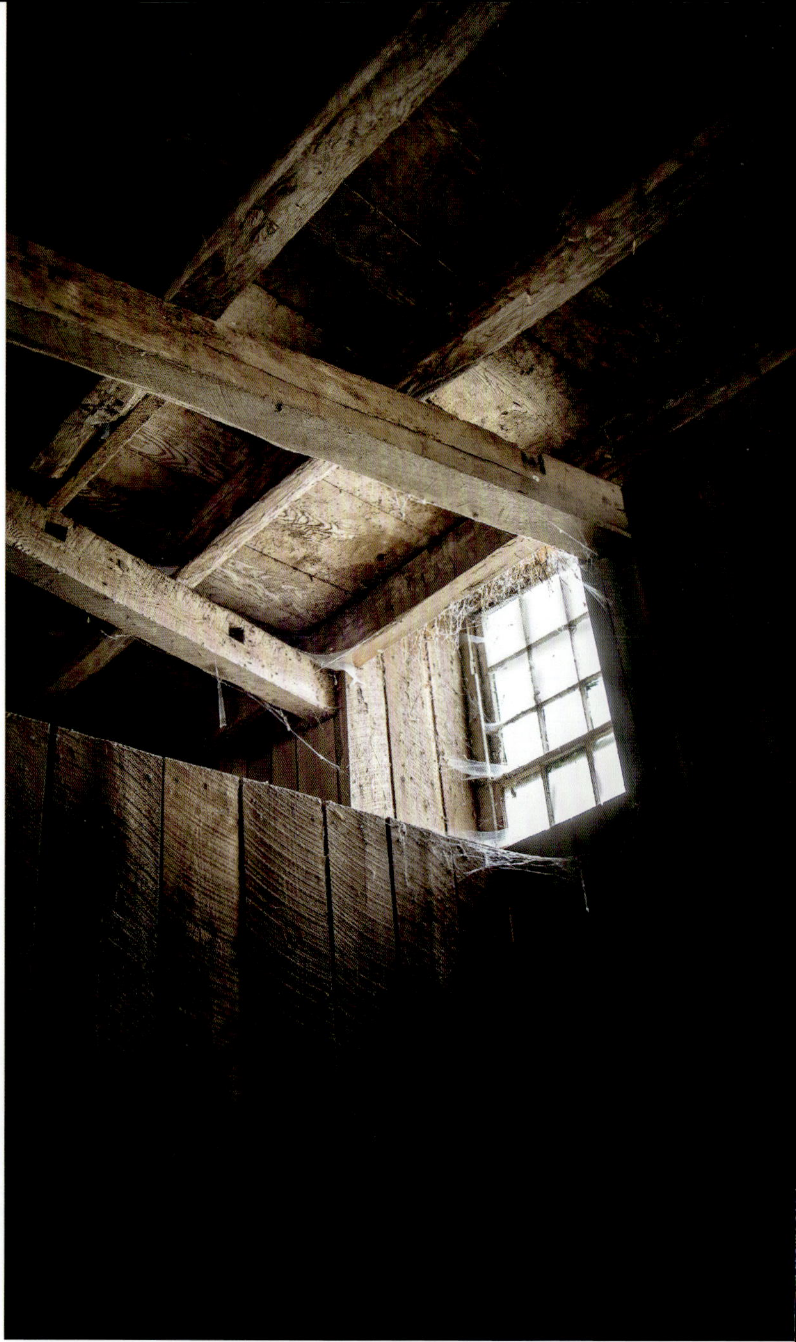

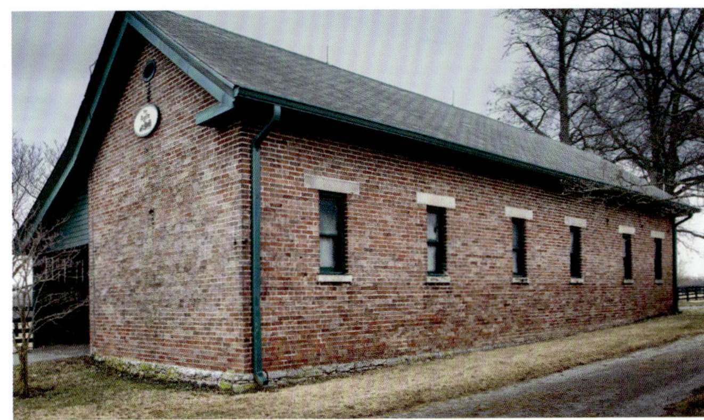

Bourbon County.

Runnymede Farm uses this brick barn for mares and foals, but in the 1850s, it was built to serve as part of a hemp factory and barn. A view of the outside still shows the character of the old barn and factory. The sandstone lintels and sills and Flemish bond brickwork remain. A recent extension has been built on the other side and stall doors installed. A look down the aisle created by the extension shows the Cooper's Run Baptist Church stone barn in the distance.

FACING PAGE

Bourbon County.

Now part of Runnymede Farm, this stone barn was built around 1803 as the meetinghouse of Cooper's Run Baptist Church. Later it was part of a hemp factory. Runnymede itself was founded in 1867 and is one of the oldest continually operating Thoroughbred breeding operations. Inside stalls have been added, but a look at the beams above show the hand-hewn work in what was once the floor of a balcony surrounding the church's interior.

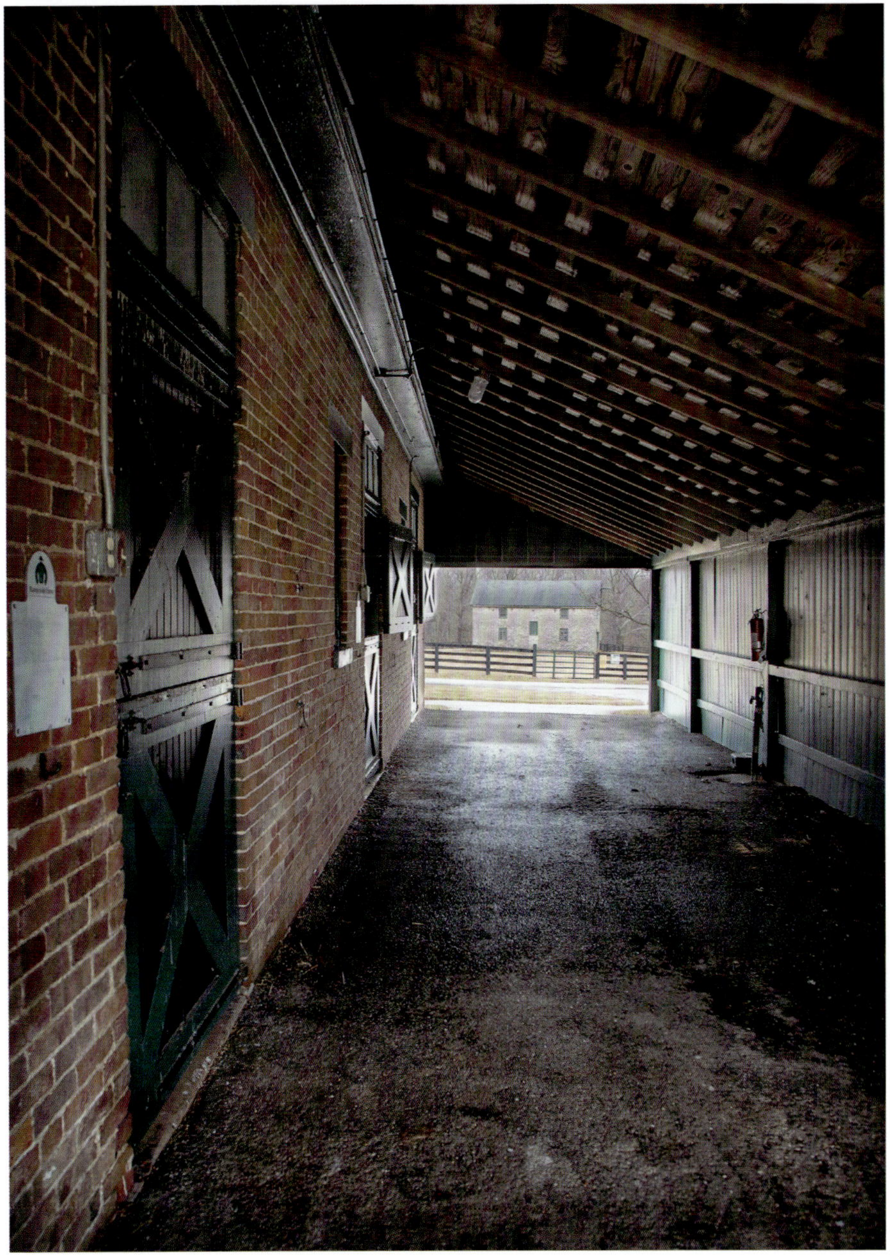

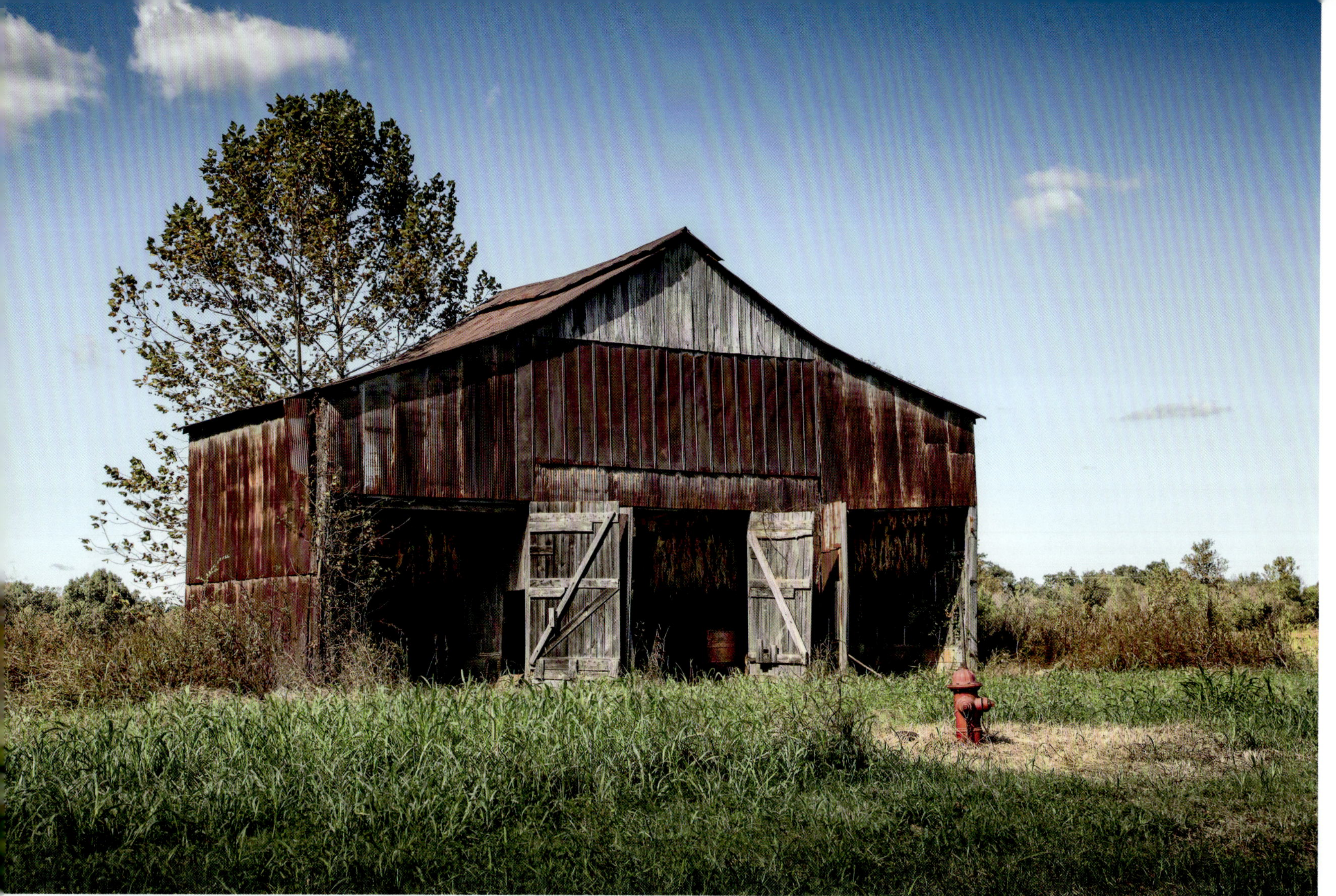

Christian County.

This roadside barn was originally intended to store equipment
or stock. It has been adapted for drying burley tobacco.

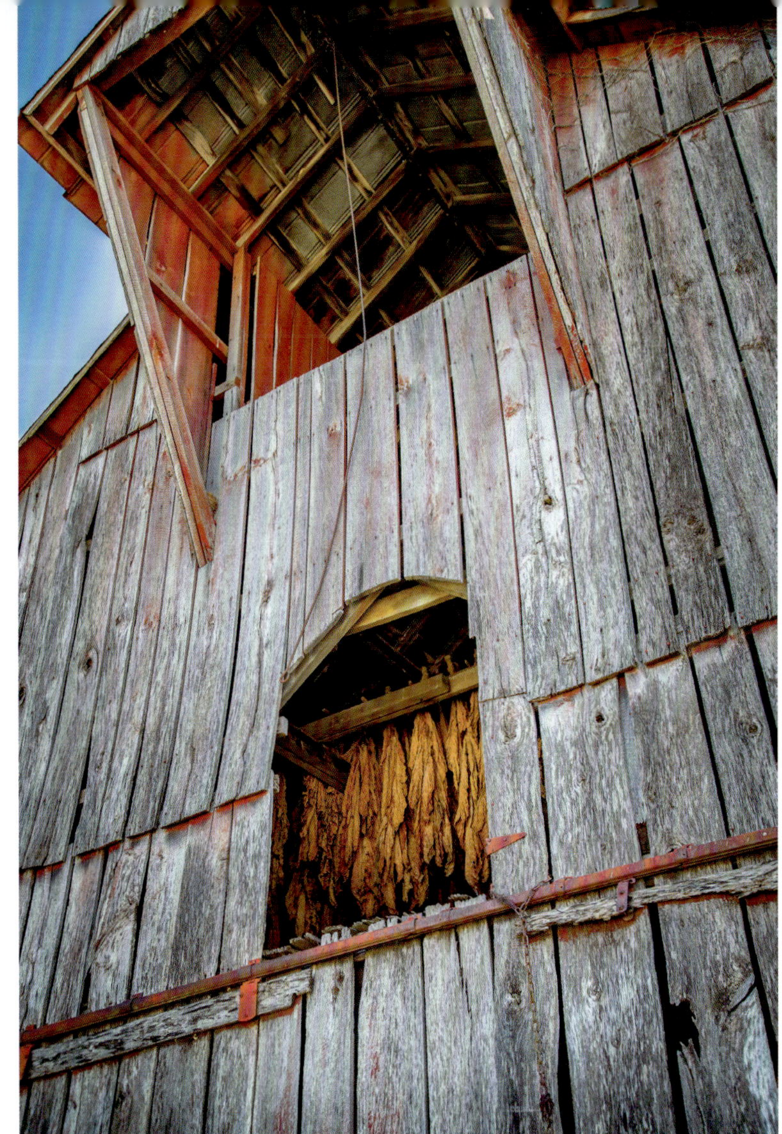

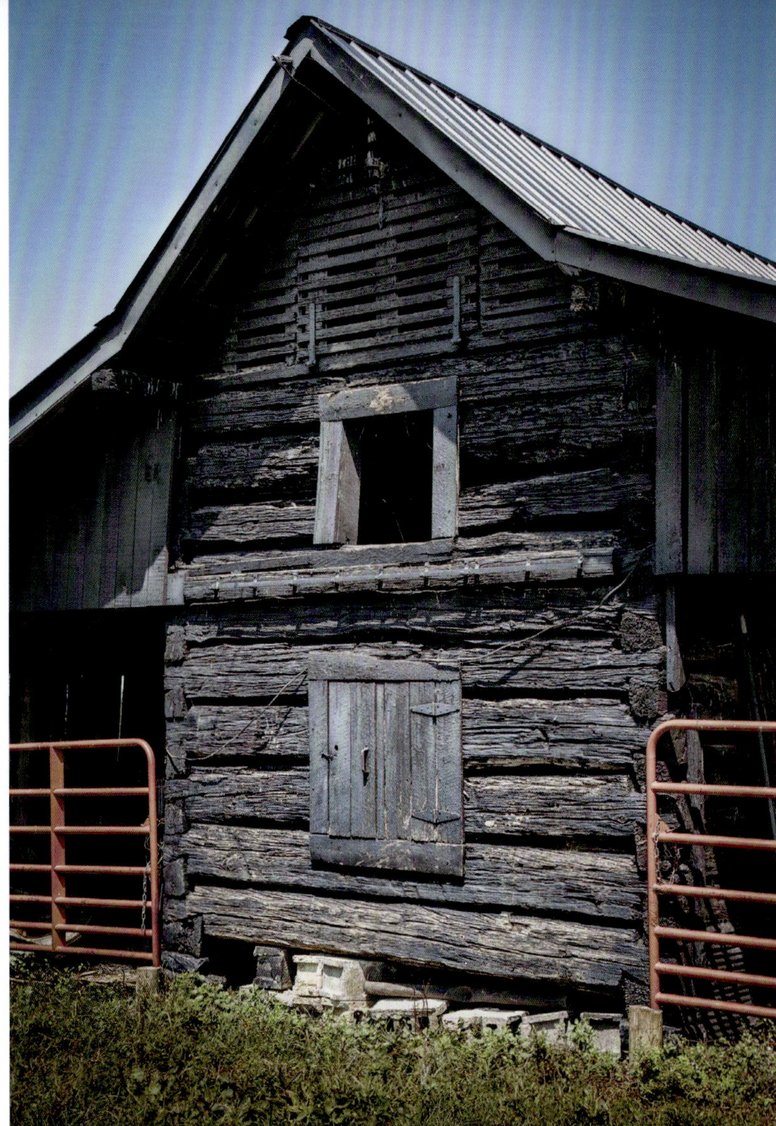

Todd County.

Inside this multipurpose barn, complete with stalls, a hayloft, and a hay hook, hangs burley tobacco.

Madison County.

The eighteenth-century log crib barn in this image was part of the original McWilliams land grant from the 1700s.

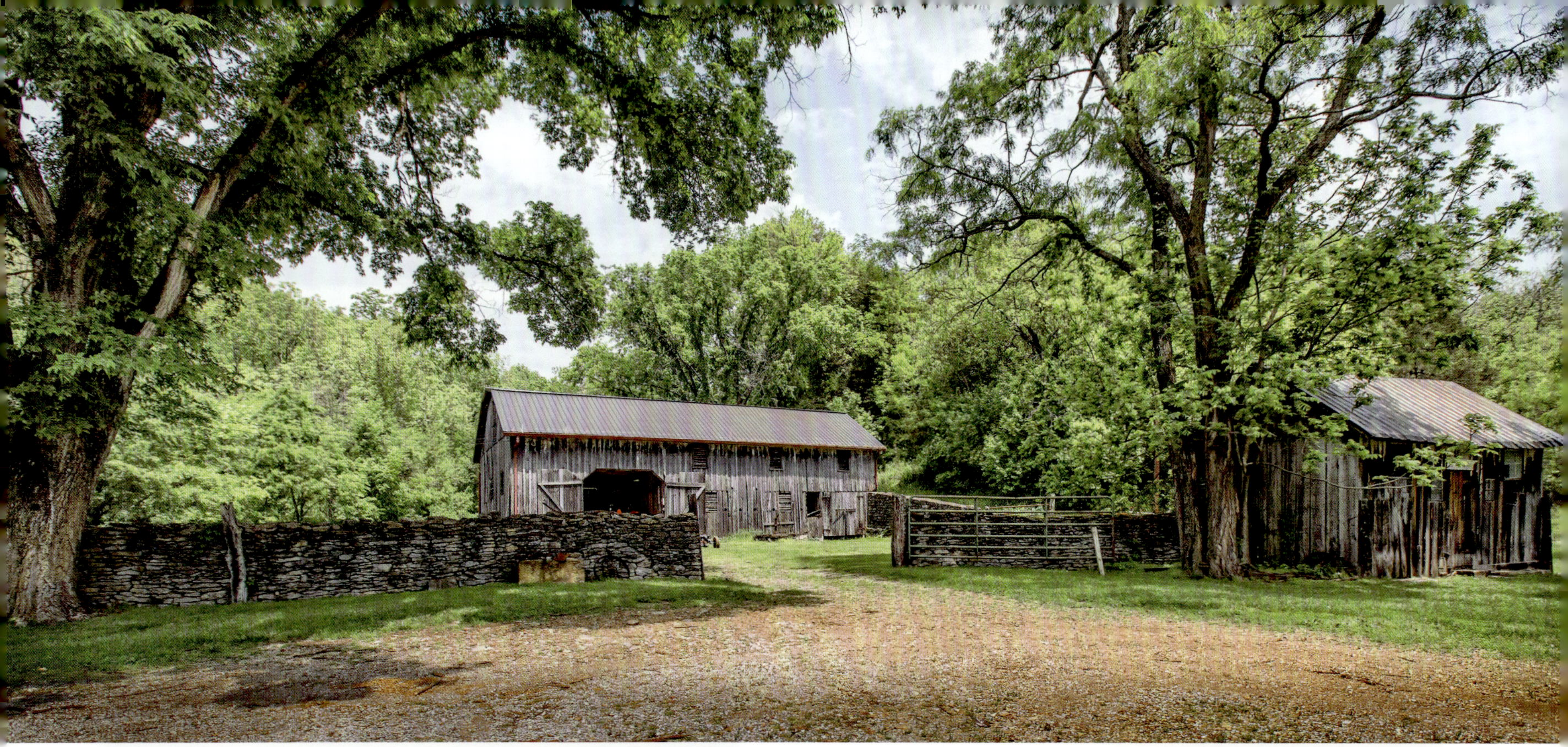

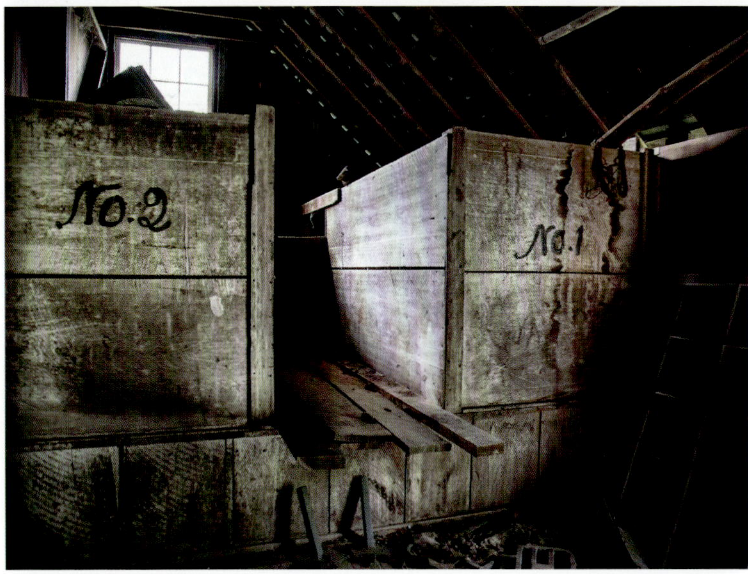
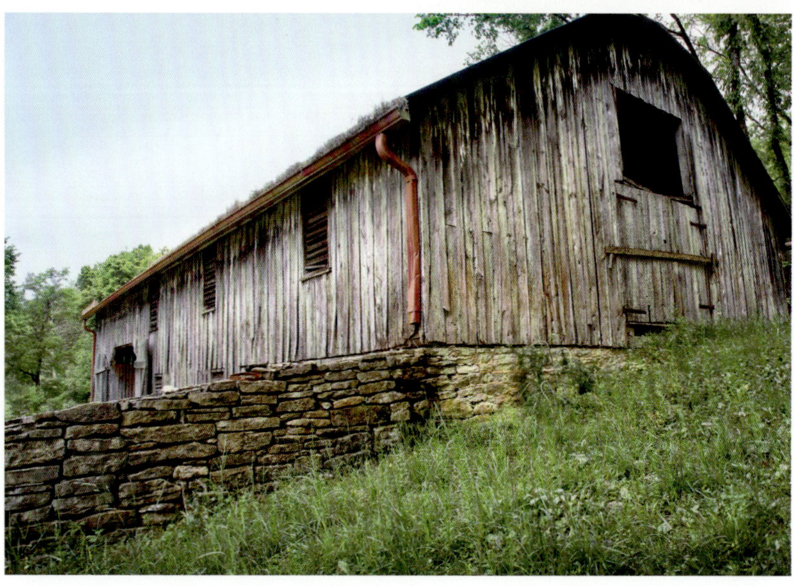

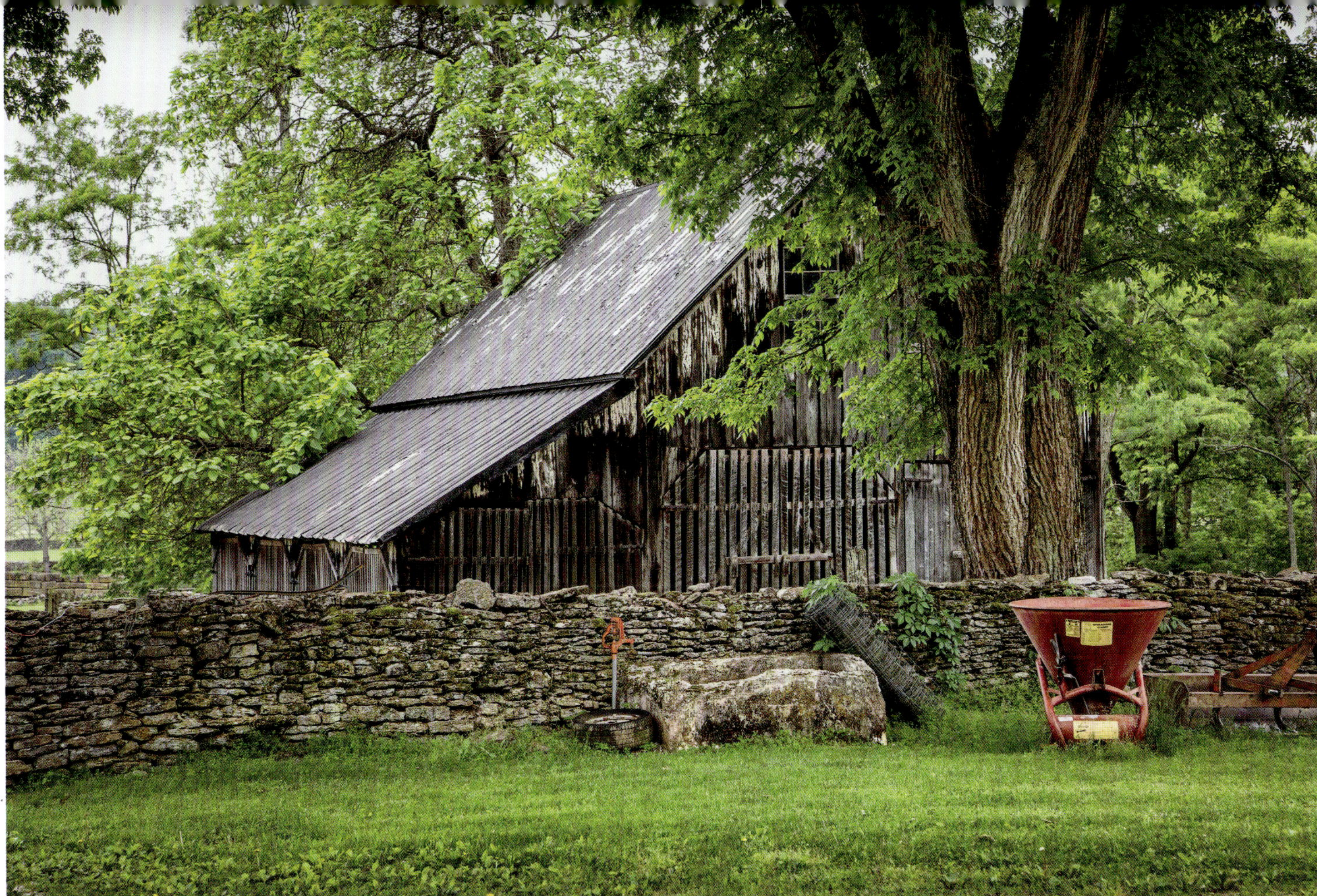

Washington County.

Hamilton Farm, also known as Parker's Landing, formed during the early settlement of Kentucky. Established in the early 1800s with a domestic building built around 1825, the farm added a dairy in 1880. This dairy and stock barn was built on a stone foundation. The interior retains original stalls and design.

Washington County.

As the agricultural life at Hamilton Farm grew, buildings proliferated. The carriage house and granary were built around 1877, and the tobacco barn with stripping room was built around 1925. All of the buildings are a short walking distance from the house.

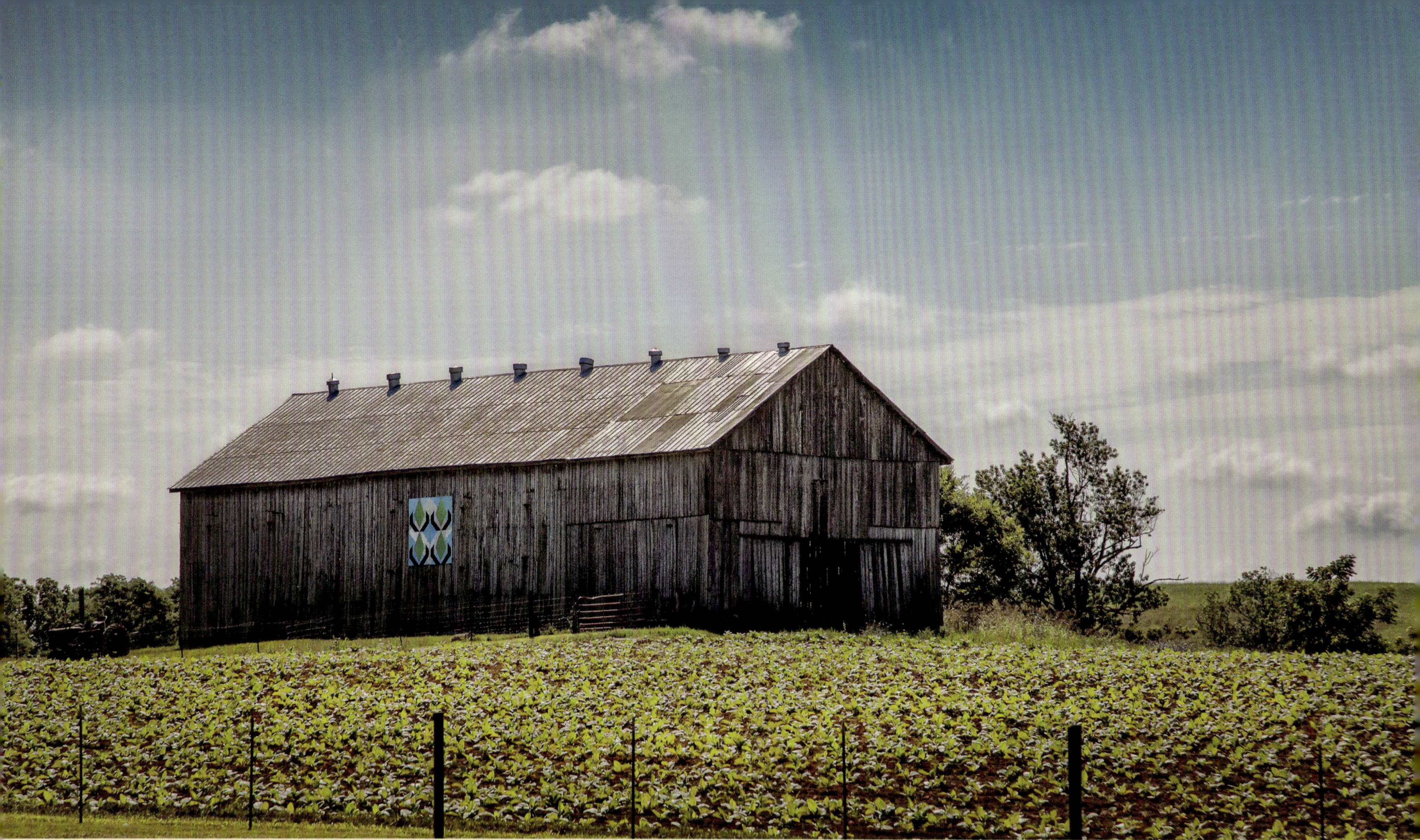

Jessamine County.

This tobacco barn sits in a field of newly planted tobacco.
The quilt motif shows a crop resembling corn or tobacco.

FACING PAGE

Rockcastle County.

This chestnut multipurpose barn sits in an area that was once one of the
biggest settlements between Shelby Branch and Reedsville. Its first half
was built circa 1925; Johnny King and his family built later additions.

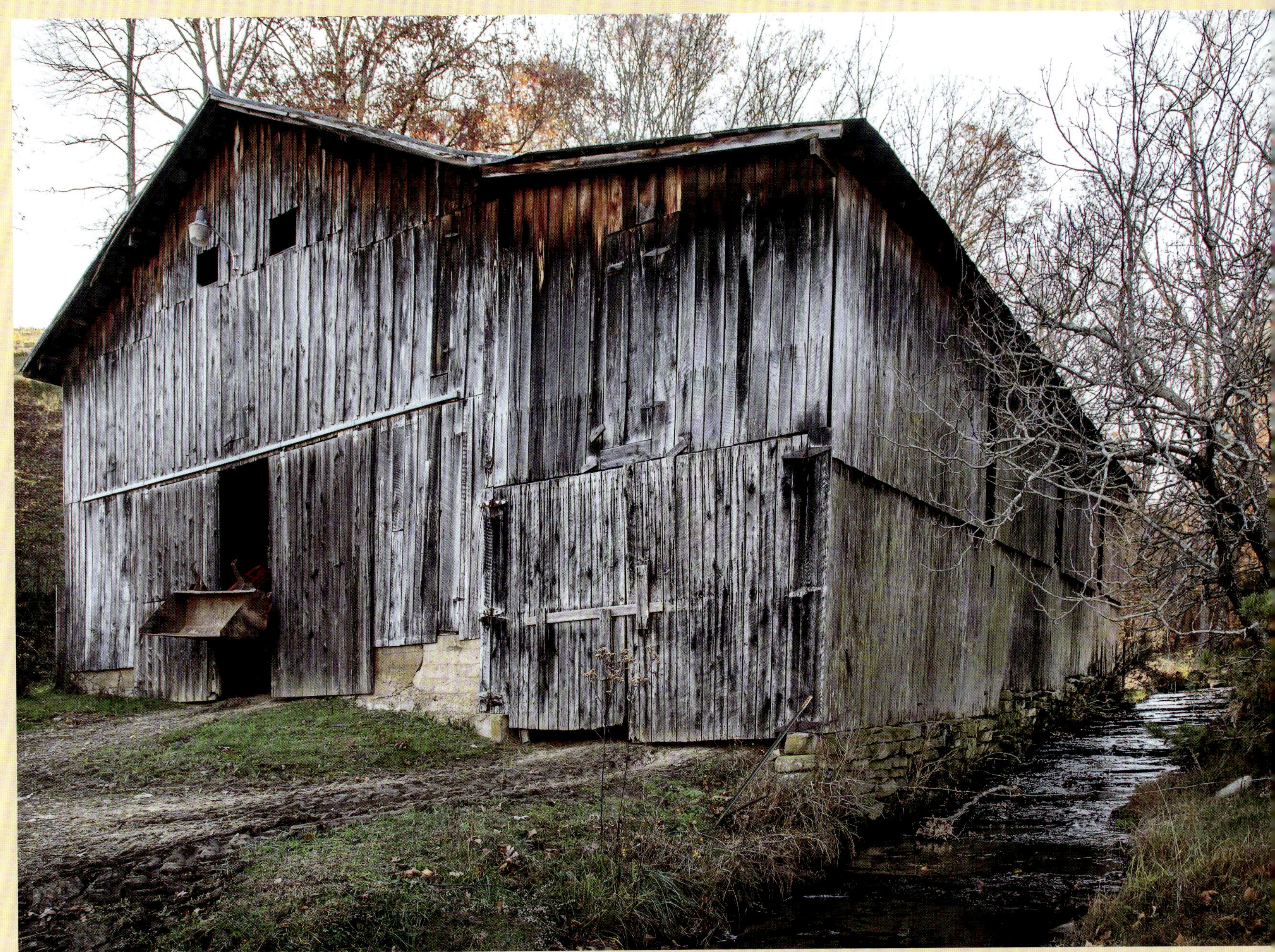

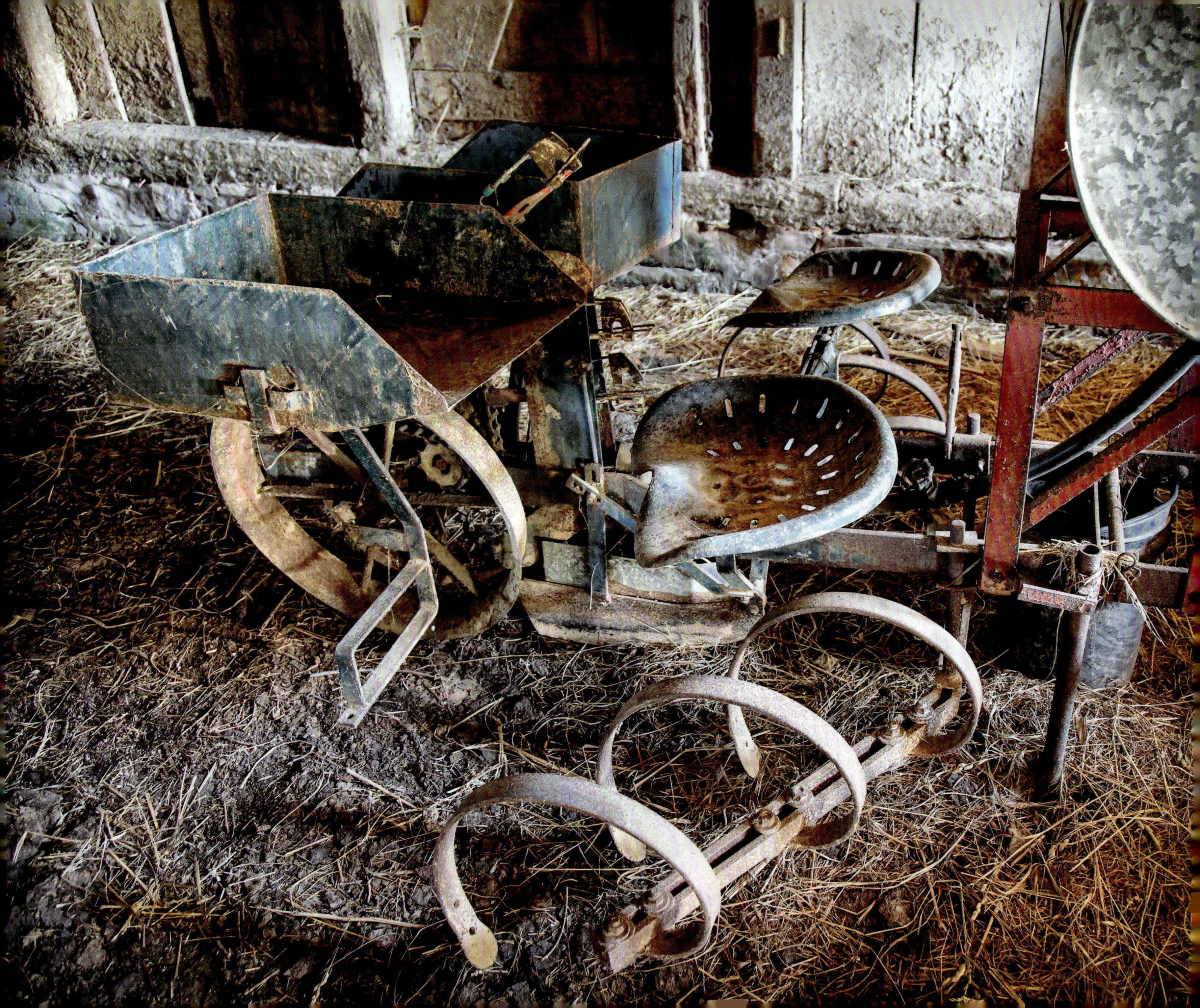

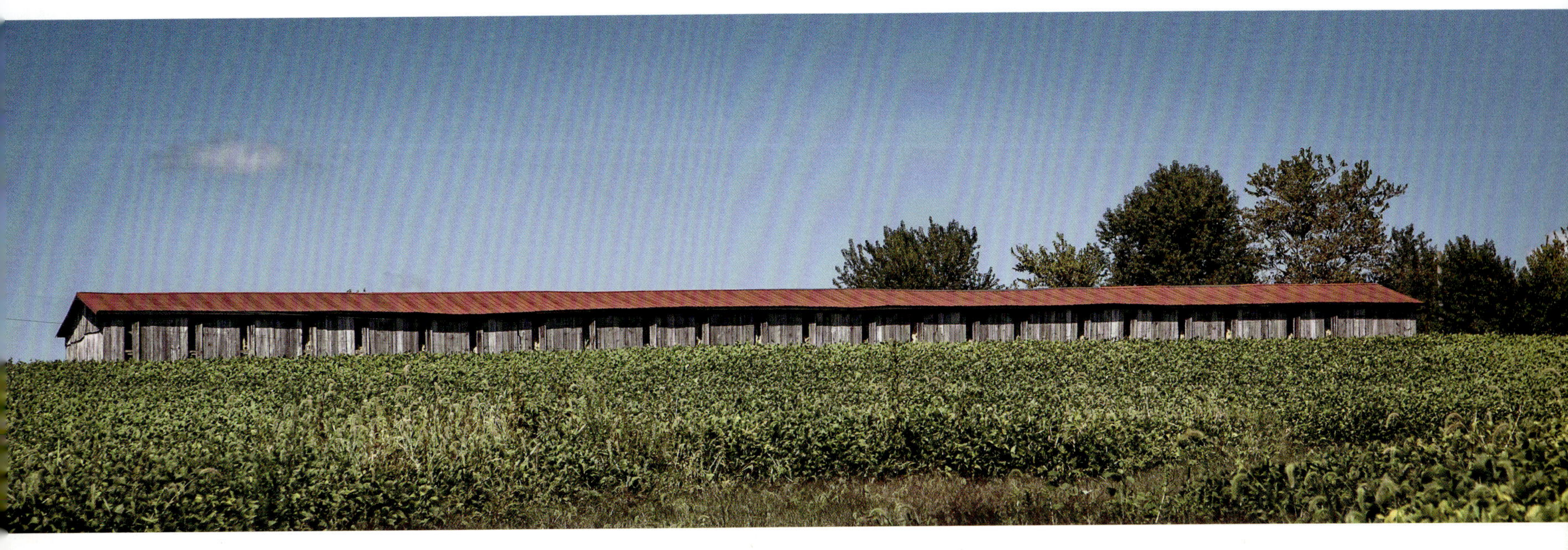

Lincoln County.

Old farm equipment, like this machine for planting tobacco, can be found in many abandoned barns.

Woodford County.

A long, low tobacco barn sits out in a soybean field. The owner of this late twentieth-century barn rents it out to tobacco growers, who use it to hang their burley for drying.

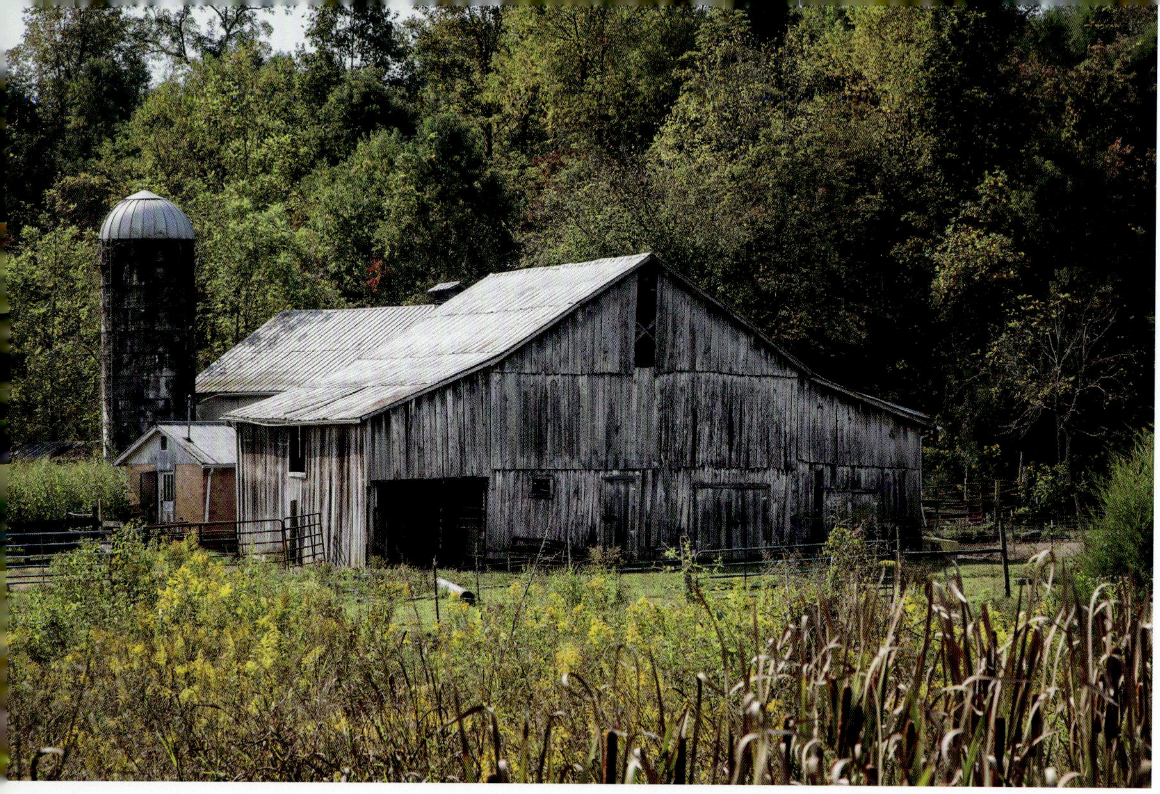

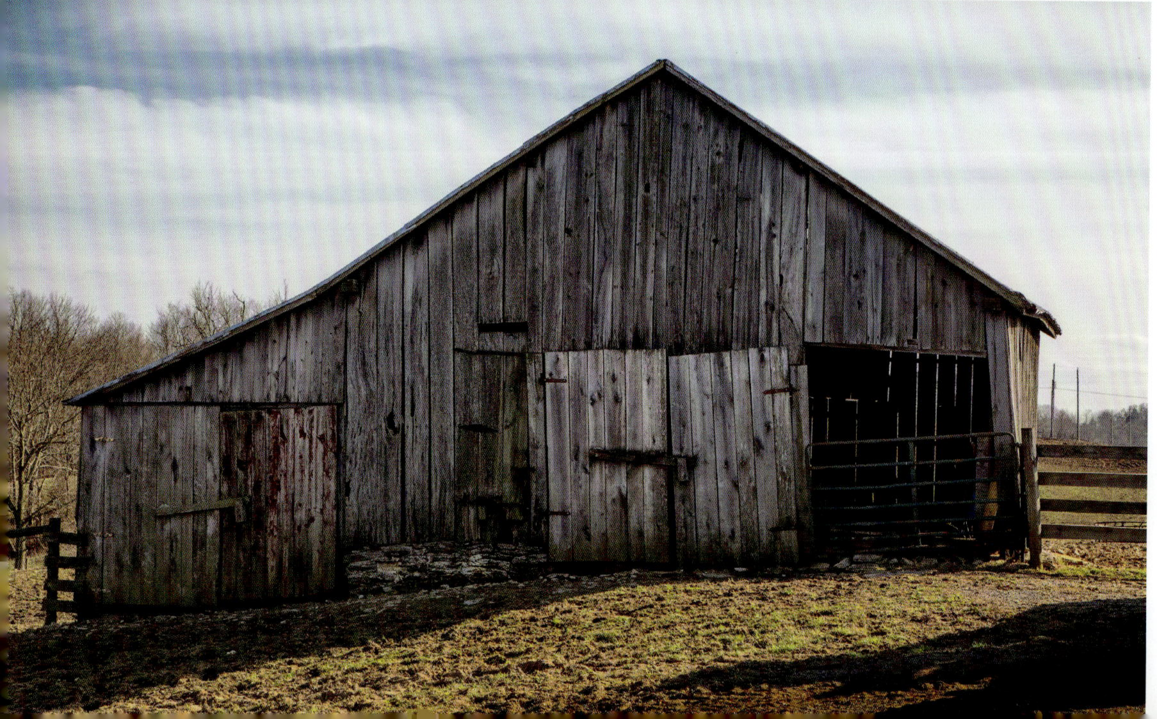

FACING PAGE

Christian County.

The Swatzell family established their farm in 1938, including this four-bay star-quilt barn with stone foundation.

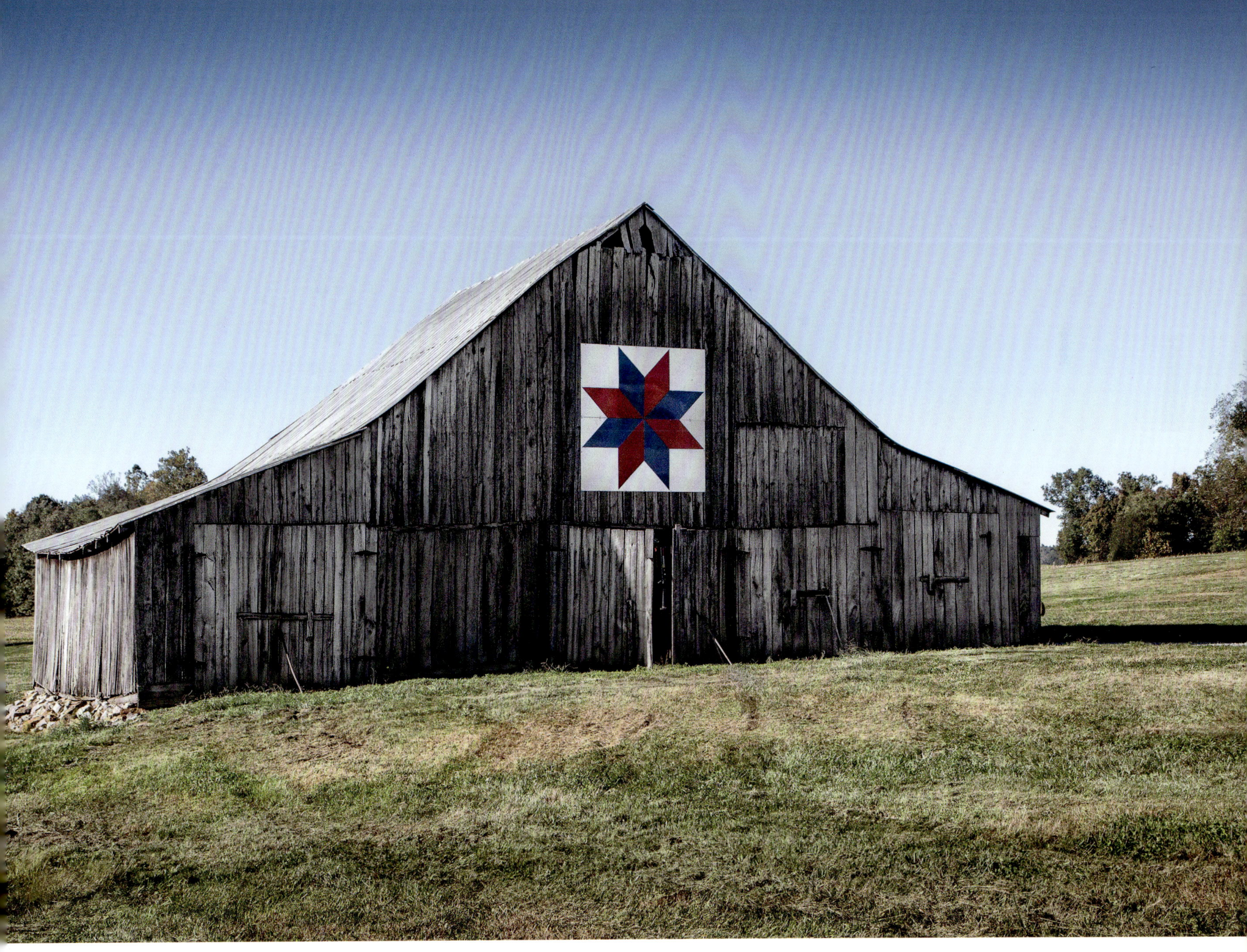

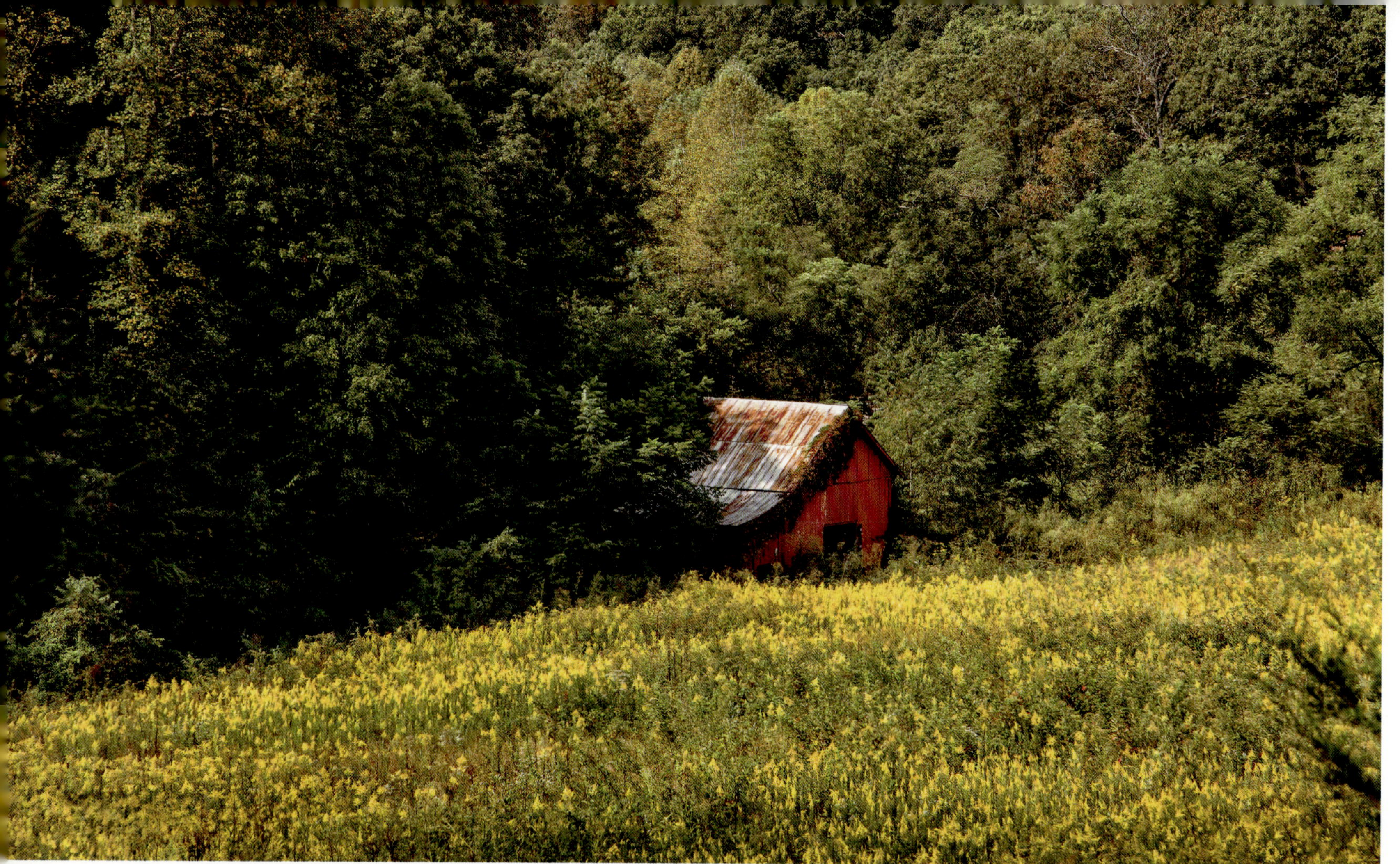

Lawrence County.

Snuggled in down a hillside, this barn is no
longer used but carries a sad beauty.

Montgomery County.

Originally built by Edward Rogers Prewitt in 1908, this barn
was used for warehousing because of its location along the C&O
Railway. The stop was known as Prewitt Station. The barn remains in
the family generations later, even after the train tracks are long gone.

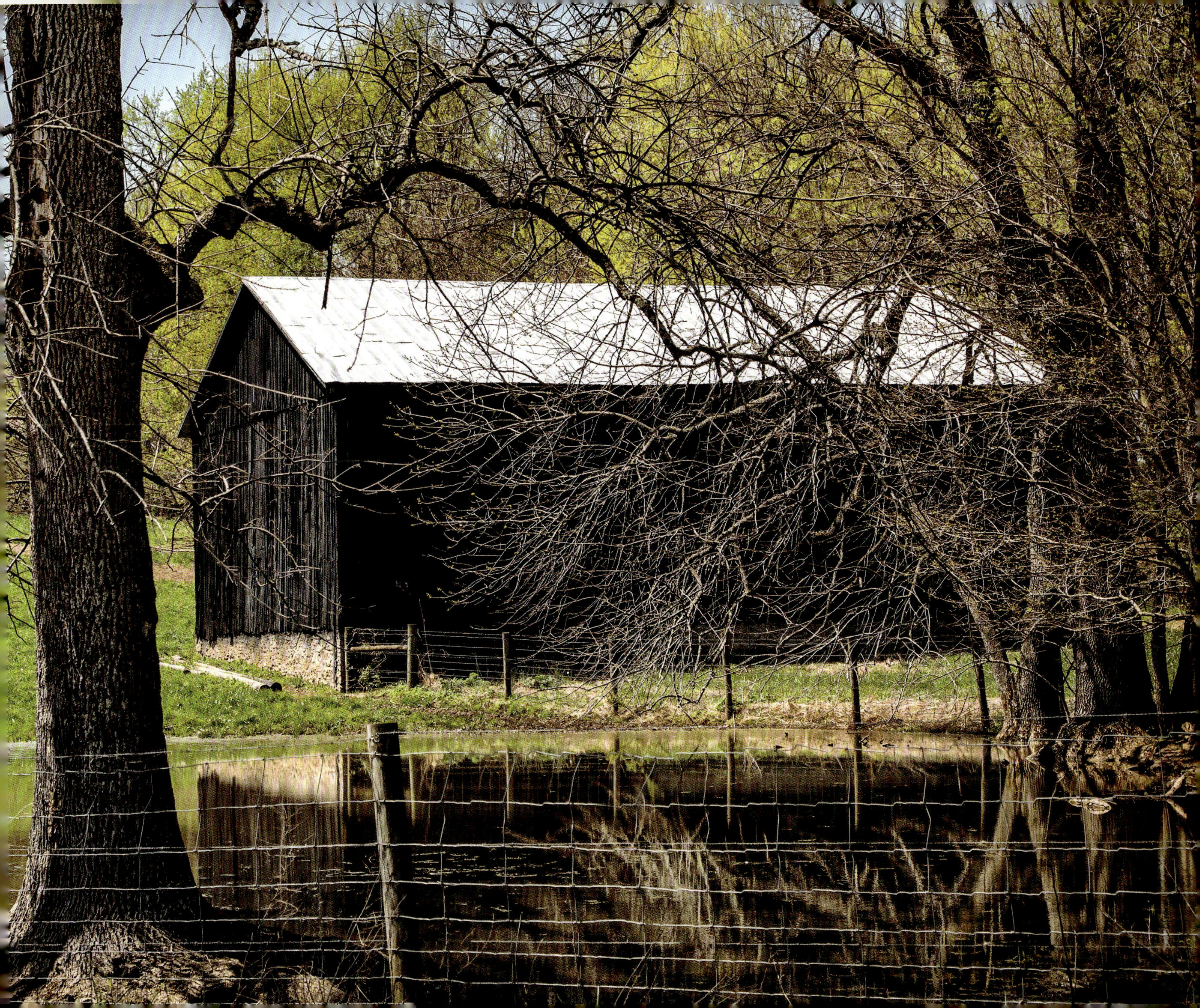

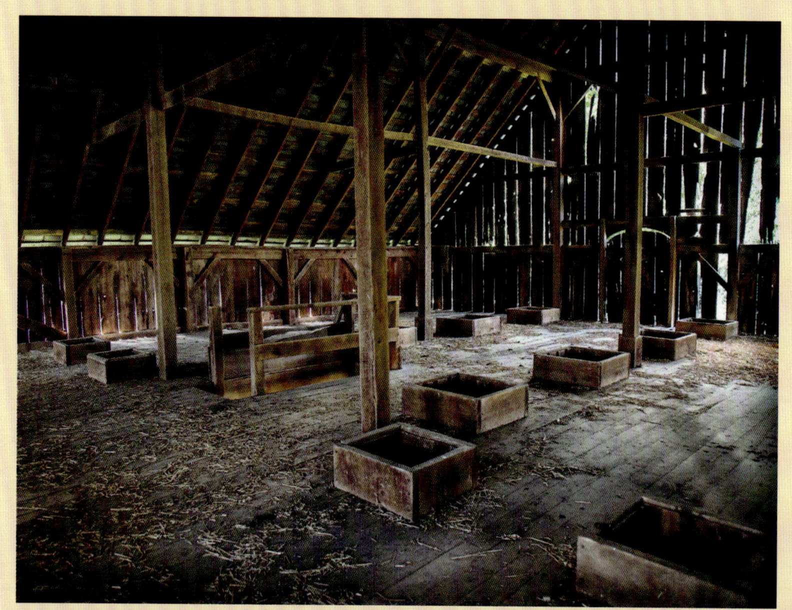
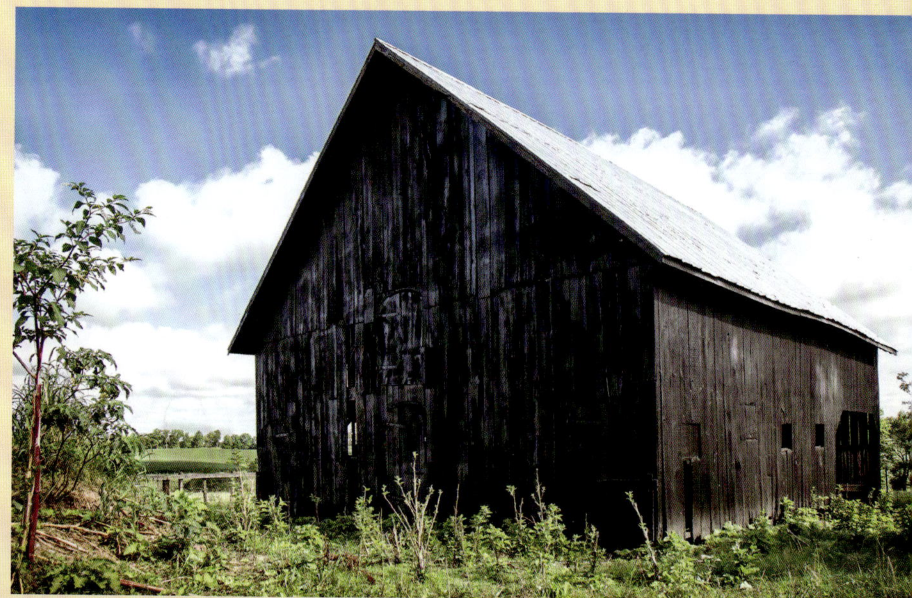
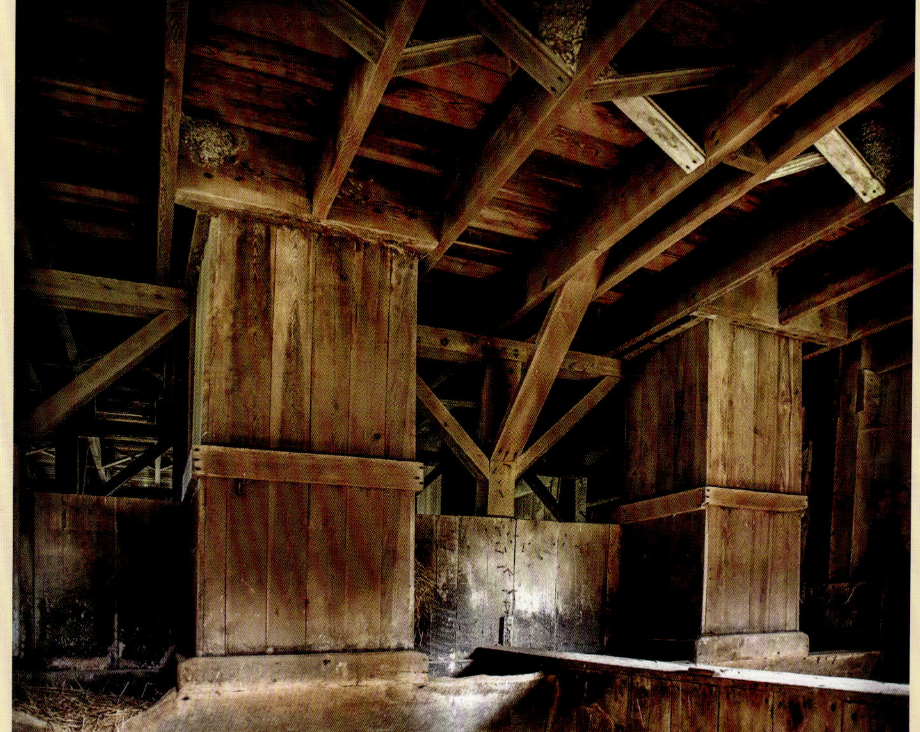
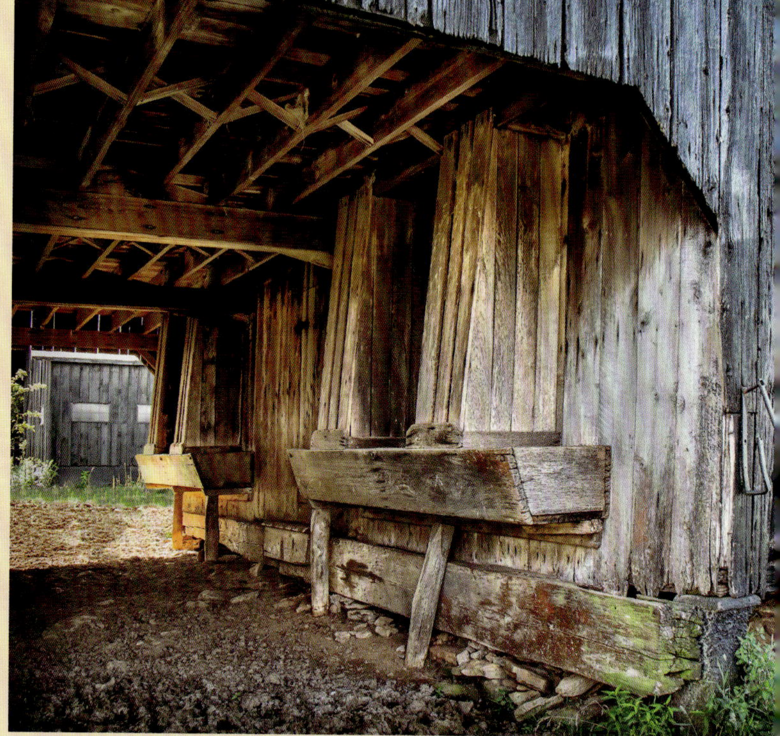

Little noticed at the Kentucky Horse Park is this crib barn on a stone foundation, situated next to another stone building. The property was originally part of the L. V. Harkness Walnut Hill Farm.

FACING PAGE

Clark County.

The interior of the mule barn in this image is made of hardwood with amazing craftsmanship and attention to detail. W. R. Sphar built the structure in the late 1800s. The box openings in the loft allow hay and grain to be dropped down the chutes into feeders, both inside and out.

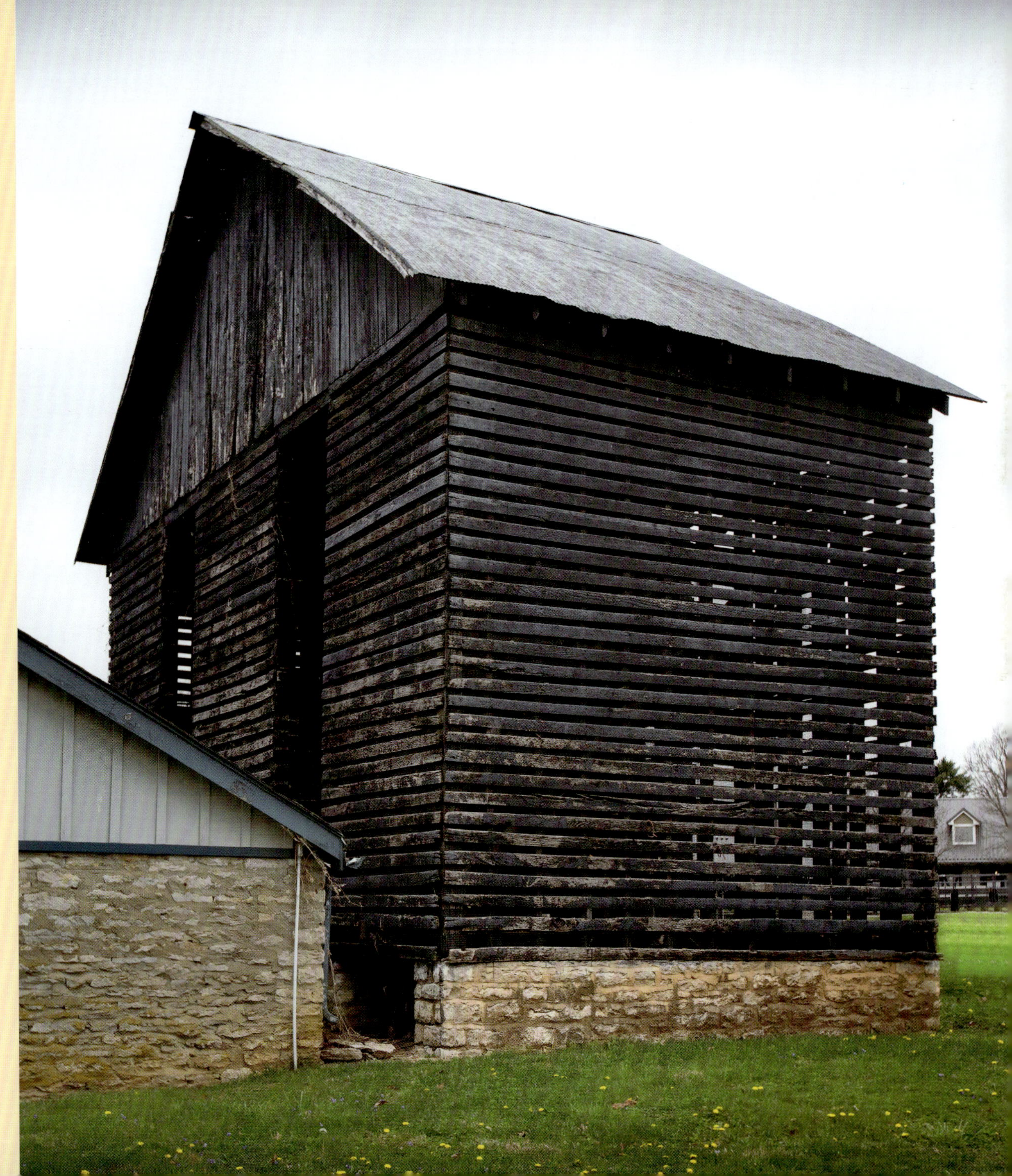

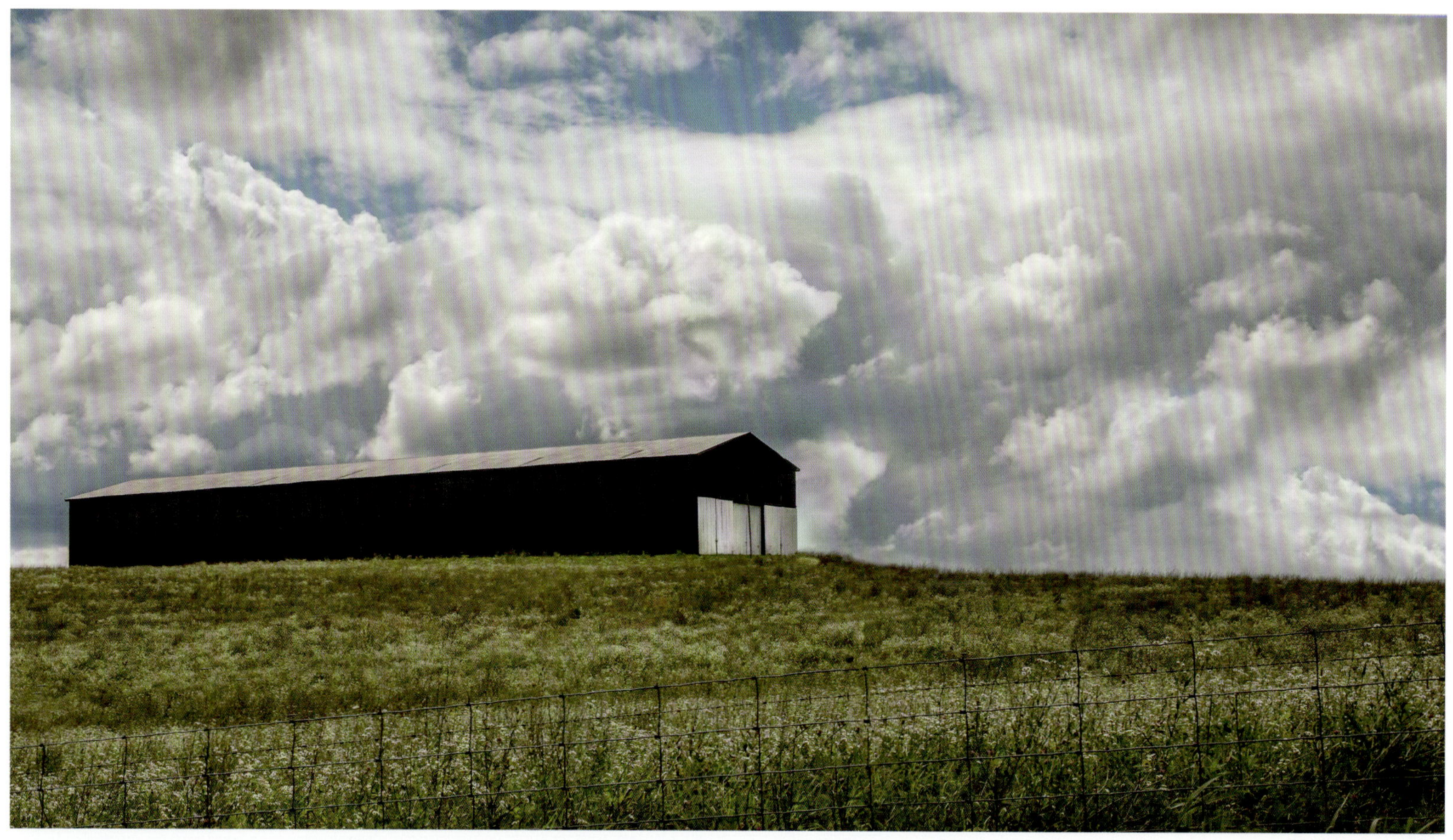

Scott County.

The tobacco barn is an iconic symbol of what was once
Kentucky's largest cash crop. It is now an endangered species.

FACING PAGE

Scott County.

Catherine Taylor has lovingly restored this beautiful twin silo barn
in memory of her late husband. Built circa 1914, the barn served
as a tobacco barn and mule stable. Two carriage openings are on
the lower level. Around the premises are implements from when
the farm was active, including an anvil and old watering trough.

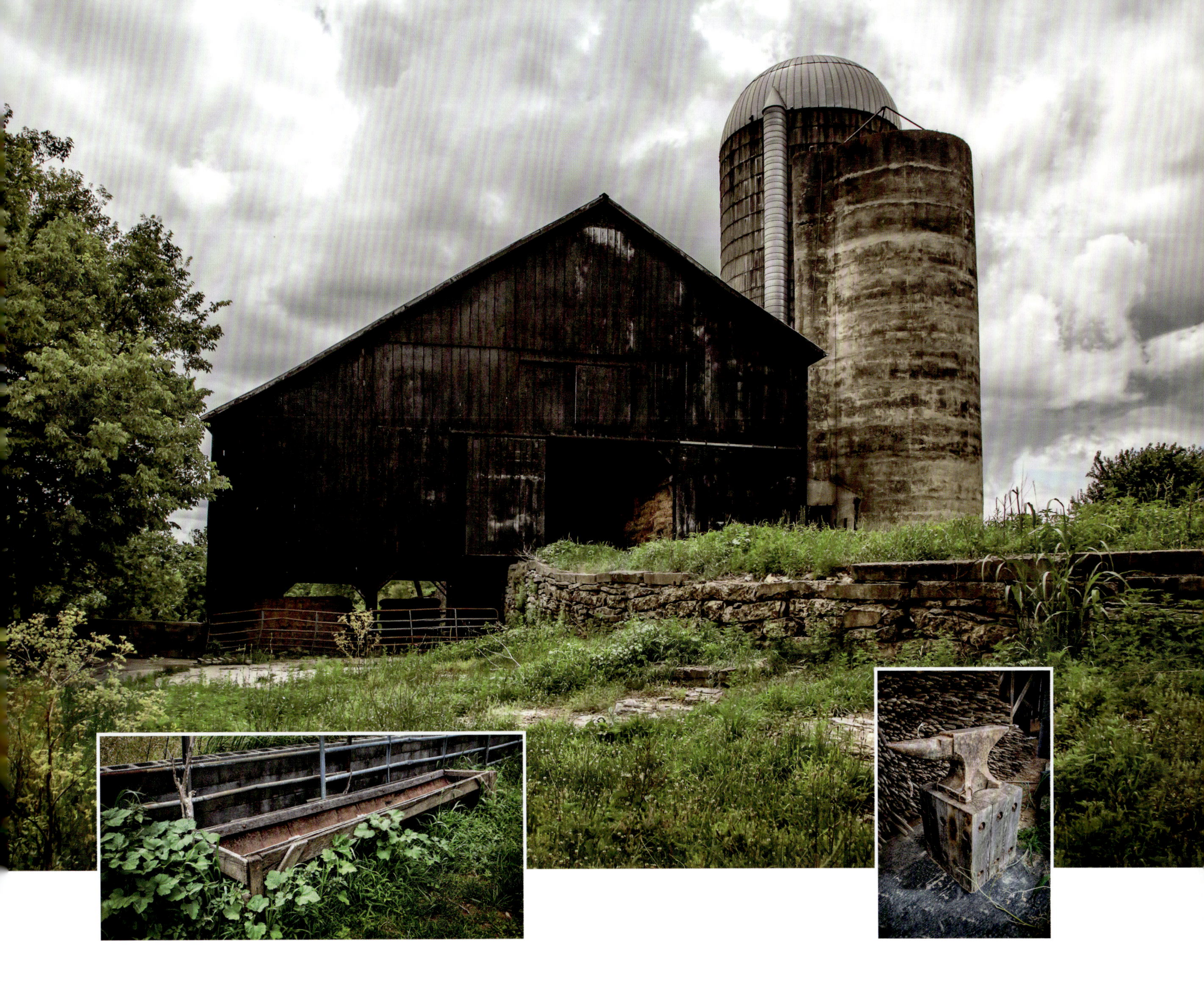

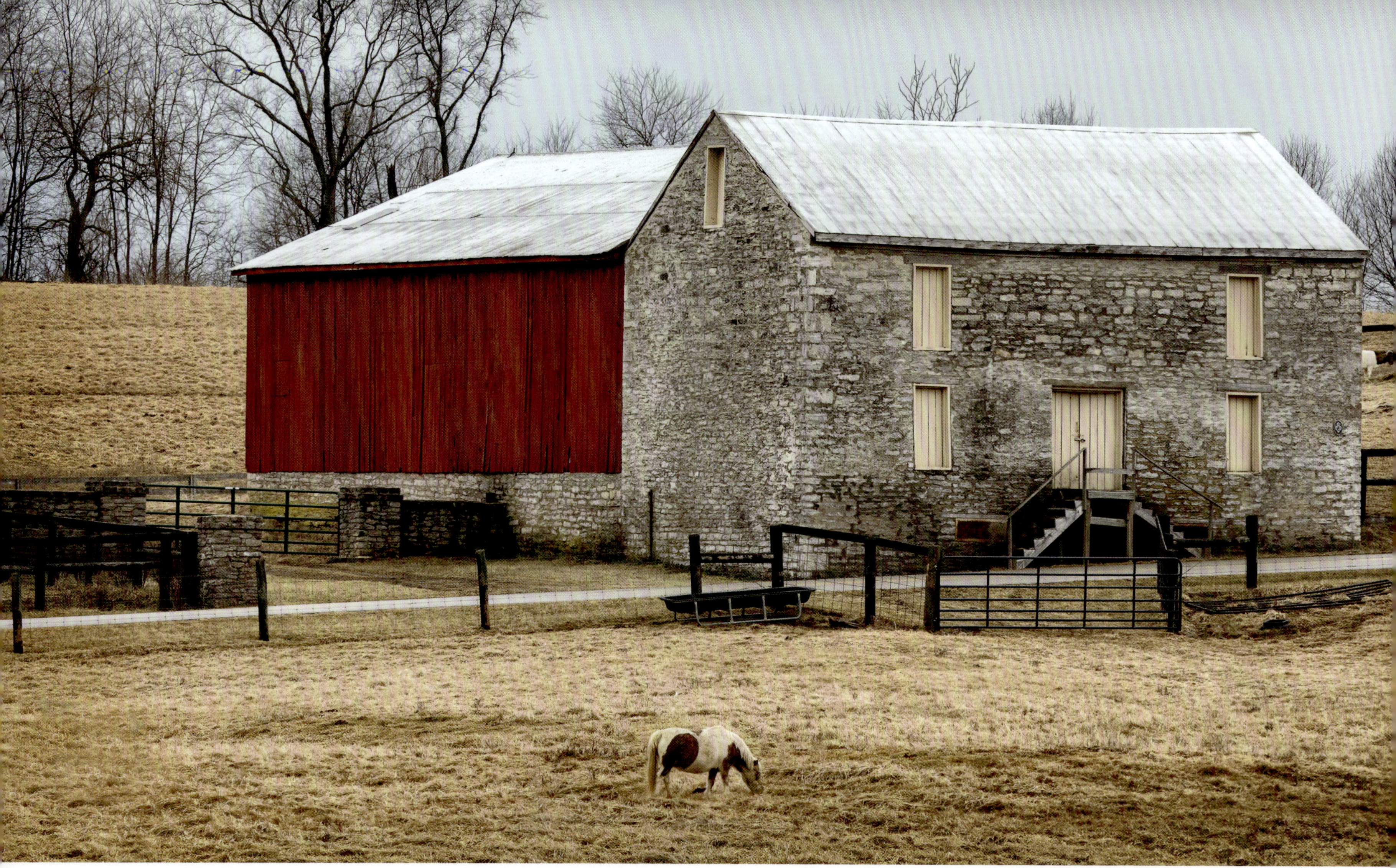

The Jacob Spears Distillery used this stone barn as a warehouse for their late eighteenth-century whiskey business. The timber addition was added later.

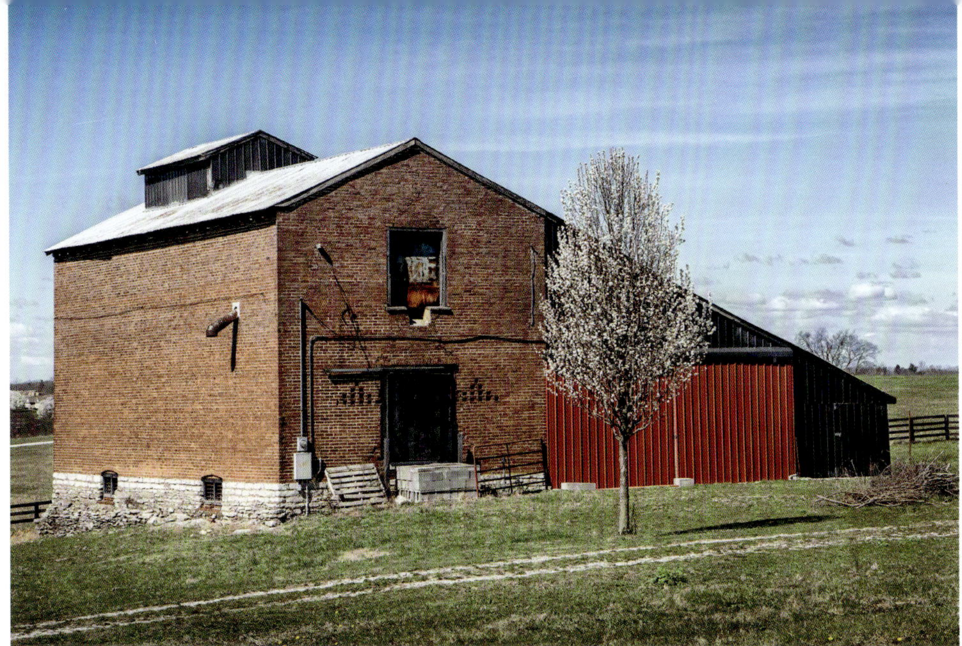

Scott County.

Newton Craig, an uncle of Elijah Craig of whiskey fame, was an early settler in Kentucky and a penitentiary keeper between 1844 and 1854. He also was winemaker, and he called his estate Spring Island. The brick building with the side addition was used as his winery; it is now used as a stable. The smaller brick and stone barn with fencing was used as a holding pen for prisoners. Their labor built most of his famous complex.

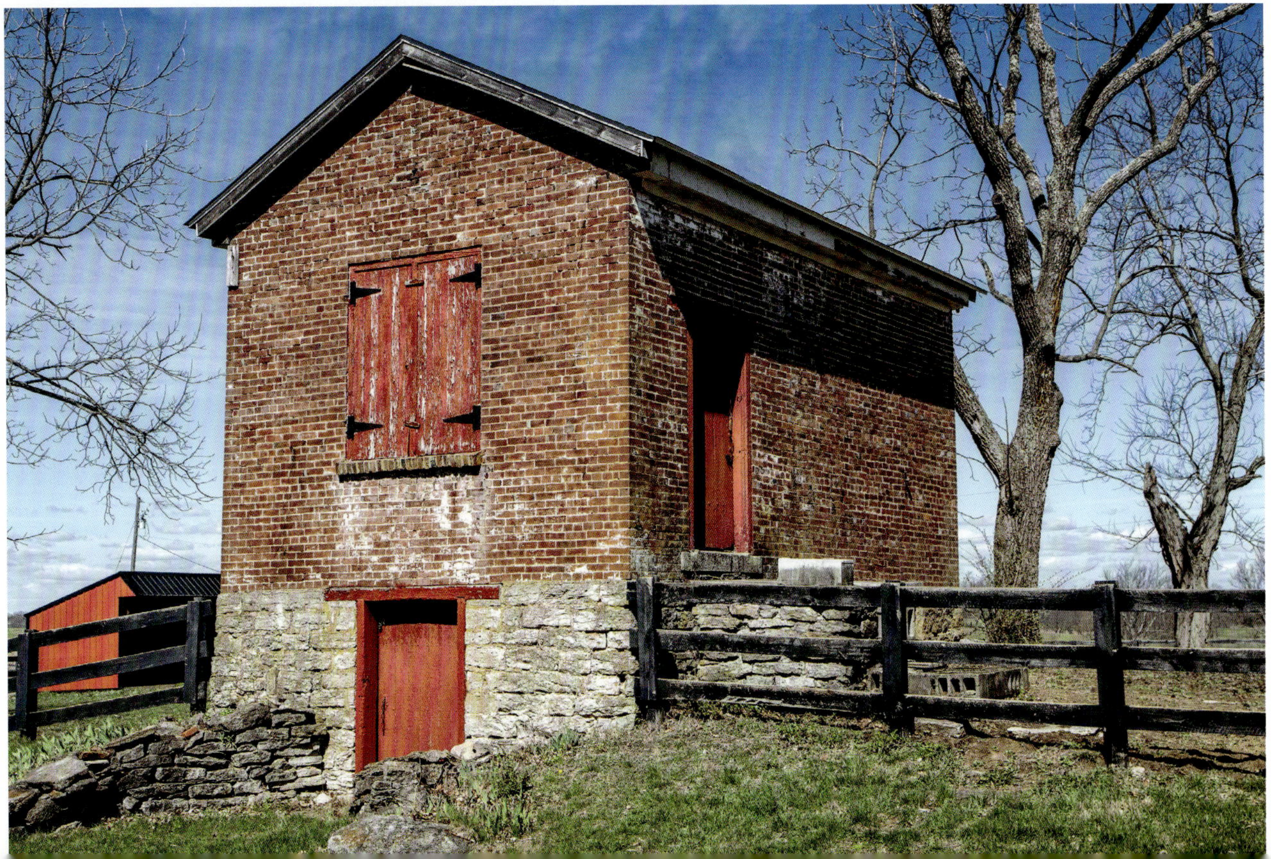

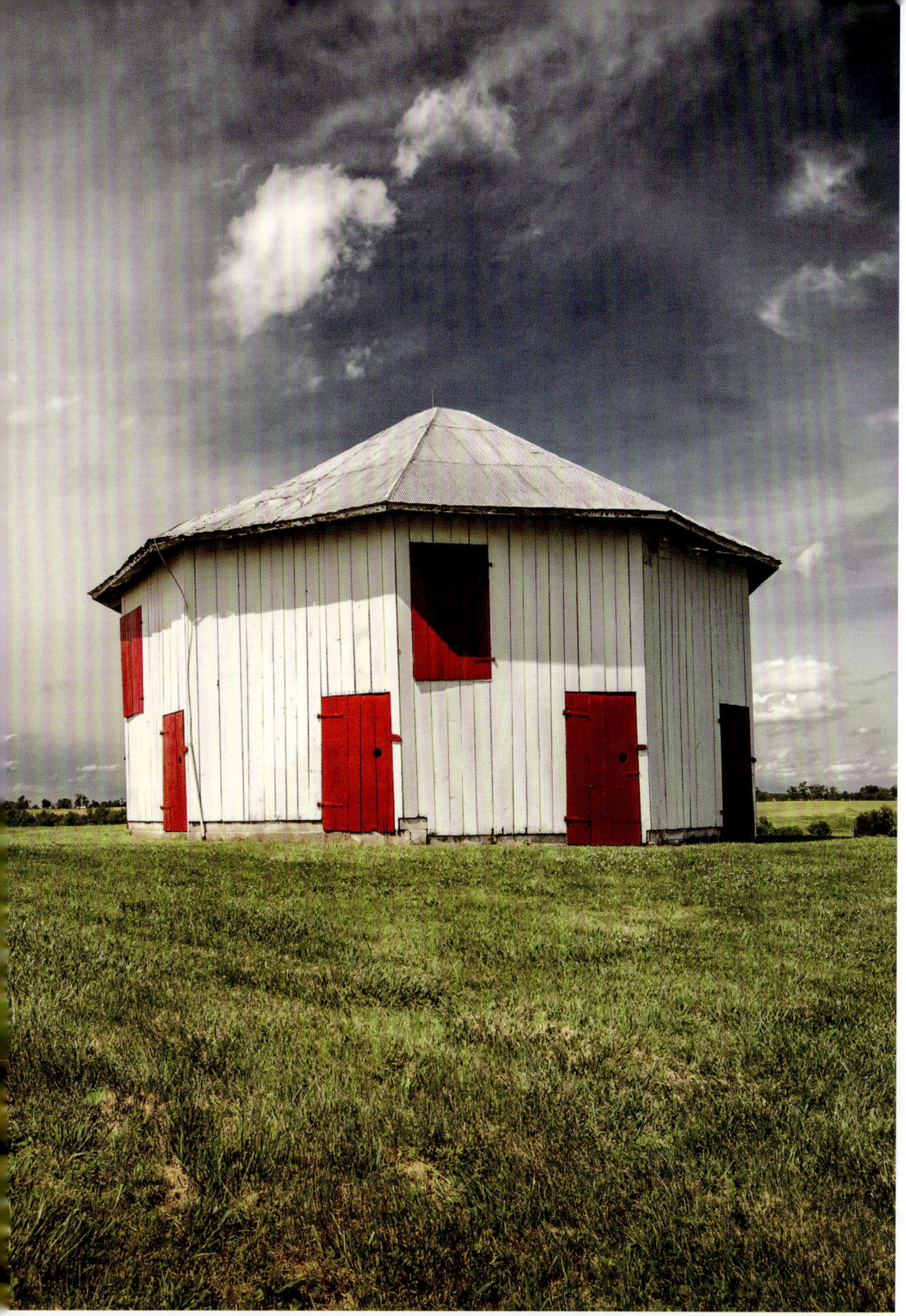

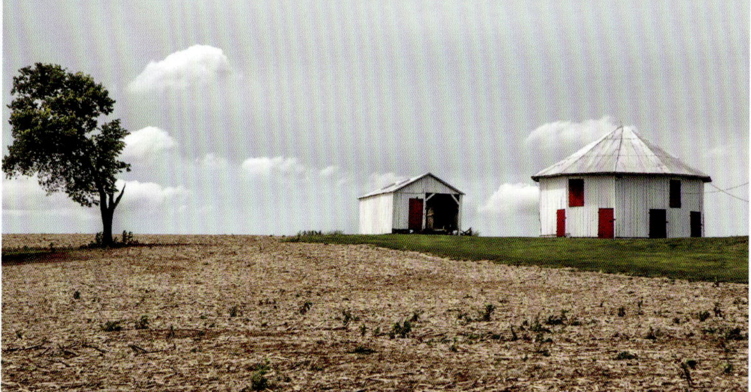

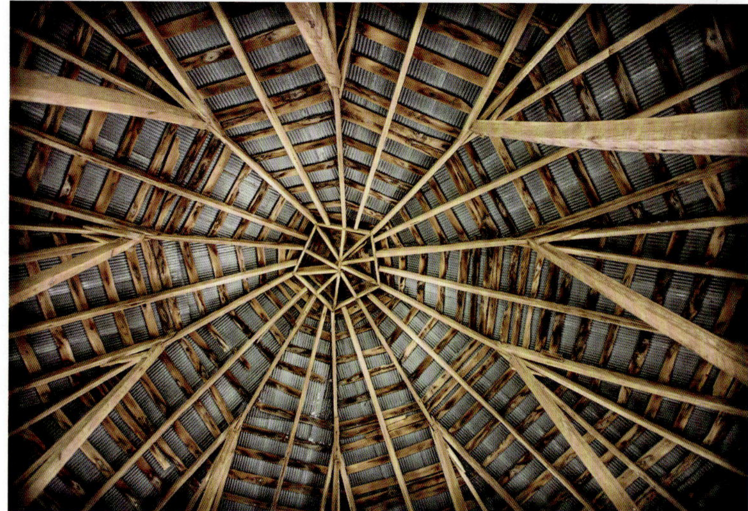

Mason County.

Some farmers built octagonal and round barns in the late nineteenth and early twentieth century as an economical use of lumber and efficient livestock management. This barn, part of the Duncan farm historically, was built by local carpenter Basil Berry. Adjacent is the Duncan farm tobacco barn.

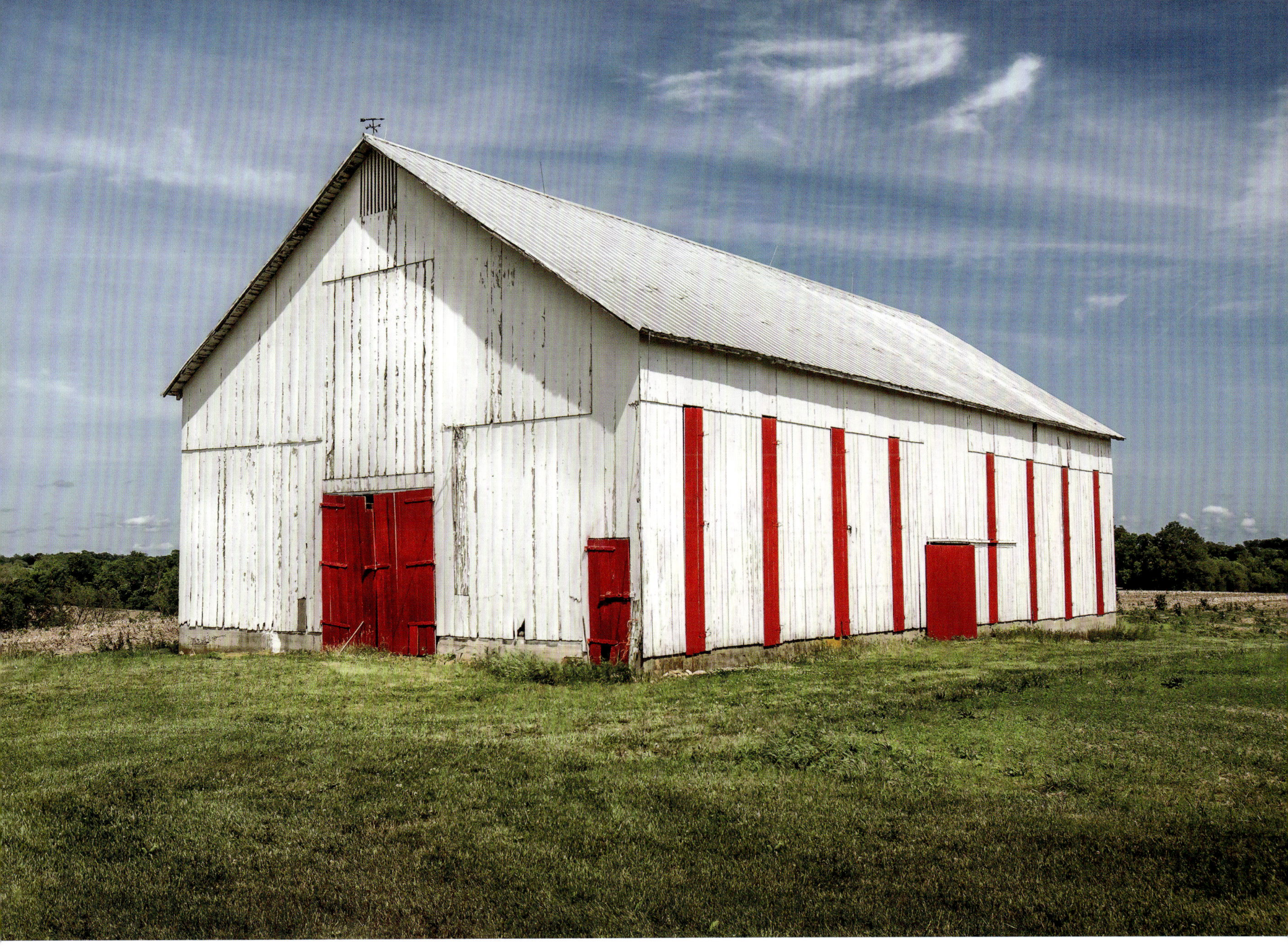

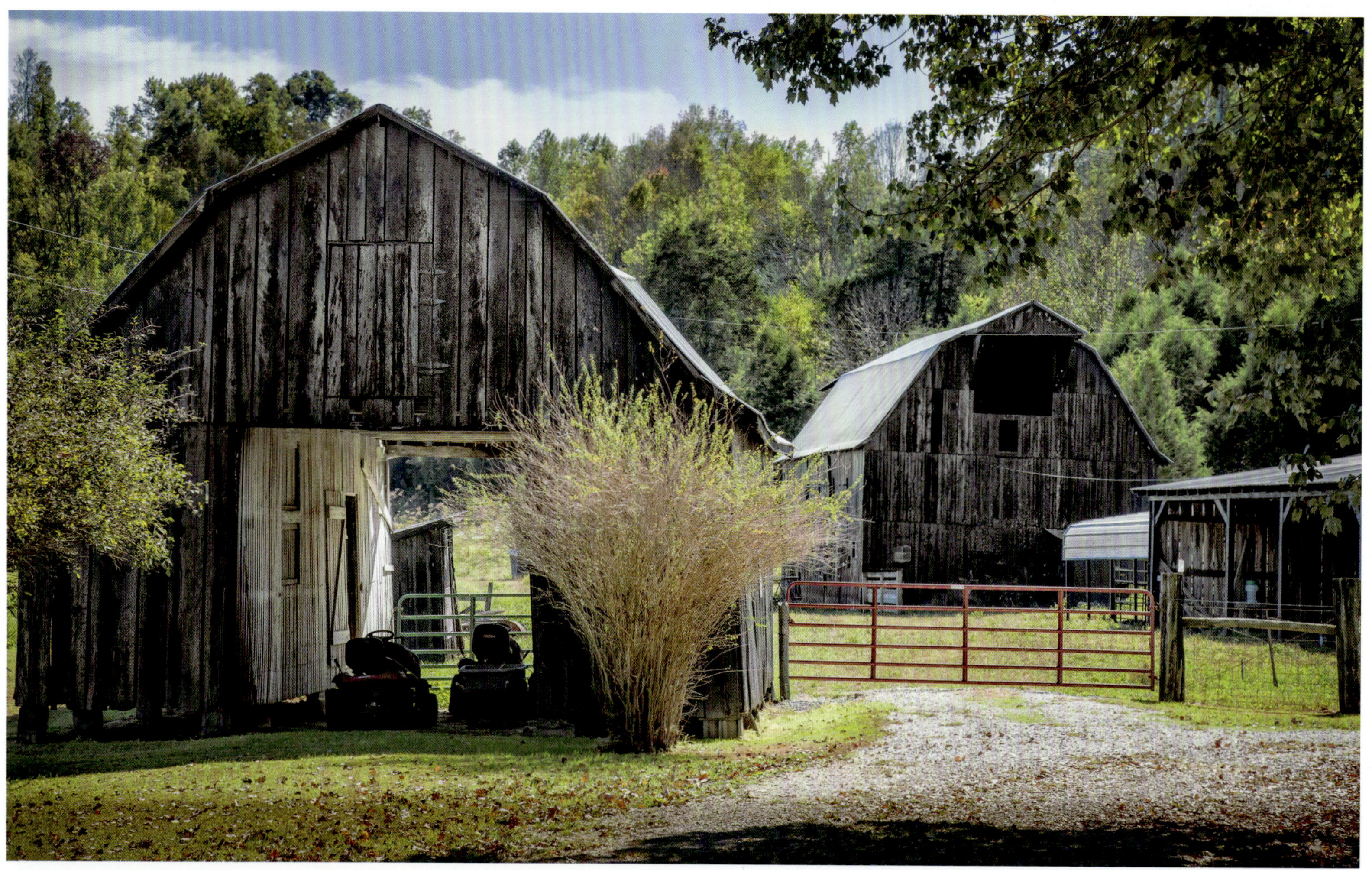

The Lemon farm is a remnant of the old Swiss settlement of Ottenheim. The carriage crib barn and stock barn with hay hood in the background were built around 1867, as was the stanchion barn, which was later converted to dry burley tobacco. Farther on the property is a stock barn with the rear converted to dry burley tobacco.

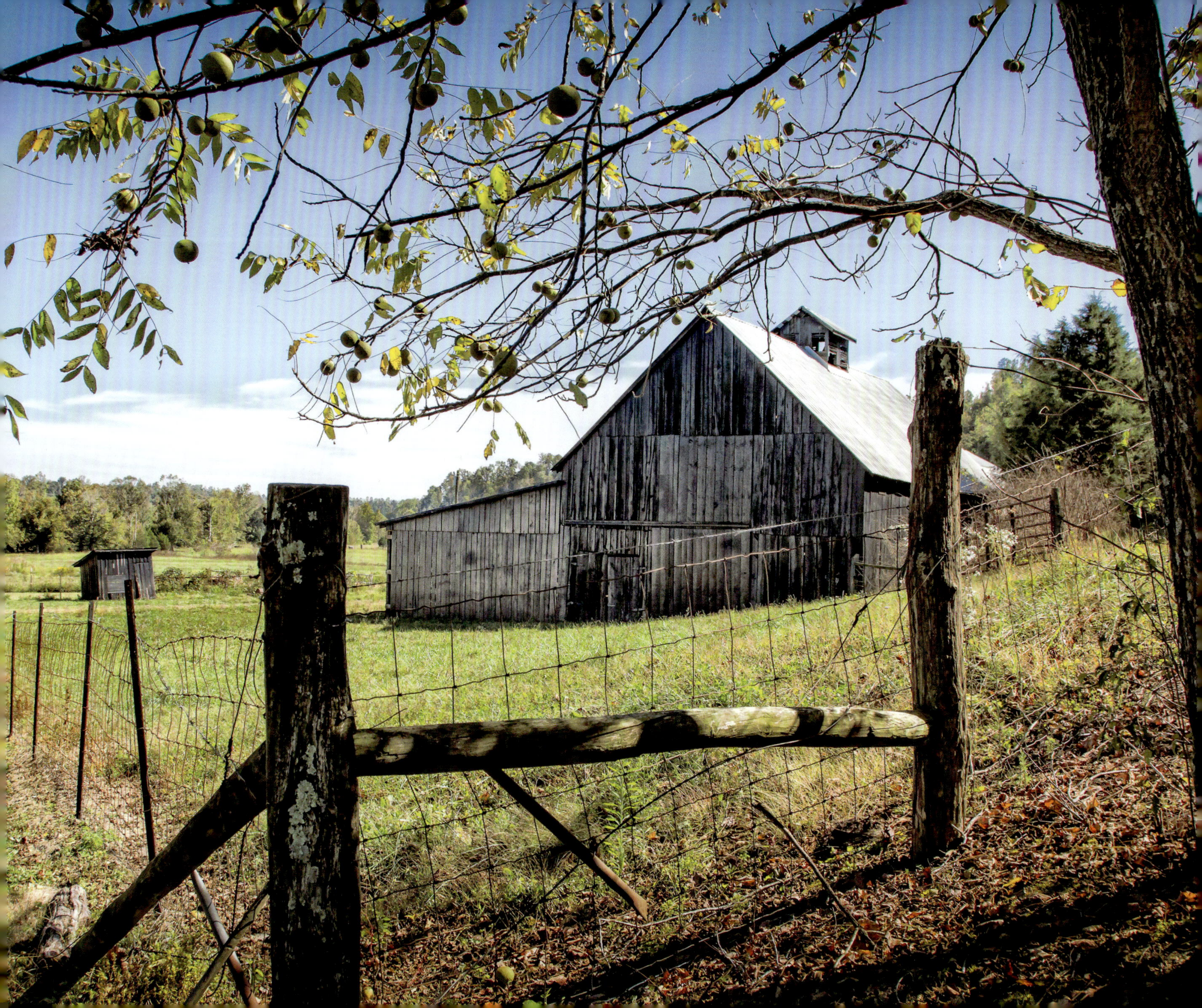

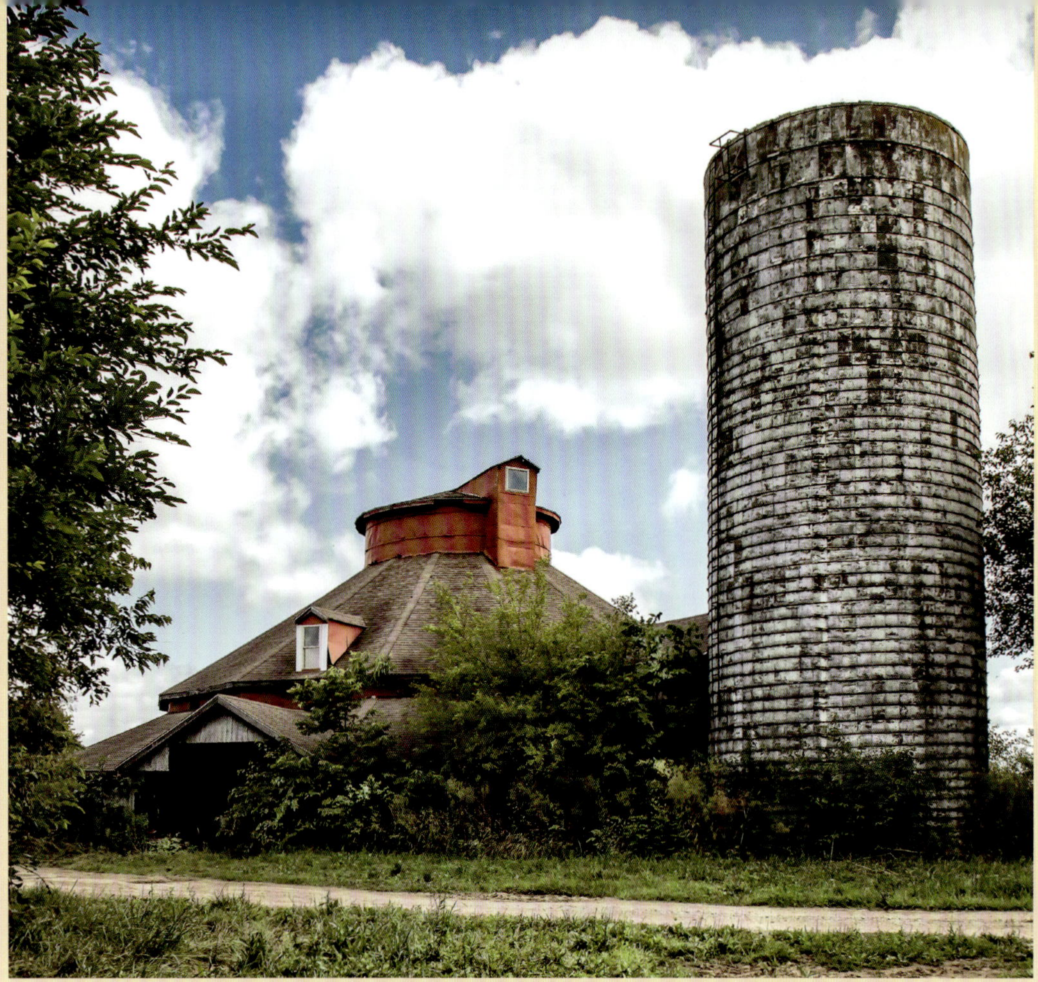

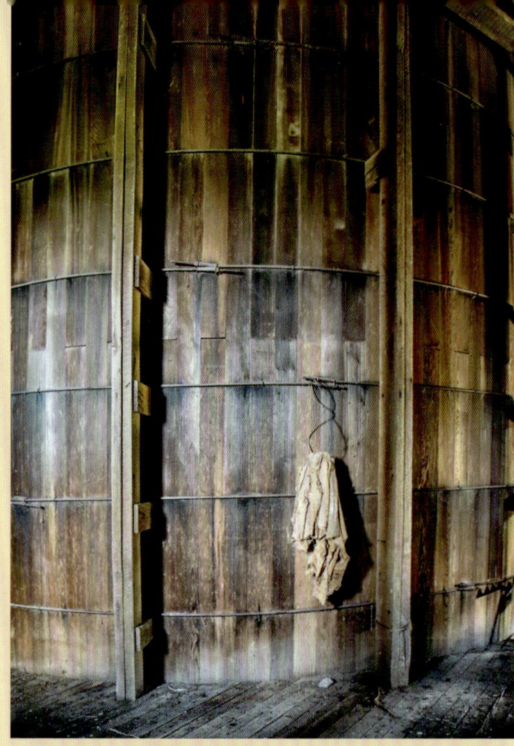

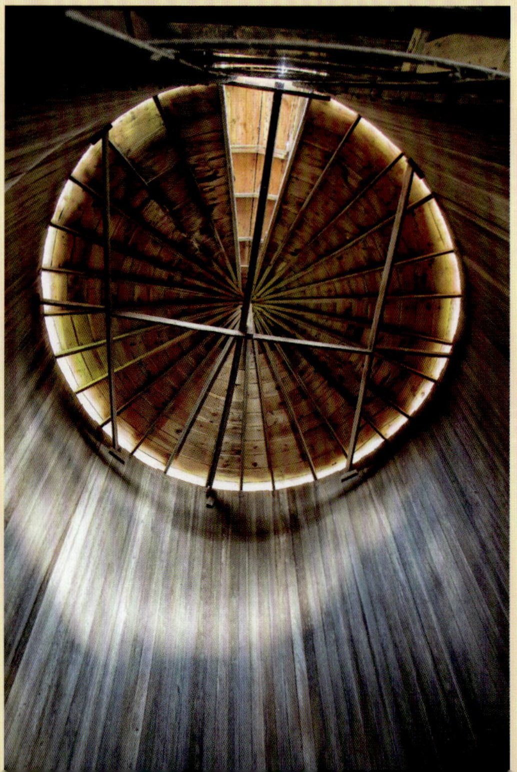

Bourbon County.

Levi Goff was a very successful businessman and banker in central Kentucky, and when he built this round livestock barn, he intended to show off his wealth. The center of the barn is a silo made of floor-quality hardwood planks. On the first level, hayracks and troughs line both outside and inside walls in a unique configuration. On the second level are dormers and ample space for hay storage, as well as a walkway to an outside cement silo.

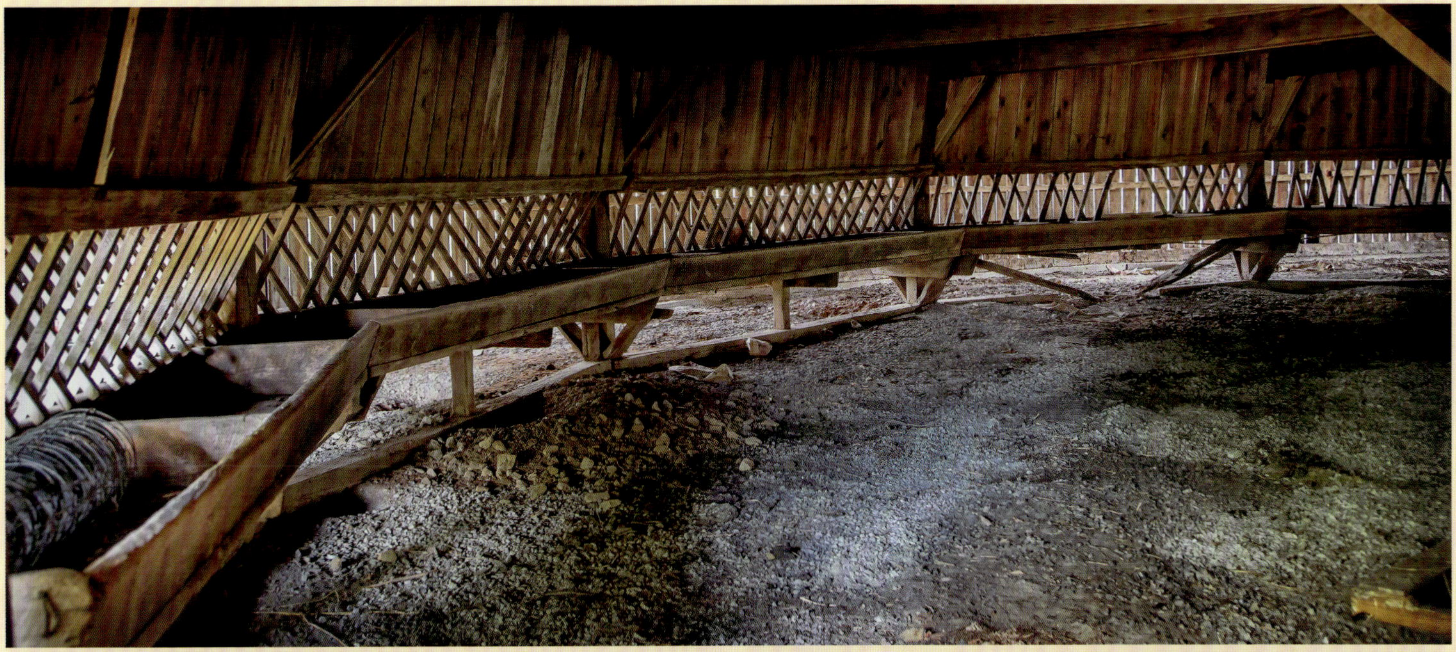

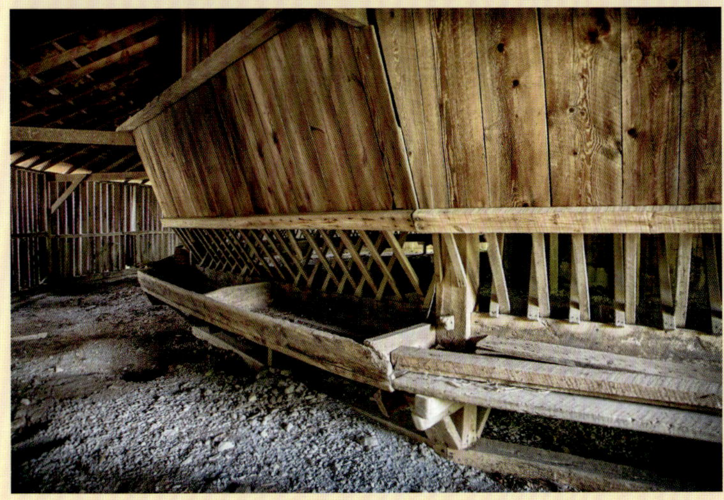

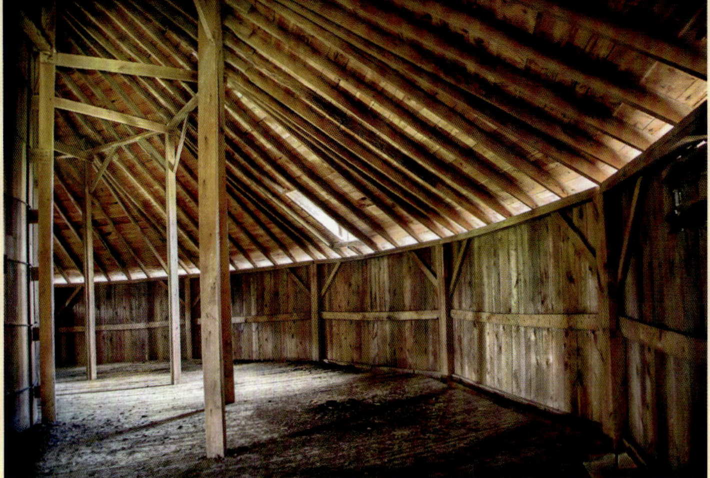

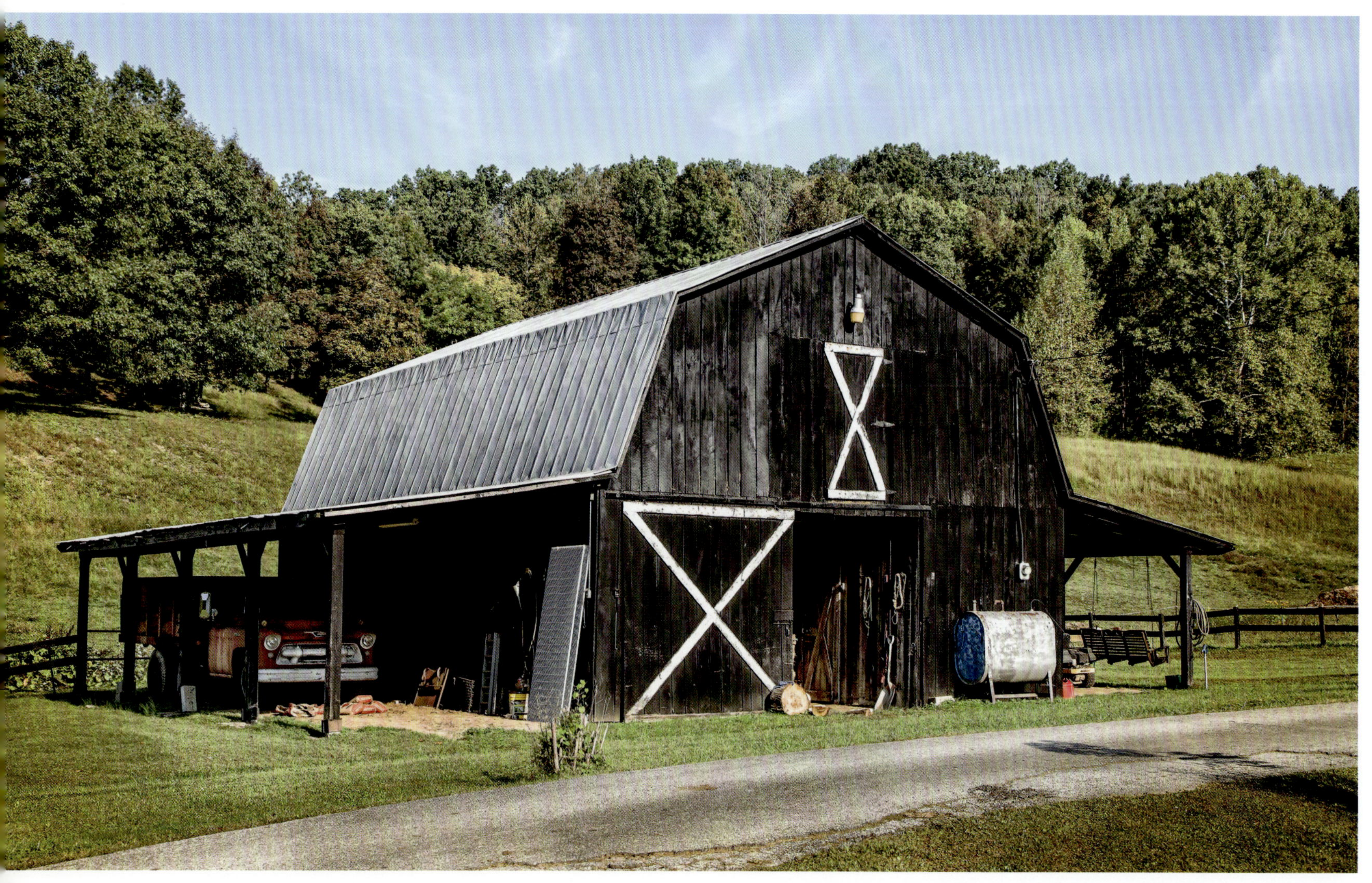

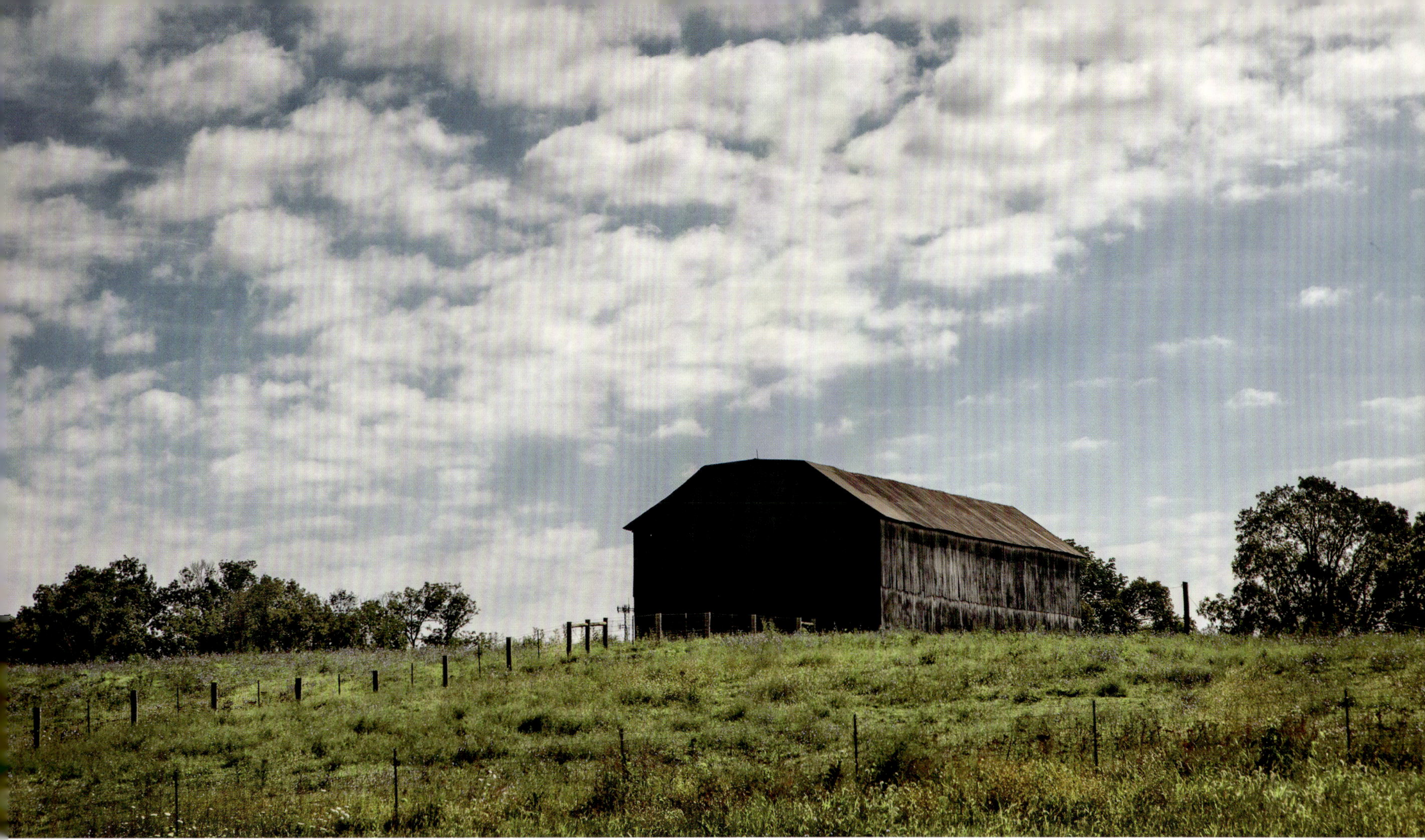

Lawrence County.

The roof on this multipurpose barn in dairy country is a classic gambrel roof design.

Madison County.

The Powell farm on the old McWilliams land grant has several historic barns, including this gambrel roof barn, built by a Mr. Venable in the early twentieth century.

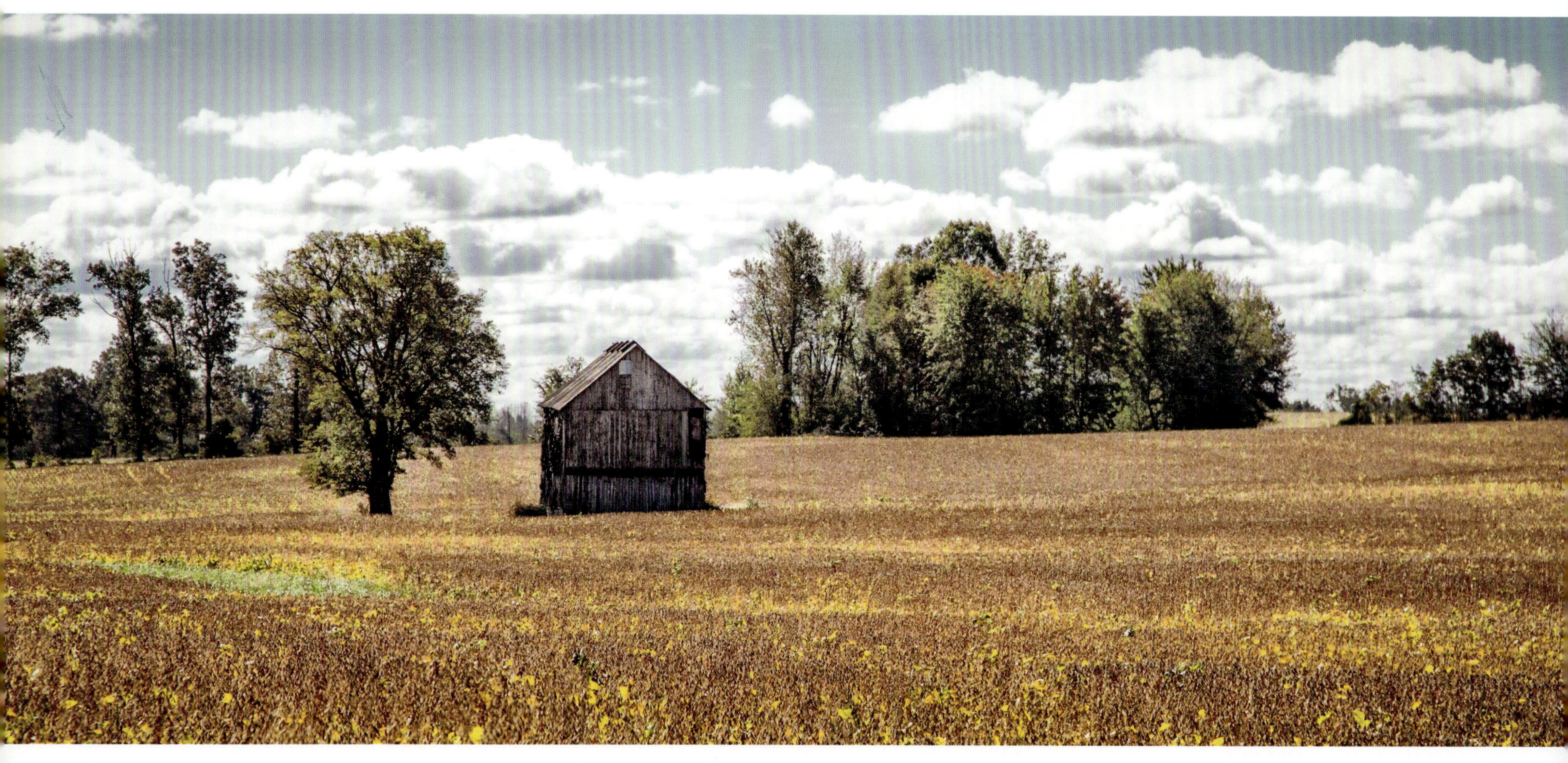

Todd County.

An old dark tobacco firing barn sits in the middle of a
soybean field, no longer needed for the new crop.

CAROL PEACHEE is an award-winning fine art photographer of historic and heritage sites. She is author and photographer of *Straight Bourbon: Distilling the Industry's Heritage* and *The Birth of Bourbon: A Photographic Tour of Early Distilleries*. She is also the photographer of *Kentucky Bourbon Country: The Essential Travel Guide*.